WRITING

TO LEARN

SCIENCE

RANDY MOORE

The University of Akron

WRITING

TO LEARN

SCIENCE

Saunders College Publishing

Harcourt Brace College Publishers

Fort Worth Philadelphia San Diego New York
Orlando Austin San Antonio Montreal
London Sydney Tokyo

Text Typeface: Garamond
Saunders College Desktop: Christine Livecchi
Executive Editor: Julie Levin Alexander
Developmental Editor: Jane Sanders Wood
Senior Project Editor: Anne Gibby
Copy Editor: Margaret Mary Anderson
Manager of Art and Design: Carol Bleistine
Cover Design: Ruth Dudero
Manager of Production: Joanne Cassetti
Marketing Manager: Sue Westmoreland

WRITING TO LEARN SCIENCE

ISBN 0-03-096488-1

Library of Congress Catalog Card Number: 96-68672

6789012345 039 987654321

PREFACE

Would you like to learn science? Begin by learning
your own language.
    — Abbe de Condilla, *On Anatomy*

It is with words that we do our reasoning, and writing is the expression of our thinking.
    — W. I. B. Beveridge, *The Art of Scientific
    Investigation*

Learn as much by writing as by reading.
    — Lord Acton

Language is the only instrument of science, and
words are but the signs of ideas.
    — Samuel Johnson

This is a book about writing to learn science. I wish it could be short and simple
like some "scientific writing" books, but I wanted to provide more than a pocket-
sized list of writing clichés such as "Be precise." Such simplistic reminders may
help experienced writers, but mean little to novices. Indeed, telling an inexperi-

enced writer to "Write concisely" is like telling a baseball player to "Hit the ball squarely." They *know* that—what they don't know is *how* to do it.

Memorizing simplistic rules will not ensure good writing. Writing well requires that you understand the *process* of writing. Only then will you understand why some writing communicates clearly while other writing does not, how word choice affects the impact of a sentence, and why much of "scholarly" writing is merely clutter that should be deleted to enhance communication. Understanding the process of writing goes beyond obeying a set of grammatical commandments, and will show you that good writing is clear thinking on paper. This is why any understanding necessary to produce good writing must be based on principles more useful than lists of anecdotes, platitudes, clichés, and self-evident truisms such as "Be specific."

But why another book about scientific writing? Most books about scientific and technical writing assume that the sole purpose of writing is to communicate. Consequently, these books seldom go beyond describing how to write a research paper and how to avoid problems of punctuation, grammar, and style. They include all of the clichés about accuracy, brevity, clarity, and cohesion, but typically provide only examples rather than an understanding of effective writing. Such one-sided accounts of writing fail to square with expert writers' advice on how to write effectively. Moreover, although these books gesture toward their audience, few help you learn to anticipate what your readers want and are looking for.

Knowing something about grammar and how to write about science is important for communicating scientific information. Consequently, these topics are included in *Writing to Learn Science*, although in a unique format. However, I contend that communication is not the *primary* goal of writing. The primary goal of writing, like reading, is to help you learn and understand; that is, writing is as much an instrument of thinking as it is a silent language that we use to pursue our thoughts. Moreover, only after you understand a subject can you make that understanding available to others in writing. Consequently, good writing should communicate *and generate* ideas. As James Van Allen said, "I am never as clear about any matter as when I have just finished writing about it."

Most books about scientific writing fail to explore writing as a way of thinking and learning. This limited approach to writing ignores *composition*, the essence of thinking. Although one function of writing is communication—the clear and direct expression of our ideas—many of our thoughts exist only when put to paper. By failing to link writing and thinking, we fall into a frustrating trap called "writer's block." This and other problems such as the obsession to "get it all on paper at one time" result from not understanding the writing process and underlie the embarrassment and self-delusion of comments such as "Well, that's not what I really meant" and "I understand the topic but just can't find the right words."

Writing can be fun and exciting. Millions of people do it for therapy and entertainment. However, many students (and professors) find writing tedious and difficult because they take a "think now and write later" approach to writing. This usually results from dwelling on the product or mental origins of writing, to the exclusion of the *activity* of writing. This approach handicaps you because it separates writing from thinking, thereby robbing you of writing's most valuable gift: its ability to help you learn.

Writing is a powerful way of thinking that can help you understand a subject, define a research problem, and hone an experiment. However, thinking an assignment completely through before writing about it pressures you to get it all done at once—to get all of the ideas onto paper in an organized way at one sitting. Such an approach is virtually impossible. A much more enjoyable and effective approach is to *view writing as thinking*—to write as you think things through. When approached this way, writing becomes a tool for communicating and for learning, because writing one sentence and paragraph suggests other sentences and paragraphs. This approach turns writing into discovery and helps you think with your pen, piece of chalk, or keyboard.

Learning About Writing

The inability of most students to enjoy writing or to write effectively attests to the failure of most university writing programs. These programs fail because they usually concentrate only on having students write without regard to understanding writing. Students already know how to write—they've cranked out all sorts of essays, book reports, and 20-page term papers for years. This hasn't improved students' writing because it has not taught them to write *effectively*. Most writing assignments have been little more than unpleasant exercises in practicing, and even perfecting, poor writing.

Although students often have an uneasy sense about what they write, they seldom know how to identify and correct the problems in their writing. Consequently, they become like drivers who know that their car won't start but have no idea about what to do to get it started. Although these people dutifully raise their car's hood, their hit-and-miss fiddling with wires and battery connections seldom starts the car and often causes even more problems. Similarly, students who rely only on simplistic rules and clichés rather than an understanding of writing usually just scramble words and sentences in hopes of improving their writing. At best, this produces frustrated students who wonder why their grades do not improve despite their having followed all of the "rules." This emphasizes an important point: Even when we don't feel anything is wrong with our writing, others often do. To prevent this problem, you must learn how to be the first critic of your writing—how to anticipate a reader's difficulties and to hear yourself as others hear you. You'll do this not when you memorize rules, but when you understand writing and readers' expectations.

Writing can be fun and educational. However, like other tools of a professional, writing is difficult to master because it requires clear, sustained thinking. Even great writers complain that writing is difficult. For example, Nathaniel Hawthorne wrote in his journal, "When we see how little we can express, it is a wonder that any man ever takes up a pen a second time." Similarly, Ernest Hemingway said, "What you ultimately remember about anything you've written is how difficult it was to write," and Flannery O'Connor wrote, "All writing is painful and if it is not painful, then it is not worth doing." Great scientists, too, have realized that good writing takes much time and is hard work, as evidenced by this quote by biologist Charles Darwin: "A naturalist's life would be a happy one if he had only to observe and never to write." If you think that writing is often hard work, you are in good company.

Writing is often difficult because it, like science, requires that you simultaneously be imaginative and critical. The imagination and discovery inherent in writing about science involve exploration, risk-taking, and leaps into the unknown. Conversely, the self-criticism necessary to revise a paper or experiment requires detachment, skepticism, and testing. This tug-of-war between discovery and criticism represents the True Believer and Doubting Thomas in each of us. Since these two mighty antagonists are not easily appeased at the same time, writing about science often seems difficult. However, so are many things that we learn to do well and come to enjoy, such as riding a bicycle, skiing, and playing a musical instrument. So take heart—writing is less a matter of mystery than a mastery of skills, many of which you already have. I hope that *Writing to Learn Science* will help take the sting out of learning to write well by helping you understand the process of writing.

Learning to write effectively doesn't require a life-long, monklike dedication. Rather, all it requires is that you want to write well and that you understand the process of writing. Understanding this process will help transform your writing from a helter-skelter "get it all on paper" chore to something you do to learn, to communicate, and even to relax.

What to Expect in This Book

If you're like most students, you probably dislike writing. Your dislike for writing may be no more than a vague feeling that you don't understand how to write, or the panic and uneasiness you feel when you're given an "essay test" or learn that you must write a term paper. Or it could result from things that are more tangible: poor grades and endless hours of frustration. Whatever its cause, your uneasiness and frustration with writing have probably caused you to ask yourself: "What's the problem here? What am I doing wrong?" Then, if ever, this book can help.

I wrote *Writing to Learn Science* to show students, scientists, and other pro-

Learning to write effectively can help you make better grades and avoid situations like the one shown above.

fessionals how writing about science can improve learning about science. You'll see that the essence of effective writing is based first on discovering your ideas, second on understanding the many decisions that writers make, and third on mastering the skills that translate those decisions into writing that communicates effectively. Contrary to what you may think, none of these decisions are hard to make, nor are the steps that turn these decisions into a well-written product. The trick is knowing what decisions to make, where those decisions will take you, and mastering the appropriate techniques. When you understand the process of writing, you'll trust the process because it will tell you what you know, what you don't know, and what you need to know. Good writers trust this process because they know it will eventually lead them to a clear and focused paper. Conversely, the inability to write effectively often typifies inexperienced writers who don't understand the process of writing and don't know where the process will lead.

Understanding writing will make writing easier than you ever thought it could be. It will also have other benefits, such as helping you write better term papers, lab reports, and answers on "essay" exams. However, give yourself time to learn and understand the process. Waiting until the last minute to start writing makes this process, and good writing, impossible. At best, you'll end up with a paper that a critic might describe as "extremely difficult, but only somewhat rewarding for the persistent and dedicated reader." Since most scientists are neither persistent nor dedicated readers, you'll probably get a poor grade on the paper or exam. Similarly, a poorly written research paper will probably be rejected or ignored.

I wrote *Writing to Learn Science* not as a writing "expert," but rather as a scientist who views writing as a useful and necessary tool for a successful career. I have tried to write a readable and accessible "how to" manual that is educational and practical. I hope that *Writing to Learn Science* will help you view writing as a process rather than as a product. I've assumed that you want to write better in a short time and without much fuss. Consequently, I've emphasized the simple and practical aspects of writing, and have tried to organize those ideas in a guide that leads you through the process of writing, shows you the purposes that writing serves, helps you improve your writing, and shows you how to use writing as a tool for learning. I stress what's immediately useful to most writers—the minimum amount of understanding needed for self-defense against common errors that impede learning and communication. This involves understanding not how sentences and paragraphs work within grammatical theories, but how to write effectively about science in the real world. I've left the abstract rules and explanations to the several excellent guides to grammar available at most bookstores (see Appendix One).

The guide listed below summarizes effective writing and will help acquaint you with this book. It is also a good place to quickly evaluate your writing skills. If you're a novice writer, you'll probably benefit most by reading the chapters in the order in which they're presented. If you're an experienced writer, use this guide to shape the first and later drafts of your paper, to rethink your ideas, and to clarify the goals and importance of your work:

Content

Subject

Have you defined your subject? See page 40.

Have you gathered your ideas? See page 35.

Audience

Have you identified your audience? See page 41.

Does your writing match that audience? See page 42.

For nonscientists, have you defined all terms and provided examples? See page 228.

For experts, have you presented enough information? See page 259.

Have you answered questions that your readers will likely ask? See page 52.

Are your conclusions supported by evidence and logic? See page 53.

Bias

Have you used inappropriate stereotypes or labels? See page 191.

Logic

Is your writing logical and coherent? See page 179.

Did you have a plan? See page 50.

Did you follow your plan? See page 52.

Interest

Did you emphasize important conclusions? See page 183.

Did you vary the length of your sentences? See page 184.

Did you write effective paragraphs? See page 180.

Did you make the reader want to read on? See page 185.

Precision and Clarity

Did you use simple words and sentences? See page 87.

Did you provide details to support generalizations? See page 183.

Did you use specific rather than vague words to ensure precision? See page 134.

Will readers understand all of the pronouns you've used? See page 145.

Have you punctuated your writing to help readers grasp your ideas? See page 193.

Have you used familiar words? See page 106.

Have you organized your thoughts logically? See page 51.

Grammar

Have you removed dangling modifiers? See page 142.

Are modifiers close to the words they modify? See page 143.

Have you expressed similar ideas in similar ways? See page 149.

Style

Usage

Have you used the right word? See page 134.

Have you avoided clichés? See page 145.

Have you used active voice? See page 156.

Have you avoided doublespeak? See page 136.

Have you avoided excessive hedging? See page 189.

Conciseness

Have you eliminated all unnecessary words? See page 93.

Have you eliminated all of the "fuzz" and filler from your paper? See page 87.

Readers' Expectations

Have you followed the subject as soon as possible with its verb? See page 151.

Have you used a strong verb to express the action of every clause and sentence? See page 154.

Have you put in the stress position the material that you want the reader to emphasize? See page 162.

Have you put familiar information at the beginning of the sentence? See page 163.

Mechanics

Have you defined all of the abbreviations that you've used? See page 147.

Have you checked the spelling of all words? See page 281.

Have you used numbers correctly? See page 302.

Are all of the tables and figures necessary? See page 311.

If so, are they numbered and well-designed? See page 313.

Have you proofread your paper to catch mistakes? See page 89.

Have you asked a colleague to read your paper? See page 199.

For students: Many of you may dread using this book, primarily because you equate writing with a long list of rules called "grammar." Your experiences with writing were probably not too pleasant, nor was sitting through those seemingly endless lectures on grammar and "rules of writing." Many of your English teachers probably taught you how to search for hidden meanings in poetry or to read literature instead of how to write effectively. Moreover, you've probably been told to strive for a "scholarly" style of writing rather than a strong and unpretentious style that will carry the thoughts of the world you live in. Consequently, by now you may be somewhat cynical and distrustful of any kind of writing or "writing book."

I went through the same kind of training—uninspiring classes and countless 20-page term-papers in which the most important criteria were the paper's length and format rather than its message. Therefore, I don't blame you for being cynical about the goals of this book. Considering your experiences, it probably has not occurred to you that writing can be a fun and useful, if not critical, tool that could make or break your career.

I hope that *Writing to Learn Science* will change your attitude by showing you that writing can be exciting, enjoyable, and easily understood. I also hope that this book will help you overcome the often overwhelming and largely unnecessary emphasis on "do's and don't's" associated with good writing. This book does not teach writing as a set of rigid rules—such an approach bores most people and has little, if any, lasting effect. Indeed, if all it took to write well was a list of rules, all scientists would write well. This is certainly not the case. Moreover, obeying all of the rules included in many writing books can produce wretched prose. That's why to many of the most lucid and precise writers, many "rules" have no standing whatsoever. Rather than concentrate on rules, *Writing to Learn Science* stresses practical writing—the kind you need to discover your ideas, to get them onto paper, and to efficiently convert those ideas into concise, well-written sentences and paragraphs. To do this, you must be able to diagnose problems and correct them, not just list rules.

Once you understand how to write well, you'll like writing for the same reasons you like science. They're both exciting, creative, lively, and inspiring; they are full of action and verifiable, defined facts; they require precision and argumentation. Moreover, learning to write well will make you a better scientist because it will show you how scientists think and work. Specifically, it will show you the nature of science, help you design better experiments, help you write better term papers, and move you away from blind acceptance to real data, interpretation, and argumentation.

This book does not pretend to teach creative writing, and I know of no magic formula to convert everyone into a Hemingway. However, this book will teach you useful, effective writing—the kind of writing that will produce papers that are clear, forceful, precise, well-organized, and easy to read and understand.

Writing to Learn Science includes exercises to help you write more effectively. It also stresses the importance of revision—preparing several drafts of a paper—to effective writing and learning. This approach to writing will reduce your anxiety about writing and show you that effective writing can be learned by understanding and applying a few principles of writing.

In *Writing To Learn Science* you'll read several of the classic papers and essays about science. These papers have diverse styles and purposes. For example, Rachel Carson's writing tries to persuade you to act, whereas Stephen Jay Gould's writing describes riddles of evolution, Albert Einstein discusses his famous equation $E=mc^2$, and James Watson's writing describes the thrill of scientific research. Despite their diversity, all of these papers will make you think. Moreover, all will show you how writing can help you learn about science.

The Stranger Side of Writing About Science

Every year we spend billions of dollars on scientific research. While many of these studies have described important discoveries such as cures for diseases, others have described leaps into the absurd. For example, a prestigious medical journal included this description of bringing mice back from the dead:

> The extract was injected into several mice, which died within two to six minutes. Given smaller doses, the mice recovered.

Another article described how another group of mice were transformed into a drug:

> The mice were treated with penicillin, then changed to erythromycin.

The scientific literature is filled with all sorts of unusual scientific studies. For example, scientists have published papers documenting that alcohol can help cure premature ejaculation in neurotic dogs, that rats are more attracted to other rats than to tennis balls, and that holy water does not affect the growth of radishes. Other studies have determined whether goldfish have hangovers, while still others have involved feeding the blood of schizophrenics to spiders.

As you might guess, sex is a favorite subject of much scientific research. One unusual paper described kinky sex practices that improve a male goat's performance, while another entitled "Paroxysmal sneezing following orgasm," published in the *Journal of the American Medical Association*, suggests that it might be a good idea for some folks to keep a box of hankies next to their beds.[1] Here are some more of my favorites:

[1]In case you're wondering, the recommended treatment for this problem is applying a long-acting nasal decongestant to each nostril about 30 minutes before intercourse. For more laughs about science and scientists, read the *Journal of Irreproducible Results.* Instead of tediously asking "Is it scientific?," this parody of scientific journals asks "Is it funny?" My favorites in the journal include "Bovinity," "The Inheritance Pattern of Death," "Reading Education for Zoo Animals: A Critical Need," and "A Drastic Cost Saving Approach to Using Your Neighbor's Electron Microscope."

In this book you'll also read some bloopers written by students, faculty, and others. For example,

A super-saturated solution is one that holds more than it can hold.

We get our temperature three different ways. Either by fairinheit, callcius, or centipede.

Pavolv studied the salvation of dogs.

The Effects of Chewing Gum Stick Size and Duration of Chewing on Salivary Flow Rate and Sucrose and Bicarbonate Concentrations (published in *Archives of Oral Biology*)

Azoreductase Activity in Bacteria Associated with the Greening of Instant Chocolate Puddings (published in *Applied and Environmental Microbiology*)

Injuries due to Falling Coconuts (published in *The Journal of Trauma*)

The Deterioration and Conservation of Chocolate from Museum Collections (published in *Studies in Conservation*)

A Classification of Pure Malt Scotch Whiskies (published in *Applied Statistics*)

Dissociation between the Calcium-induced and Voltage-driven Motility in Cochlear Outer Hair Cells from the Waltzing Guinea Pig (published in *Journal of Cell Science*)

Three Cases of Disputed Paternity in Dogs Resolved by the Use of DNA Fingerprinting (published in *New Zealand Veterinary Journal*)

Energy of the Closed Universe with Respect to the Gas of Clocks (published in *Classical and Quantum Gravity*)

Painless Jogging for 15,000 km After a Lumbrosacral Stabilization with Screws and Cement (published in *The Medical Journal of Australia*)

The Mechanics of the Anal Sphincter (published in *Journal of Biomechanics*)

Seizures Induced by Thinking (published in *Annals of Neurology*)

Termination of Intractable Hiccups with Digital Rectal Massage (published in *Journal of Internal Medicine*)

Acute Management of the Zipper-Entrapped Penis (published in *The Journal of Emergency Medicine*)

However, the most intriguing title of a paper appears in *Veterinary Record*: "The erect dog penis: a paradox of flexible rigidity." Rather than spoil your fun, I'll let you look up that one for yourself.

A molecule is so small that it can't be seen by a naked observer.

After soaking in acid, I washed the glassware thoroughly.

Madman Curie discovered radio.

Gravity was invented by Isaac Walton. It is chiefly noticeable in the autumn, when the apples are falling off the trees.

Magellan circumcised the globe with a hundred foot clipper.

I hope these bloopers will teach you about writing. I also hope they make you laugh.

Whether this is your first or last course in science, *Writing to Learn Science* will help you better understand science. It is meant as a handbook, so keep it handy. Write as you use this book—use it as a writing coach and reference to guide you through your writing projects. Keep a marking pen handy to underline or highlight points that you especially want to remember. Moreover, don't wait until you've read the whole book to start practicing what you learn. Remember the saying, "What I read, I forget. What I see, I remember. What I do, I understand."

I hope you enjoy this book.

Randy Moore
Akron, Ohio
April 1996

ACKNOWLEDGMENTS

According to Spanish writer Baltasar Gracián,

> Good things, when short, are twice as good.

So I'll be brief.

With *Writing to Learn Science* I've tried to give science teachers, students, and others a practical and entertaining book about how to use writing to learn science. This book required much work and I am grateful to many people for their help. I thank:

Julie Levin Alexander and Jane Sanders Wood, my editors, for guiding me through this project. I greatly appreciate the freedom they gave me as I wrote this book.

Flo Fiehn for helping me organize my writing, my job, and my life. Everyone should have such a helpful colleague and friend.

My many students who have taught me much about science and writing.

Finally, I thank my family and friends for their support and encouragement. This book is for Mom, Dad, Kris, and Stella the Bunny.

CONTENTS

UNIT ONE

WRITING

ABOUT

SCIENCE

Reading maketh a full man, conference a ready
man, and writing an exact man.
  — Francis Bacon

. . .[W]riting comes in grades of quality in the fashion of beer and baseball games: good, better, best.
. . . Better ways can be mastered by writers who are
serious about their writing. There is nothing arcane
or mysterious about the crafting of a good sentence.
  — James Kilpatrick

Young writers often suppose that style is a garnish
for the meat of prose, a sauce by which a dull dish
is made palatable. Style has no such separate entity;
it is non-detachable, unfilterable. The beginner
should approach style warily, realizing that it is himself he is approaching, no other; and he should
begin by resolutely turning away from all devices
that are popularly believed to indicate style—all
mannerisms, tricks, adornments. The approach to
style is by way of plainness, simplicity, orderliness,
sincerity.
  — E.B. White

The scientific man is the only person who has anything new to say and who does not know how to say it.
— Sir James M. Barrie

If thought corrupts language, language can also corrupt thought.
— George Orwell

Everything that can be thought at all can be thought clearly. Everything that can be said can be said clearly.
— Ludwig Wittgenstein

CHAPTER ONE

What is

Effective Writing?

It is impossible to dissociate language from science
or science from language, because every natural science always involves three things: the sequence of
phenomena on which the science is based; the
abstract concepts which call these phenomena to
mind; and the words in which the concepts are
expressed. To call forth a concept a word is needed.
— Antoine Lavoisier

A word to the wise is not sufficient if it doesn't make
any sense.
— James Thurber

But just because people work for an institution they
don't have to write like one. Institutions can be
warmed up. Administrators and executives can be
turned into human beings. Information can be
imparted clearly and without pompous verbosity.
— William Zinsser, *On Writing Well*

The more students write, the more active they
become in creating their own education: writing frequently . . . helps students discover, rehearse,
express and defend their own ideas.
— Toby Fulwiler

I always wanted to write, but I always figured it'd be
no good unless somehow the hand just took the
pen and started moving without me really having
anything to do with it. Like automatic writing. But it
just never happened.
— Jim Morrison

Put it before them briefly so they will read it, clearly
so they will appreciate it, picturesquely so they will
remember it and, above all, accurately so they will
be guided by its light.
— Joseph Pulitzer

Good writing does not come from fancy word
processors or expensive typewriters or special pencils
or hand-crafted quill pens. Good writing comes from
good thinking.
— Ann Loring

Science is a process that helps us learn. Scientists design experiments to test
hypotheses, and use the results of experiments to learn about our world. In
doing so, scientists face a dual challenge—to understand what they're doing and to
communicate this understanding to others. Scientists must create meaningful
descriptions of their ideas so that readers can easily re-create, and therefore understand, their ideas. These challenges mean that a scientist must know how to do and
write about science.

Many scientists equate effective writing with "correct" writing—the kind of writing that breaks none of the rules they memorized in their English classes. To show
the fallacy of this logic, consider the following two papers:

**Watson, J.D. and F.H.C. Crick. 1953. Molecular structure of nucleic acids:
a structure for deoxyribose nucleic acid.** *Nature* **171: 737–738.**   This
paper—the paper that many scientists say gave birth to molecular biology—
sketched the double-helical structure that Watson and Crick had deduced for
DNA. Although the paper reports no experimental data, it had a tremendous
and immediate impact, not only because it was accurate, but also because it
was written effectively.

Avery, O.T., C.M. MacLeod, and M. McCarty. 1944. Studies on the chemical nature of the substance inducing transformation of pneumococcal types. *Journal of Experimental Medicine* **79: 137-158**. This was the first published demonstration that DNA is the hereditary material. Avery *et al.* studied *Streptococcus pneumoniae*, a bacterium common in people having pneumonia. Avery *et al.* extracted DNA from a virulent strain of the bacterium and mixed it with cells of a less virulent strain. When the DNA-treated bacteria were grown into colonies, some of the bacteria had characteristics of the virulent strain. That is, these bacteria had been transformed. Avery *et al.* showed that the substance that caused the transformation was "a deoxyribose-containing nucleic acid."

The paper by Avery *et al.* was important because it overturned a widespread assumption and paved the way for Watson's and Crick's subsequent paper about DNA. Yet, despite the paper's novel conclusions and technical strength, it was not widely appreciated or accepted when it was published. Indeed, some scientists claimed that the discovery by Avery *et al.* that DNA is the genetic substance, like Mendel's discovery of the gene in 1865, was "premature." As you'll see in the following exercise, this perception and the paper's lack of immediate impact was due largely to its writing style.

Confidence of the Authors

Avery *et al.* were hesitant to make conclusions. Watson and Crick were extremely confident.

Length

Avery *et al.* were verbose; their paper was about 7,500 words long. Watson's and Crick's paper was much more concise; it was only about 900 words long.[1]

Organization

Avery *et al.* did not state their thesis in their opening paragraph, and did not mention DNA until about halfway through their paper. Watson and Crick state their thesis in the opening sentence of their paper.

---

[1] Other great biologists have used concise writing to enhance the impact of their writing. For example, Arthur Kornberg, who produced the first good evidence that the helices of DNA run in opposite directions, described the enzymatic synthesis of DNA with only 430 words. Cournand and Ranges described the first catheterization of a human heart with only 950 words, and Fritz Lipmann described coenzyme A with only 250 words and a table. For comparison, the U.S. Department of Agriculture's directive for pricing cabbage contains 15,629 words. Also see Moore, R. 1995. *Writing to Learn Botany.* Dubuque, Iowa: William C. Brown Communications, Inc.

Claims of Importance

Avery *et al.* make no claims that their work is important. Rather, they introduce the paper as merely a "more detailed analysis" of an already well-known phenomenon (transformation). Watson and Crick boldly claim that their work is important. As Watson wrote to a colleague one month before publicly announcing the structure of DNA, "It is a strange model and embodies several unusual features. However, since DNA is an unusual substance, we are not hesitant in being bold."

Writing Style

Avery *et al.* used an abstract, impersonal style of writing that includes much passive voice; they even refer to themselves abstractly as "the writers." Watson and Crick used first person and active voice to enhance the impact of their paper and to claim the work and model as their own. Similarly, the dull presentation of Avery *et al.* is based on the belief that "facts" and data can speak for themselves. Watson and Crick reject this notion, and consistently show that meaning is created by human activities ("We wish to suggest . . .", "We wish to put forward . . .", "We have postulated . . .").

If you're still not convinced that the impact of what you write will depend largely on how you write it, read the following essays without pausing too much, and then consider your impressions of the quality of each writer as a scientist:

Brown's version

In the first experiment of the series using mice it was discovered that total removal of the adrenal glands effects reduction of aggressiveness and that aggressiveness in adrenalectomized mice is restorable to the level of intact mice by treatment with corticosterone. These results point to the indispensability of the adrenals for the full expression of aggression. Nevertheless, since adrenalectomy is followed by an increase in the release of adrenocorticotrophic hormone (ACTH), and since ACTH has been reported (Brain, 1972), to decrease the aggressiveness of intact mice, it is possible that the effects of adrenalectomy on aggressiveness are a function of the concurrent increased levels of ACTH. However, high levels of ACTH, in addition to causing increases in glucocorticoids (which possibly accounts for the depression of aggression in intact mice by ACTH), also result in decreased androgen levels. In view of the fact that animals with low androgen levels are characterized by decreased aggressiveness the possibility exists that adrenalectomy, rather than affecting aggression directly, has the effect of reducing aggressiveness by producing an ACTH-mediated condition of decreased androgen levels.

Smith's version

The first experiment in our series with mice showed that the total removal of the adrenal glands reduces aggressiveness. Moreover, when treated with corticosterone, mice that had their adrenals taken out became as aggressive as intact animals again. These findings suggest that the adrenals are necessary for animals to show full aggressiveness.

But removal of the adrenals raises the levels of adrenocorticotrophic hormone (ACTH), and Brain (1972) found that ACTH lowers the aggressiveness of intact mice. Thus the reduction of aggressiveness after this operation might be due to the higher levels of ACTH which accompany it.

However, high levels of ACTH have two effects. First, the levels of glucocorticoids rise, which might account for Brain's results. Second, the levels of androgen fall. Since animals with lower levels of androgen are less aggressive, it is possible that removal of the adrenals reduces aggressiveness only indirectly: by raising the levels of ACTH it causes androgen levels to drop.[2]

Both essays present the same information in the same order and use the same technical words. Both essays are also "correct"—they differ only in their use of ordinary language. However, Smith's essay is more readable because it avoids unfamiliar words, avoids inflated roundabout phrases, and uses shorter, more direct sentences. Conversely, Brown's essay is hard to read because it contains long sentences, big words, and convoluted constructions. These differences have a tremendous impact on other scientists. Indeed, almost 70 percent of the 1,580 scientists who read these essays judged Smith's essay as more stimulating, more interesting, more impressive, and more credible than Brown's essay. Readers also thought Smith was more helpful, dynamic, and intelligent than was Brown. Most importantly, when asked to judge Smith and Brown's competence—specifically, which scientist seemed to have a better organized mind—almost 80 percent chose Smith. Heed the message here: although both of the essays are *correct*, only one is *effective*.

It's not enough to write a "correct" essay or paper—such writing often fails to advance your argument or accomplish the goal of your writing. Rather, strive for *effective* writing—writing that is clear, simple, precise, accurate, and concise. If your writing is effective, other people will not only enjoy reading your writing, but they will also think that you have a better organized mind, are more competent, and, in this case, are a better scientist than someone who does not write well. If readers can easily understand what you're saying, they are more likely to be impressed with and learn from your ideas.

---

[2]From Christopher Turk and John Kirkman. 1989. *Effective Writing*. 2nd edition. London: E. & F.N. Spon. For more information about this study, see Barbell, Ewa. 1978. "Does style influence credibility and esteem?" *The Communicator of Scientific and Technical Information* 35:4-7; Turk, C.C.R. 1978. "Do you write impressively?" *Bulletin of the British Ecological Society* ix(3) 5-10; Wales, LaRae H. 1979. *Technical Writing Style: Attitudes Towards Scientists and Their Writing*, Burlington, Vermont: University of Vermont Agricultural Experiment Station.

Myths About Writing

Writing is judged by how easily it conveys ideas and helps us learn, not by its adherence to grammatical rules. To communicate well, you must understand what you did, what you want to say, and why it is important. You'll communicate effectively only when you have something to say, say what you mean, and logically support your statements with evidence. This requires no inspiration, wit, or rhythm. All you must do is make sure that your ideas are obvious to readers and never make readers guess or wonder if you have anything important to say.

Many students assume that they are good writers, yet begrudgingly admit that they cannot distinguish a subject from a verb, or a pronoun from a preposition. Others believe that they write well because they studied Latin, diagrammed hundreds of sentences, and can distinguish grammatical contraptions such as subjunctive pluperfect progressive and retained objects. Still others believe that good writing results from sincerity, from writing like they speak, or from just being "gifted." The best evidence indicates that none of these things has much to do with one's ability to write well.

The first step toward improving one's writing involves understanding what writing is and what it isn't. Most conceptions about writing are, in fact, misconceptions.

Writing requires inspiration. Contrary to folklore, writing isn't sitting for hours in front of a blank piece of paper or computer screen, waiting to be "inspired." Writing never happens that way, not even for the best of writers. Writing only *looks* like it happens that way because many of the decisions that writers make are invisible to those who do not understand writing.

Writing is less an art than a craft. Like any craft, writing involves a series of decisions that, when done in the correct order and with the proper attention to detail, can produce a decent and acceptable paper. That paper may not read like it was written by Hemingway, but it will communicate your thoughts effectively and help you learn about and understand your subject.

Writing to Learn Science teaches practical and effective writing—writing that will communicate your message clearly and quickly. To help you understand what I mean by this, consider this sentence from the back of a tube of *Crest* toothpaste: "For best results, squeeze the tube from the bottom and flatten it as you go up." No one would claim that this sentence is overly creative or that it represents great literature. However, it is perfect for its function because it efficiently does its job: It contains no excess words and its message cannot be misunderstood. Scientific writing, like the sentence on the back of a tube of toothpaste, is measured not by the pleasure that it gives, but by *how well it does a job*. Learning to write simple, functional sentences will help you learn about your subject and, in the process, make you a better student and a more productive professional. If you refuse to write simple, effective sentences, much of your education will be wasted because what you write will not be understood, and therefore will be ignored.

The myth that effective writing occurs only when a writer is inspired is based on the notion that writing is something that happens to you when you're inspired. This is not true. Writing is something that you do—a process that helps you communicate and learn. Waiting to be inspired ensures failure because it postpones learning and justifies a writer's worst enemy: procrastination. Writing to learn requires only a few guidelines and a logical approach that are closer to common sense than inspiration.

Just write the way you talk—after all, both are means of communication. The fallacy of this approach is obvious when you read transcripts of conversations. Such transcripts seem disjointed and confusing without the gestures, pauses, facial expressions, and emphases that accompany talking. Read the transcript of a conversation and you'll see that you talk more loosely and informally than you would want to have recorded in writing. This is because talking is all that effective writing isn't—natural, informal, habitual, and relaxed. You've also practiced talking more than you have writing. Indeed, you speak more words in a month than you'll write during your lifetime. For example, a person speaks an average of about 12,000 words per day; this translates into an 84,000-word novel every week, and about 4 million words per year. Even Shakespeare, one of history's most prolific writers, did not write this much during his lifetime.

Good writers can quickly and effortlessly produce a perfect paper on their first try. This misconception typifies amateurs and poor writers. People who believe that good writing comes naturally on their first try are either incredibly talented or have low standards. For example, although the folksy stories of Andy Rooney and the homespun musings of Erma Bombeck seem so effortless and easy to understand, they require hours of hard work to write. Few people can let words flow without having them sound spilled. As a philosopher said, "What is written without effort is read without pleasure."

Good writers know that good writing is hard work and that their first drafts are usually unsatisfactory. To write well, you must learn to revise well. As a famous writer once said, "There is no good writing, only good rewriting." *Writing to Learn Science* will teach you how to revise your work—not as a means of merely correcting mistakes, but as a way of rethinking and learning more about your ideas, your subject, and your research.

Simple writing is not scholarly. Students who use this excuse confuse simple writing with simplistic writing, and think that simple writing is not impressive or important. These students try to dignify their writing by choosing words such as "endeavor" instead of "try" because they think that endeavor is so much more, well, *sophisticated*. Such words are not sophisticated at all. They merely clutter your writing and bog readers down.

Contrary to what some students think, verbose and fancy writing usually hides shallow thought. More importantly, the research involving Smith's and Brown's essays (see pages 6 and 7) shows that a simple, straightforward style

Science Headlines

Newspaper headlines are notorious for their double meanings. Articles about science are no exception, as shown by headlines such as "Milk Drinkers Are Turning To Powder." Headlines such as "Genetic Engineering Splits Scientists" might make people think twice about a career in molecular biology, while other headlines describe astounding evolutionary tales. For example, an article entitled "Lung Cancer In Women Mushrooms" apparently describes some renegade female fungi that have evolved lungs and have started puffing on cigarettes. However, few will ever beat this gem that appeared in the travel section of a local newspaper: "Canada's Virgin Forests: Where The Hand Of Man Has Never Set Foot."

impresses readers more than the long-winded "scholarly" writing style typical of many professionals. Heed the advice of the survey: You'll impress other people most not if you write scholarly, but rather if you write so they can quickly and clearly understand what you're saying.

Good writing is an art.   Effective writing is neither a science nor an art. Rather, it is a craft that can be learned by understanding and applying a few principles—not rules—of writing.

Effective writing is as simple as following a set of rules.   Rules such as those for punctuation, spelling, and grammar are important to scientists and other professionals because they help us communicate our ideas. However, principles of grammar are not rigid rules, but rather are tools that help us detect and eliminate flaws in our writing that inhibit learning and communication. Although poorly written sentences and paragraphs are easy to identify because they sound awkward or confusing, correcting the problem is another matter entirely. This is best done by understanding the principles of writing, not by rattling off a list of grammatical rules. In many instances, a good ear will serve you better than a rigid rule.

Some rules are important. For example, a disregard for spelling and punctuation will make you seem uncaring, illiterate, or lazy. Other rules are useless to good writers (see "What do I need to know about grammar?", p. 160). Writers who rely only on rules may write "correctly," but not effectively. Rather than use writing as a tool to learn and communicate, these writers dwell on simplistic rules such as always using passive voice, always stringing modifiers together, never splitting an infinitive, always repeating ideas, never using personal pronouns, and never beginning a sentence with a forbidden word. Such rules are panaceas that, without an understanding of writing, seldom improve one's writing or learning.

As shown by the essays of Smith and Brown (see pp. 6–7), merely avoiding mistakes by following rigid rules does not ensure good writing. Moreover,

---

The Dangers of Writing

Most people do not consider writing to be a dangerous activity. After all, what's the worst thing that can happen? A paper cut?

The U.S. Consumer Product Safety Commission estimates that there were 32,197 pen and pencil-related injuries in the United States in 1992. To put this in perspective, the sport of fencing produced too few accidents to report. Apparently, the pen is not only mightier than the sword; it is also riskier.

Medical journals have described a variety of writing-related ailments. For example, since the 1830s physicians have written about *scrivener's palsy*, better known as writer's cramp. As one study mentioned, writer's cramp occurs when "the pen does not move quite as intended." If you've ever sat down and *intended* to write like Einstein, but got different results, then you may be afflicted with scrivener's palsy.

Not all desk-based problems can be blamed on pens and pencils. For example, an article in *Indiana Medicine* titled "Paper Clip Stab Wounds: Four Case Reports" described some unusual wounds. Similarly, the highly regarded *Annals of Internal Medicine* reported a condition dubbed *photocopier's phalanx*, which results from "the repetitive motion of forcibly applying the original document to the glass plate and, to insure adequate contact, exerting pressure with an adducted, extended thumb and adducted index finger." Finally, there's an article in *Lancet* that describes *photocopier's papillitis*, an inflamation of the tongue caused by repeated finger-wetting to count pages coated with toner.

---

pedantically correct writing is often dull and colorless. Rules for writing are valuable only if they improve communication. Use them as guidelines, not as substitutes for thought.

To write well you must have a large vocabulary.   Words alone do not create meaning and communication. If writing were that simple, then only a thesaurus or dictionary would be essential for becoming a good writer and the size of one's vocabulary would determine one's writing skills. You don't need a huge vocabulary to write well. Indeed, a vocabulary of only about 1,000 words covers about 85 percent of a writer's needs to describe ordinary subjects. Thus, buying books, tapes, and videos with titles like "Nine Billion Gigantic Words That You Should Know" and "Super Duper Word Power" that promise to make you a verbal Charlie Atlas won't improve your writing nearly as much as working to ensure that you write so that you cannot be misunderstood. Indeed, the words that you choose *not* to use are often more important than those that you use.

Reject the notion that long and unusual words are interesting and elegant. Elegance may or may not be a by-product, but it can never be an intention of your writing. As you write, try to learn and communicate rather than be clever or elegant. Take Albert Einstein's advice: "When you're out to describe the truth, leave elegance to the tailor."

Writing is less important than it used to be. Top scientists spend more than one-fourth of their working-day writing. Most of these scientists claim that their ability to write effectively helped advance their careers. Similarly, *Fortune* magazine recently asked successful executives what students should learn to help them prepare for careers. Their answer: *Learn to write well.* Editors and professors who must sift through heaps of poorly written papers would yell, "Amen."

Language determines how effectively you communicate. Although professionals who write poorly may fool readers for a paragraph or two, they eventually lose their readers.

Scientists' Excuses for Writing Poorly

Although scientists spend much of their time writing and may be highly educated, many express themselves wordily and obscurely, and therefore ineffectually. Listen to the laments of these two editors of scientific journals:

> [Published papers] not only want of rhetorical finish (a slight blemish, comparatively speaking), but of all regard to correctness or appropriateness of language . . . An inexcusable defect in composition, for the reason that it is so easily avoided, is the commonplace, inaccurate, in short, illiterate, language suffered to find its way into our journals.

> The majority of articles submitted for publication could be cut down one-half, and not a thought be eliminated in so doing. The repetition of well-known facts, padding with abstracts from text-books, and words, words, words, too often constitute the papers that appear as "original" in medical journals. And if the editor presumes to use the blue pencil in the least, the majority of authors consider it an insult.

Many scientists are curiously eager to perpetuate the poor writing described by these editors. Rather than take responsibility for becoming a better writer and, therefore, a better scientist, these scientists rely on several excuses for their poor writing:

"Writing belongs in English classes, not science classes." This misconception underlies the notions that writing and science are unrelated, that words and science don't mix, and that writing is something done well only by people with names like Hemingway and Twain. All of these notions are incorrect.

Words and science are intimately linked, as are writing and thinking. Writing about science forces you to *think* about science—that's why writing is a such a powerful tool for learning.

"Great scientists write poorly."   This excuse is ridiculous. Albert Einstein, Marie Curie, Richard Feynman, and James Watson are just of a few of the famous scientists who have used personal, direct, and forceful writing to describe their brilliant ideas. In *Writing to Learn Science*, you'll read the work of these and other scientists. Unfortunately, many other scientists' descriptions of research are as impersonal, dull, and lifeless as a phone-book or weather report.

Clear writing reveals how a clear mind attacks and solves a problem. Since one's writing reflects one's thinking, poor writers cannot pretend to be clear thinkers, much less effective scientists. In science, your ideas are only as good as your ability to express them to others.

"I just wasn't born a good writer. I'm a good scientist, not a good writer." This "blame nature" excuse typifies people who think that good science and effective writing are mutually exclusive. These individuals, who view scientific writing as a mystic art rather than a learnable craft, often become frustrated when their poor writing discredits their work. Contrary to what these people think, effective scientific writing is a craft that can be learned, not an effortless outpouring by a genius.

Poor writing reflects poor thinking—it is shifty and unpredictable, much like Brownian movement. Thus, most scientists who hide behind the "I'm a good scientist, not a good writer" excuse are usually only half right—right about not being a good writer, but usually wrong about being a good scientist.

"Simple writing is not scientific."   These scientists, who confuse simple writing with simplistic writing, do not want to communicate clearly. They resist clear, simple writing because they know that clear, simple writing reveals problems and faulty logic, just as it can also reveal genius. Wrap their work in flowery, vague language, they reason, and others will ignore deficiencies in their work. Papers written by these scientists are usually ignored because other scientists cannot understand their message.

"The science is all that's important—if the science is good, then blemishes in the writing are irrelevant."   Scientists with this "grammar don't matter" attitude either have an inflated self-opinion or understand little about science. Writing is an integral part of thinking, learning, and science. Data cannot stand alone—rather, they must be interpreted and incorporated into an argument supporting or refuting a hypothesis. "Blemishes" such as ambiguity, poor grammar, unnecessary jargon, and inaccuracy typify shoddy thinking and carelessness. These blemishes produce weak arguments that discredit your work.

Scientists who think that what matters is what you say, not how you say it, are half right. Trite, shallow, or illogical ideas do not magically become brilliant ideas when they're written well: An embroidered sow's ear remains a sow's ear, and style—the order and movement you put into your ideas—is no substitute for substance. However, neither can substance substitute for a lack of style. Ideas buried in confusing writing aren't worth much because they're nothing more than confusion. Knowing how to use writing to learn will help identify your weak ideas, thereby allowing you to strengthen them as you learn more about your ideas and your subject.

Scientific writers establish their credibility in two ways: (1) with facts, evidence, and logic, and (2) with their writing style. Although poor writing won't camouflage a lack of substance, good ideas are useless unless they're explained well. The hallmark of important research papers is that they're readily understandable by other scientists. Scientists naive enough to believe that writing is irrelevant to science are often equally naive about research.

"Don't worry about the details." The effectiveness of writing depends on details which, if ignored, can damage your credibility. For example, one student submitted a paper describing a non-existent and apparently satanic chemical having the formula $C_6H_{12}O_{666}$. Similarly, I recently received an advertisement for a seminar entitled "Effective Scientific Writting." As best I could tell, the only consolation was that the course was non-credit.

Using words carelessly can confuse readers, as can mistakes created by "details" such as misplaced decimals, incorrect measurements, or grammatical mistakes such as stacked modifiers. Scientists who ignore these details are often ineffective because their work is either incomprehensible or unrepeatable, and is therefore ignored. Sloppy writing can have even more serious implications. For example, the "Code for Communicators" of the Society for Technical Communication states that precision, use of simple and direct language, responsibility for how readers understand your message, and respect for readers' need for information (rather than your need for self-expression) show a technical writer's commitment to professional excellence *and ethical behavior.* "Details" that produce vague, misleading papers damage your credibility and hinder science.

"They'll know what I mean." This "they can read my mind" excuse for poor writing tries to shift the responsibility for communicating to the reader by forcing the reader to do your thinking for you. This approach fails because *it is always the writer's job to communicate with the reader.* Papers must communicate, not merely "make information available to others," and scientific papers must be read *and understood.* Merely converting your data and observations to sentences is irrelevant—it matters only that readers accurately perceive what you had in mind. This requires no fancy words, grammatical tricks, or gimmicks. Rather than impress readers, these ornaments usually only confuse readers. Effective scientific writing has no place for fanciful leaps or implied truths. Facts and deductions are the rule.

Regardless of its difficulty or sophistication, a complex subject can be made as accessible as a simple subject by an effective writer. Never mind what you think your writing is *supposed* to mean—all that matters is what it *says.* Readers will know what you say, but only if you write effectively will they know what you *mean.* How you write about science *is* the science. Similarly, your ideas alone, no matter how brilliant, will be irrelevant if you can't describe them effectively. As Vladimir Nabokov wrote, "Style and structure are the essence of a book; great ideas are hogwash."

"Scientists have always written in scholarly prose." This "we've always done it this way" excuse is used by many poor writers. These writers, if

describing atomic structure, would write something like this:

> It is hypothesized by this author that, in essence, the initial material exis-
> tence of physically manifest substances was relatively dense, massive in
> weight, durable, and particulate in form; the extreme manifestation of
> hardness being categorically displayed by resistance to diminution in size
> due to to abrasive processes and by counter fragmentation systems.

Although such writing is common in many scientific journals, scientists have
not always written like this. The sentence you just read is, in fact, a "modern"
version of a sentence written by renowned physicist Isaac Newton almost 300
years ago:

> It seems probable to me, that God in the beginning formed matter in solid,
> massy, hard, impenetrable, moveable particles; . . . even so very hard, as
> never to wear or break in pieces.[3]

The scientific literature contains many other examples of poetic and aesthetic
responses to scientific experiments. For example, Roman philosopher Lucretius
wrote *De Rerum Natura* (On the Nature of Things), which summarized scientific thought in his time, as a poem. In 1665 Robert Hooke wrote, "These pleasing and lovely colours have I also sometimes with pleasure observ'd even in
Muscles and Tendons." Similarly, you quickly sense Malpighi's excitement
when he quotes Homer in a letter written after he saw capillaries for the first
time: "I see with my own eyes a certain great thing."[4]

"Scholarly" writing often impedes communication because it unnecessarily
forces readers to wade through clutter such as *it is hypothesized by this author*.
Here's a scientific example of "scholarly" writing:

> The following describes the activities of five immature mammals of the
> family of nonruminant artiodactyl ungulates. All five of these may be
> described as being of less than average magnitude; however, no informa-
> tion is given as to the relative size of one with respect to another. Available
> evidence indicates that the first of the group proceeded in the direction of
> an area previously established for the purpose of commerce. Data on the
> second of the group clearly show that, at least during the time period
> under consideration, it remained within the confines of its own place of
> residence. Reports received on the activities of the third member of the
> group seem to show conclusively that it possessed an unknown quantity
> of the flesh of a bovine animal, prepared for consumption by exposure to

[3]Newton, Sir Isaac. Optiks London (1704), quoted by Pyke, Magnus (1960) in "This Scientific Babel." *The
Listener*, **LXIV** (1641)(8th September), 380.

[4]Marcello Malpighi was an Italian physician who founded microscopic anatomy, demonstrating that blood
flows to tissues through tiny capillaries too small to be seen with the naked eye. William Harvey, a famous English
physician who discovered the circulation of blood, had inferred that these capillaries must exist, but he had never
seen them. To learn more about this subject, see Foster, M. 1901. *Lectures on the History of Physiology during the
Sixteenth, Seventeenth and Eighteenth Centuries.* Cambridge: Cambridge University Press.

dry heat. The only information available on the fourth member of the group is of a wholly negative nature, namely, that its possessions did not include any of the type previously described as having been in the possession of its predecessor in this discussion. As to the fifth and last member of the group, fairly conclusive evidence points to its having made, during the entire course of a movement in the direction of its place of residence, a noise described as "wee, wee, wee."[5]

As you discovered at the end of this paragraph, this is a "scholarly" version of "This Little Pig Went to Market." Although you enjoyed hearing it as a child while your smiling mother or father pulled at your toes, you probably found this scholarly version of the story annoying—annoying because its most noticeable trait is its pompous writing, not its message.[6]

Good writing is invisible because it draws attention not to itself, but to the writer's ideas. Readers don't automatically think "This sure is good grammar" when they read a well-written paper. Rather, they quickly and clearly understand the author's message. Similarly, the mistakes that typify poor writing do not announce themselves with titles such as "dangling modifier" or "passive voice." Readers know the writing is poor because it doesn't communicate clearly or quickly.

Reject excuses for poor writing. Do not let them betray your ideas and ability to think.

Why is Writing Important?

Speech is the representation of the mind, and writing is the representation of speech.
— Aristotle

If any man wishes to write a clear style, let him first be clear in his thoughts.
— Johann W. von Goethe

A good narrative style does not attract undue attention to itself. Its job is to

[5]From Mancuso, Joseph C. 1990. *Mastering Technical Writing*. Reading, MA: Addison-Wesley Publishing Co.

[6]Tired, prefabricated phrases such as *at this point in time*, *pursuant to our agreement*, and *your assistance in this matter will be greatly appreciated*, common in "professional" writing, are similar to the dreaded cousin who tells the same stories over and over again regardless of the occasion. The many people who consider such phrases critical to sounding "professional" should reconsider. Indeed, the *Harvard Business Review* studied 800 letters written by the most prominent business executives in the United States, and concluded that all of these letters were concise and clear—almost the exact opposite of what many call "professional."

keep the reader's mind on the story, on what's happening, the event, and not the writer.
— Leon Surmelian

Communication is the root of scientific progress.
— Joshua Lederberg

This is a book about how to use writing to learn science. Chances are you've probably not considered that science and writing could be related. After all, science involves experiments and verifiable "facts," whereas writing involves rhetoric and grammatical rules. Moreover, the beauty of science is in the science, not the language used to describe it. How, then, is writing important in science?

Effective writing is extremely rewarding, but it can also be hard work. Words seldom flow and writing rarely comes easily because effective writing requires sustained thought. Clear thinking, like clear writing, is a conscious act that scientists and other professionals must force upon themselves. It's also critical to your career, both as a student and as a professional:

- Learning to write effectively will improve your grades. Much research shows that students who write well achieve better grades than do students who write poorly.

- Effective writing saves time and money. Boring writing is tossed aside unread, a waste of the investment made to produce it. Similarly, costly research must be repeated when scientists do not clearly describe their methods and results.

- Regardless of whether you become a scientist, businessperson, lawyer, or other professional, your ability to write will help determine the impact of your ideas and work. Many brilliant ideas have died in obscurity, their discoverers unknown because they wrote leaden prose that no one could or wanted to understand. Since these ideas were not communicated well, they were overlooked and ignored. Conversely, the recognition that accompanies well-written papers can lead to promotions, awards, and salary raises. Heed the advice of Sir William Osler, "In science the credit goes to the man who convinces the world, not to the man to whom the idea first comes." To do yourself justice, you must think and write well.

- Writing and thinking exert a back-pressure on each other that greatly enhances learning. Writing about a subject forces you to be precise. In doing so, it uncovers faulty logic that betrays your search for truth and knowledge.

- Careful writing helps you develop ideas, and therefore is an important tool for helping you learn (such as by "taking notes"), observe (such as with a

Keeping a Personal Journal

I had, also, during many years followed a golden rule, namely, that whenever a published record, new observation or thought came across me, which was opposed to my general results, to make a memorandum of it without fail and at once; for I had found by experience that such facts and thoughts were far more apt to escape from the memory than favourable ones.— Charles Darwin, *Life and Letters*

In my journal, anyone can make a fool of himself.— Rudolph Virchow

Many serious thinkers, including many scientists, keep an informal personal journal to capture their thoughts. Indeed, the journals of naturalists and scientists such as Thoreau, Darwin, Freud, and Einstein mapped the development of their revolutionary ideas. This highlights a journal's primary value: *discovery*. Such informal writing helps you sort your ideas and theories and, in the process, discover new ideas. As we'll discuss in Chapter 2, such free-writing often takes you where you never intended to go—not by reflecting your ideas, but by *leading your thoughts*. There are few better ways to learn about yourself and your thoughts than by keeping a journal, because writing about what you don't know is an excellent way to know. In pausing to write, you'll listen to yourself think.

Journals are a cross between a student notebook and writer's diary. Journals describe topics such as "What happened in class today?" and "Why do I believe this?" They may include observations about a field site, speculation about data, and syntheses of your ideas. Personal journals are also excellent places to voice questions and express your doubts. This is important because writing questions often helps sharpen the questions so that they can be answered.

Keep a personal journal. Since no one will see the journal but you, feel free to write about whatever you please. After a while, you'll treasure the time you set aside for writing in your journal. It will become a relaxing time of discovery.

sketch), plan, show relationships, review, organize, communicate, remember, clarify, and discover what you know, what you don't know, and what you need to know. We think and capture our thoughts with words. Therefore, consistently poor writing betrays one's inability or unwillingness to think clearly. It is more than stylistic inelegance—it's an outward and visible sign of inner confusion.

- Writing and thinking are related tasks—that's why nothing helps you learn about and clarify a topic better than writing. To show this, consider the following paragraph questioning if a law requiring a five-cent deposit on bottles and cans reduces litter:

The law wants people to return the bottles for five cents, instead of littering them. But I don't think five cents is enough nowadays to get people to bother. But wait, it isn't just five cents at a blow, because people can accumulate cases of bottles or bags of cans in their basements and take them back all at once, so probably they would do that. Still, those probably aren't the bottles and cans that get littered anyway; it's the people out on picnics or kids hanging around the streets and parts that litter bottles and cans, and they sure wouldn't bother to return them for a nickel. But someone else might—boy scout and girl scout troops and other community organizations very likely would collect the bottles and cans as a combined community service and fund-raising venture; I know they do that sort of thing. So litter would be reduced.[8]

Besides illustrating everyday reasoning, this is an excellent example of the close relationship of thinking and writing. Each sentence challenges the preceding one and, in the process, advances the writer's argument.

- Writing is not a dull, irrelevant task to be endured. Rather, it will enhance your career by helping you learn and communicate. Many of your best ideas will be lost to you if you do not write well.

- In many professions, writing is the primary means by which you will become known (or remain unknown). This is especially true in science. Data cannot speak for themselves; rather, they must be interpreted and used to make an argument supporting or rejecting a hypothesis. Effective writing helps you advance your argument, and is therefore the primary way by which you'll establish your reputation. You must use written evidence to convince your peers that your hypotheses are significant and that your ideas are good. Indeed, neither dexterity in lab, innate knowledge, familiarity with the literature, brilliant ideas, creativity, nor personal charm will overcome the frustration and poor reputation established by poor writing. Poor writing breeds distrust and prompts readers to wonder if language is the writer's only area of incompetence. Don't let poor writing hide the importance of your ideas.

- Science is a collective enterprise based on the free exchange of ideas. Consequently, writing is critical to science because it extends the history of science and is the vehicle for sharing scientific knowledge. Scientists must publish their work in a permanent and retrievable form so that others can examine, test, and build on your data and ideas. Publications are the cornerstone of science because they confirm or add to our knowledge.

- The goal of scientific research is to discover and communicate new infor-

[8]Perkins, David N., R. Allen, and J. Hafner. 1983 "Difficulties in everyday reasoning." In *Thinking: The Expanding Frontier*, ed. W. Maxwell. Hillsdale, NJ: Lawrence Erlbaum Associates.

mation. Communicating new information requires clear, concise writing to convince colleagues of the importance and validity of your discovery. Consequently, your ability to write will largely determine your influence as a scientist. Indeed, merely "doing research," having a breakthrough idea, designing a brilliant experiment, or making an important discovery will do you little good if you cannot communicate its brilliance or importance to others. Scientists must do *and write about* science. Research is incomplete until it is published and made available for other scientists to test and build on. Writing that is verbose, pretentious, and dull communicates poorly, delays or prevents publication, and therefore delays scientific progress.

- Scientists at most "major" universities must publish their research in peer-reviewed journals to keep their jobs. This "publish or perish" requirement means that scientists unable to write effectively seldom keep their jobs (see "Publish or Perish," p. 264). Similarly, most scientists must obtain money from agencies such as the National Science Foundation to support their research. Getting this money requires that scientists write research proposals. The inability of a scientist to write convincingly almost always means that he or she will get no money for their work, and therefore that they cannot do the research. This results in no data for publication which, in turn, often causes the scientist to be fired.

  Scientists who write well often progress faster in their careers than do those who write poorly. Thus, knowing how to write effectively will help you get and keep a job, and is therefore an investment in your future. Also, since the printed word is indelible, you must write well to protect your future reputation.

- Finally, scientists must communicate their discoveries to the public. Einstein said it best:

  > It is of great importance that the general public be given the opportunity to experience—consciously and intelligently—the efforts and results of scientific research. It is not sufficient that each result be taken up, elaborated, and applied by a few specialists in the field. Restricting the body of knowledge to a small group deadens the philosophical spirit of a people and leads to spiritual poverty.

Your ability to write effectively will, in part, determine your success as a professional. Despite the importance of writing, most students are trained only to obtain information rather than to communicate it. Consequently, most students studying science learn "scientific writing" by imitating their mentors or colleagues. Indeed, scientific writing is a learned "skill," as described by physician and writer Michael Crichton in the *New England Journal of Medicine*:[9]

[9]Michael Crichton is a physician and professional writer. His first novel, the suspenseful *The Andromeda Strain* centered on complex scientific issues and became a best-seller, prompting him to devote increasingly more time to writing. His other books include *Sphere* and *Jurassic Park.*

It now appears that obligatory obfuscation is a firm tradition within the medical profession. . . [Medical writing] is a highly skilled, calculated attempt to confuse the reader. . . . A doctor feels he might get passed over for an assistant professorship because he wrote his papers too clearly—because he made his ideas seem too simple.

Consequently, students often perpetuate the pompous, boring, and ineffective writing that has characterized many scientific journals for decades. In many instances, it sounds as if scientists are trying to keep secrets rather than communicate.

In the following chapters, you'll learn about the process of writing and the tools that good writers use to discover and communicate their ideas. However, knowing about these tools is about as useful as listing just the ingredients of a great recipe and then expecting someone to make the dish. Just as knowing the ingredients and how to use them distinguishes reading cookbooks from cooking, how you use what you learn in this book will distinguish reading this book about writing from writing. Practice what you learn.

Isaac Asimov

The Next Frontiers for Science

The late Isaac Asimov (1920–1992) was a biochemist and world authority on science and medicine. He was world-famous for his science-fiction books and articles. He was incredibly prolific—he wrote more than 450 books.

His essay entitled *"The Next Frontiers For Science"* was published in 1991 in several popular magazines, including *Money* and *Fortune.* In the essay, Asimov makes a controversial claim—that the only scientific research that deserves major funding is research to save our planet.

We are living in an age where many scientists are thinking big. There is the supercollider, a new unprecedentedly powerful particle accelerator which may give us an answer at last to the final details of the structure of the universe, its beginning, and its end.

There is the genome project, which will attempt to pinpoint every last gene in the human cells and learn just exactly how the chemistry of human life (and of inborn disease) is organized.

There is the space station, which will attempt, at last, to allow us to organize the exploitation of near space by human beings.

All these things, and others of the sort, are highly dramatic and will be, at least potentially, highly useful. All are also highly expensive, something of great importance in a shrinking economy. Worse yet, all are, at the moment, highly irrelevant.

What is relevant is that we are destroying our planet.

A steadily increasing population is placing ever-higher demands on Earth's resources, is forcing the conversion of more and more land to human needs and is wiping out the wilderness and the ecological balance of the planet, something on which we all depend.

A steadily rising use of fossil fuels for energy (at a rate that is increasing more rapidly than the population) is choking Earth's atmosphere with gases that are slowly poisoning it. In addition, it allows the atmosphere to conserve heat more efficiently, so that the planet is experiencing a greenhouse effect that may have catastrophic impact.

A steadily increasing production of chemical substances that are highly toxic, or that cannot be recycled by biological processes, or both, is poisoning the soil and water of the Earth, is destroying the ozone layer and is converting much of the planetary surface into a garbage heap. Since there can be nothing on Earth, simply nothing, that is more important than saving the planet, our coming priorities must be to reverse these destructive tendencies. And America must lead. It is, for instance, foolish, absolutely foolish, to put the study of reproductive physiology to work on test-tube babies and on producing babies after menopause; we must not increase the number of babies, but decrease them.

We must find alternate sources of energy, long-lasting and non-polluting. We must continue the search for nuclear fusion, in the hope that it will be a far richer and safer source than nuclear fission. We must develop wind-power, wave-power, the use of Earth's internal heat and, most of all, the direct use of solar power. All these things are highly practical, but cost more money than oil and coal, so the challenge is to make them cheaper. (The fact that we can destroy our planet so cheaply, by the way, does not mean we ought to destroy it.)

We must find ways of detoxifying toxic products produced by industrial plants. We must find substitutes for packaging, substitutes that are recyclable. We must find substitutes for chemicals that destroy the ozone layer.

We must find methods of saving our forests, of saving threatened species, of maintaining a healthy ecological balance on Earth.

If there is any spare effort left over from these absolute necessities of scientific advance, we can put them into other projects—otherwise not.

I regret this, for I am emotionally on the side of the big projects, all of them, but necessity is a hard task-master, and necessity is now in the saddle and holds the whip.

Understanding What You've Read

Do you agree with Asimov? Write a one-page essay that describes your opinion. Substantiate your opinion with evidence, and develop your ideas with logic.

How does Asimov present his argument? What are his biases and assumptions?

Albert Einstein

E=mc²

Albert Einstein (1879–1955) did more to change our view of nature than did any other scientist. He made several major discoveries, including Brownian movement (based on his study of the motion of atoms) and the photoelectric effect, for which he received the Nobel prize in physics in 1921. However, Einstein is most famous for his Theory of Relativity, which revolutionized modern thought about the nature of time and space and provided a theoretical basis for the exploitation for atomic energy.

The theory of relativity is actually two theories: the Special Theory (1905) and the General Theory (1916). The Special Theory, published when Einstein was only 26 years old, is "special" because it refers only to bodies moving at a constant speed—that is, bodies in uniform (unaccelerated) motion with respect to each other. The Special Theory shows that, contrary to what Newton had assumed, there can be no single absolute system of measuring time and space for all moving bodies in the universe; that is, any description of a given body's dynamics must be relative to an external frame of reference. Thus, the fact that your classroom seems motionless is relative to the earth on which the classroom sits. But relative to the earth's axis, the classroom's speed is about 1,000 miles per hour. Relative to the sun, the classroom's speed is about 10,000 miles per hour.

The Special Theory of Relativity predicts that light (or radio) waves moving from one planet to another will be bent off their path by a massive body such as the sun, slowing the light down for a fraction of a second. The Special Theory was confirmed on May 29, 1919 by English astronomer Arthur Eddington, who photographed a solar

Albert Einstein, *The General Theory of Relativity*, "E=mc²" is reprinted with permission of Philosophical Library, Inc., New York.

eclipse on the island of Principe off West Africa. Einstein predicted that a ray of light just grazing the surface of the sun would be bent by 1.745 arcs of light—twice the gravitational deflection estimated by Newtonian theory. Einstein was right.

Einstein's General Theory addresses gravitational attractions in the universe. Unlike other scientists, who viewed gravity as "force at a distance," Einstein viewed gravity not as a force but as a field that distorts the space that it inhabits. This distortion produces an acceleration in passing bodies, conveniently labeled "gravitational attraction."

Einstein's famous formula $E=mc^2$ was announced in a paper entitled "The Electrodynamics of Moving Bodies." According to that formula, mass is equivalent to energy when multiplied by the square of the speed of light.

In order to understand the law of the equivalence of mass and energy, we must go back to two conservation or "balance" principles which, independent of each other, held a high place in pre-relativity physics. These were the principle of the conservation of energy and the principle of the conservation of mass. The first of these, advanced by Leibnitz as long ago as the seventeenth century, was developed in the nineteenth century essentially as a corollary of a principle of mechanics.

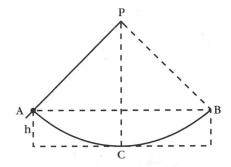

Consider, for example, a pendulum whose mass swings back and forth between the Points A and B. At these points the mass m is higher by the amount h than it is at C, the lowest point of the path (see drawing). At C, on the other hand, the lifting height has disappeared and instead of it the mass has a velocity v. It is as though the lifting height could be converted entirely into velocity, and vice versa. The exact relation would be expressed as $mgh=\frac{m}{2}v^2$, with g representing the acceleration of gravity. What is interesting here is that this relation is independent of both the length of the pendulum and the form of the path through which the mass moves.

The significance is that something remains constant throughout the process, and that something is energy. At A and at B it is an energy of position, or "potential" energy; at C it is an energy of motion, or "kinetic" energy. If this concept is correct, then the sum $mgh + m\frac{v^2}{s}$ must have the same value for any position of the pendulum, if h is understood to represent the height above C, and v the velocity at that point in the pendulum's path. And such is found to be actually the case. The generalization of this principle gives us the law of the conversation of mechanical energy. But what happens when friction stops the pendulum?

The answer to that was found in the study of heat phenomena. This study, based on the assumption that heat is an indestructible substance which flows from a warmer to a colder object, seemed to give us a principle of the "conservation of heat." On the other hand, from time immemorial it has been known that heat could be produced by friction, as in the fire-making drills of the Indians. The physicists were for long unable to account for this kind of heat "production." Their difficulties were overcome only when it was successfully established that, for any given amount of heat produced by friction, an exactly proportional amount of energy had to be expended. Thus did we arrive at a principle of the "equivalence of work and heat." With our pendulum, for example, mechanical energy is gradually converted by friction into heat.

In such fashion the principles of the conversation of mechanical and thermal energies were merged into one. The physicists were thereupon persuaded that the conservation principle could be further extended to take in chemical and electromagnetic processes—in short, could be applied to all fields. It appeared that in our physical system there was a sum total of energies that remained constant through all changes that might occur.

Now for the principle of the conservation of mass. Mass is defined by the resistance that a body opposes to its acceleration (inert mass). It is also measured by the weight of the body (heavy mass). That these two radically different definitions lead to the same value for the mass of a body is, in itself, an astonishing fact. According to the principle—namely, that masses remain unchanged under any physical or chemical changes—the mass appeared to be the essential (because unvarying) quality of matter. Heating, melting, vaporization, or combining into chemical compounds would not change the total mass.

Physicists accepted this principle up to a few decades ago. But it proved inadequate in the face of the special theory of relativity. It was therefore merged with the energy principle—just as, about 60 years before, the principle of the conservation of mechanical energy had been combined with the principle of the conservation of heat. We might say that the principle of the conservation of energy, having previously swallowed up that of the conservation of heat, now proceeded to swallow that of the conservation of mass—and holds the field alone.

It is customary to express the equivalence of mass and energy (though somewhat inexactly) by the formula $E=mc^2$ in which c represents the velocity of light, about 186,000 miles per second. E is the energy that is contained in a stationary body; m is its mass. The energy that belongs to the mass m is equal to this mass, multiplied by the square of the enormous speed of light—which is to say, a vast amount of energy for every unit of mass.

But if every gram of material contains this tremendous energy, why did it go so long unnoticed? The answer is simple enough: so long as none of the energy is given off externally, it cannot be observed. It is as though a man who is fabulously rich should never spend or give away a cent; no one could tell how rich he was.

Now we can reverse the relation and say that an increase of E in the amount of energy must be accompanied by an increase of $\frac{m}{c^2}$ in the mass. I can easily supply energy to the mass—for instance, if I heat it by 10 degrees. So why not measure the mass increase, or weight increase, connected with this change? The trouble here is that in the mass increase the enormous factor c^2 occurs in the denominator of the fraction. In such a case the increase is too small to be measured directly; even with the most sensitive balance.

For a mass increase to be measurable, the change of energy per mass unit must be enormously large. We know of only one sphere in which such amounts of energy per mass unit are released: namely, radioactive disintegration. Schematically, the process goes like this: An atom of the mass M splits into two atoms of the M' and M'', which separate with tremendous kinetic energy. If we imagine these two masses as brought to rest—that is, if we take this energy of motion from them—then, considered together, they are essentially poorer in energy than was the original atom. According to the equivalence principle, the mass sum $M' + M''$ of the disintegration products must also be somewhat smaller than the original mass M of the disintegrating atom—in contradiction to the old principle of the conservation of mass. The relative difference of the two is on the order of 1/10 of one percent.

Now, we cannot actually weigh the atoms individually. However, there are indirect methods for measuring their weights exactly. We can likewise determine the kinetic energies that are transferred to be disintegration products M' and M''. Thus it has become possible to test and confirm the equivalence formula. Also, the law permits us to calculate in advance, from precisely determined atom weights, just how much energy will be released with any atom disintegration we have in mind. The law says nothing, of course, as to whether—or how—the disintegration reaction can be brought about.

What takes place can be illustrated with the help of our rich man. The atom M is a rich miser who, during his life, gives away no money (energy).

But in his will he bequeaths his fortune to his sons M′ and M″, on condition that they give to the community a small amount, less than one thousandth of the whole estate (energy or mass). The sons together have somewhat less than the father had (*the mass sum M′ + M″ is somewhat smaller than the mass M of the radioactive atom*). But the part given to the community, though relatively small, is still so enormously large (*considered as kinetic energy*) that it brings with it a great threat of evil. Averting that threat has become the most urgent problem of our time.

Understanding What You've Read

What analogies did Einstein use to make his points in this essay?

What are some implications of $E = mc^2$?

Why is it possible to equate gravity with inertia?

What usually tells us we are experiencing acceleration rather than gravity?

In 1934, Nazi officials confiscated his property and revoked Einstein's German citizenship because he was Jewish. What other important scientists' lives have been affected so directly by politics?

Einstein wrote that "Physical objects are not in space, but spatially extended. In this way the concept of 'empty space' loses its meaning." Explain.

Like Isaac Newton and many other great scientists, Einstein became involved in politics late in his life. For example, although an ardent pacifist, Einstein urged President Franklin Roosevelt to study the possible use of atomic energy in bombs. Does the pursuit of political goals hurt or enhance science? Explain your answer.

Converting a tiny amount of mass can lead to the release of a huge amount of energy. Explain.

On January 30, 1929, Einstein announced in six pages of equations that he had formed a "unified gravitational field theory." Go to the library and read about this theory. Was Einstein correct?

Einstein said that "God does not play dice." Renowned physicist Stephen William Hawking (b. 1942) disagrees, saying that "God not only plays dice, he also sometimes throws the dice where they cannot be seen." What are Einstein and Hawking talking about? Who is correct? Explain your answer.

Exercises

1. Science has changed our lives, but the public knows little about scientists. Indeed, many publications stereotype scientists as soulless, eccentric hermits who study things that they can't explain in anything except gibberish. Similarly, movies such as *Back to the Future III, Gremlins 2: The New Batch, Die Hard 2, Total Recall,* and *E.T.* depict scientists as heartless and morally blind—determined to dissect or otherwise abuse lovable creatures such as a beautiful mermaid, E.T., a baby dinosaur, a revived neolithic man, and intelligent chimpanzees.

   What are your conceptions of scientists? Nobel laureate James Watson said, "One could not be a successful scientist without realizing that, in contrast to the popular conception supported by newspapers and mothers of scientists, a goodly number of scientists are not only narrow-minded and dull, but also just stupid." Do you agree with this statement? Why or why not?

2. Several years ago the U.S. Supreme Court upheld scientists' right to patent new forms of life. The vote on this case was 5–4, indicating that the justices' opinions were divided. How would you have ruled on the case? Write an essay explaining your position. Your argument is directed at people who hold the opposing view, and your purpose is to persuade them that your opinion is correct.

3. Read a science-related article in a recent issue of *Time* magazine. Summarize the article for a friend who has not read it. Do not evaluate the article or give your opinion about it; simply inform your friend of its contents.

4. List what you feel are the 20 most important ideas in science. Include a brief discussion of each idea and of your ranking. Compare your ranking with that of Hazen and Trefil in *Science Matters: Achieving Science Literacy.*

   The universe is regular and predictable.

   One set of laws describes all motion.

   Energy is conserved.

   Energy always goes from more useful to less useful forms.

   Electricity and magnetism are two aspects of the same force.

   Everything is made of atoms.

   Everything—particles, energy, the rate of electron spin—comes in discrete units, and you can't measure anything without changing it.

   Atoms are bound together by electron "glue."

   The way a material behaves depends on how its atoms are arranged.

   Nuclear energy comes from the conversion of mass.

   Everything is really made of quarks and leptons.

   Stars live and die like everything else.

The universe was born at a specific time in the past, and it has been expanding ever since.

Every observer sees the same laws of nature.

The surface of the earth is constantly changing, and no feature on the earth is permanent.

Everything on the earth operates in cycles.

All living things are made from cells, the chemical factories of life.

All life is based on the same genetic code.

All forms of life evolved by natural selection.

All life is connected.

Do you agree with their ranking? Why or why not?

5. Find and read a short article in a journal or magazine that you find easy to read and understand. Study the writing. Why is the article easy to read? How many of the sentences are simple? Are there any grammatical mistakes? Spelling mistakes? How do the paragraphs lead you to the conclusion?

6. How does science affect your life? Write a short essay about your favorite topic in science. What questions did you think of as you wrote?

7. On the desks of Anthony Trollope and John Updike (a two-time Pulitzer Prize winner) are framed copies of a quotation of Pliny the Elder: "Nulla dies sine linea"— "Not a day without a line." Why do you think these writers have this quotation on their desks? How does this relate to the myth of writers needing "inspiration" to write effectively?

8. Do you agree with the idea that clear writing leads to clear thinking? Cite examples of how putting your thoughts on paper have helped clarify your ideas.

9. Support or refute this statement:

    Having knowledge is not what makes a scientist. For example, if all of the scientific information known today were to be accepted with no new work, there would be no science. Statements such as $E=mc^2$ may look scientific, but they are not science if the only reason you believe them is because you read them or because your professor said them.

10. Many people base their lives on blind faith in things such as astrology, fortune tellers, and religious dogma. Science has also had its share of blind faith—for example, the Scholastics of the late Middle Ages accepted Aristotle's conclusions merely because Aristotle had said them, not because the conclusions were supported by data. Does blind faith have a place in science? If so, when and where? If not, why?

11. Write a letter to the editor of your local newspaper expressing your view or concern about a scientific topic.

12. Many famous scientists have communicated their ideas by ascribing human chara-teristics to their subjects. For example, Charles Darwin wrote that a tree "tried to raise its head . . . and . . . failed." This type of writing is called anthropomorphic writing and is rejected by many scientists. Why do you think that great scientists such as Darwin and others have used it to describe their work? Do you think that anthropomorphic writing is effective? Why or why not?

13. Why do you write? What kinds of writing do you do? What is the purpose of each kind of writing? What kind do you like best? What kinds do you dislike? Why?

14. What kinds of publications do you read? What has this reading taught you in the last week? What effects does it have on you and what you do?

15. Write a one-page essay about what you think of science and your science class.

16. What is effective writing? Do you write well? What evidence do you have to support your answer?

17. Discuss, support, or refute the ideas included in these quotations:

> It is important that students bring a certain ragamuffin, barefoot irreverence to their studies; they are not here to worship what is known, but to question it. — Jacob Bronowski

> The main object of all science is the freedom and happiness of man.— Thomas Jefferson

> It is not what the man of science believes that distinguishes him, but how and why he believes it.— Bertrand Russell

> Circumstantial evidence can be overwhelming. We have never seen an atom, but we nevertheless know that it must exist. — Isaac Asimov

> It is the easiest thing in the world to deny a fact. People do it all the time. Yet it remains a fact just the same.— Isaac Asimov

18. Write a short essay describing how the *The Far Side* cartoon on page 32 relates to science.

THE FAR SIDE By GARY LARSON

"Hold it right there, young lady! Before you go out, you take off some of that makeup and wash off that gallon of pheromones!"

Figure 1–1
The Far Side. © 1988. UNIVERSAL PRESS SYNDICATE. Reprinted with permission.

CHAPTER TWO

Preparing a First Draft:

Writing to Learn

Writing and thinking and learning (are) the same process.
— William Zinsser

The improvement of understanding is for two ends: first our own increase of knowledge; secondly to enable us to deliver that knowledge to others.
— John Locke

Have something to say, and say it as clearly as you can. That is the only secret of style.
— Matthew Arnold

He has half the deed done who has made a beginning.
— Horace

Well begun is half done.
— Anonymous

Writing comes more easily if you have something to say.
— Sholem Asch

Thinking is the activity I love best, and writing to me is simply thinking through my fingers.
— Isaac Asimov

If a man will begin with certainties, he shall end in doubts, but if he will be content to begin with doubts, he shall end in certainties.
— Francis Bacon

[You must] convince yourself that you are working in clay and not marble, on paper and not eternal bronze; let that first sentence be as stupid as it wishes. No one will rush out and print it as it stands. Just put it down; then another.
— Jacques Barzun

Brainstorming

> The finest thought runs the risk of being irretrievably forgotten if it is not written down.
> — Arthur Schopenhauer

> Thoughts, like fleas, jump from man to man. But they don't bite everybody.
> — Stanislaw Lec

Writing is making meaning, a process that begins with brainstorming. Brainstorming is the most critical part of writing because it helps you generate ideas and gather information as you simultaneously define your subject and audience. Later, you can use brainstorming to identify gaps in your knowledge or to show where more research needs to be done.

Brainstorming is critical because it's the first time you transfer thoughts from your head onto paper. Brainstorming will help you find and identify your thoughts —not just the ones you already have, but also some new ones. Indeed, you'll hardly know many of your thoughts until you see them in writing. This is why Sartre quit writing when he lost his sight: He could no longer see his words, the symbols of his thoughts.

Brainstorming has led many scientists to brilliant ideas and major discoveries, and it can do the same for you. Here are some good ways to brainstorm:

Do **not** restrict yourself by trying to think of everything before you start. If you do, you will lose some of your thoughts because you'll be worrying about where they fit in. Similarly, do **not** write an outline and stick to it. All of these things come later. If you do them first, you'll only force yourself into thinking that you must completely understand everything before writing about it. This ignores the ability of writing to help you understand your work. Do not try to predict what you will write.

Try **not** to formulate everything in advance. Rather, generate your ideas in any way that you please, such as with notes, lists, terms, sketches, questions, or reminders. Write in whole sentences or phrases with diagrams; this will help you discover, identify, and preserve your thoughts. This is a favorite technique of many scientists, including Charles Darwin:

I have as much difficulty as ever in expressing myself clearly and concisely; and this difficulty has caused me a very great loss of time; but it has had the compensating advantage of forcing me to think long and intently about every sentence, and thus I have been led to see errors in reasoning and in my own observations or those of others.

Writing with a Word Processor

> There are about as many word processing programs as there are recordings of Beethoven's Fifth — maybe more. They are all different, but some are less different than others. — Peter McWilliams, *The Word Processing Book*

Word processors have revolutionized writing by improving the speed and efficiency of writing. Although computers can't think for you, they can make it much easier and faster to type, correct, and manipulate your writing. Computers also let you bypass note cards, make quick and sweeping changes of your manuscript, change and re–examine the paper immediately, and quickly recall all parts of the paper. Also, a computer allows you to access electronic data-bases such as the *World Wide Web*, *Science Citation Index*, *Scientific Abstracts*, and *Current Contents*. Writing at a computer saves time and paper—you need no scissors, paste, ink, erasers, "white out," or trash cans when you write at a computer. Last-minute rewrites, once considered an impossible task, become routine.

If used properly, computers can do more than merely move text or make quick changes in your paper. They can also keep track of format and organize a paper, store text, justify margins, hyphenate words, build a glossary, count words and sentences, insert footnotes and headers, produce different fonts, and highlight text. Many word-processing programs include spelling checkers, while other programs include dictionaries, graphics, grammar checkers, and a thesaurus. Style-checking programs such as *Correct Grammar* can check for misused words (for example, *affect* and

---

There seems to be a sort of fatality in my mind leading me to put at first my statement or proposition in a wrong or awkward form. Formerly I used to think about my sentences before writing them down; but for several years I have found that it saves time to scribble in a vile hand whole pages as quickly as I possibly can, contracting half the words; and then correct deliberately. Sentences thus scribbled down are often better ones than I could have written deliberately.

Do **not** lose ideas while you struggle for the perfect phrase. If you get stumped for a specific word or phrase, mark the spot with a distinctive character (such as #). Later, use the search command of your word-processing program to locate that character; by then, you may have discovered the right word (see "Writing with a Word Processor"). If you have several candidates for the right word, list them all and decide about them later. The choice will be easier after you've let some time pass.

Let your writing generate ideas; think with your pencil or keyboard. Go for quantity, not quality. At this stage of writing, as in Las Vegas, never stop when you're on a roll.

effect, see Appendix 2), clichés, passive voice, and proper spacing of text. Although you don't need all of these programs to write well, you should at least know about them because they can save you much time and work and, in the process, free you to concentrate on your ideas rather than the drudgery of tasks such as formatting and hyphenation.

Although computers can greatly improve the efficiency of writing, they can also cause problems. For example, word-processing programs allow uncontrolled cutting and pasting of identical paragraphs from one paper to another. This clutters the literature with redundancies and blocks the learning and improved communication that accompany revising and rethinking what you've written.

Computers and word-processing programs are not a substitute for knowing how to write effectively. They cannot gather information or write your paper. Similarly, style-checking programs cannot define your audience, check your logic, or meaningfully evaluate your writing. That's your job. You'll still need to read your paper carefully to ensure that all words are spelled correctly and used properly. If you don't, you're likely to include misspelled words that will make you appear either uneducated or careless. Also, remember that computer programs do not necessarily save what you write; be sure to keep a back-up copy of all your work.

Learn how to use a word-processing program. However, don't write only at a computer because, in doing so, you'll see only a small, screen-sized part of your paper at a time. Print and read a "hard copy" of your paper so you can get a broad perspective by seeing several pages at once.

Refer frequently to information you gathered from the library, your lab notebook, reprints of relevant papers, and any other materials that might help you generate ideas.

Gather your ideas and understand your data, references, and subject. If you fail at this, you'll have nothing to say and you're doomed. Indeed, even educated, intelligent people write poorly when they have nothing to say or when they've not thought much about what they're writing.

Be sure it's the page, not your mind, that's blank. Chances are if you can't find the right words, you haven't yet found the right idea.

Think about the topic uncritically. Write every idea you have in whatever order and form it occurs to you. Write freely and do not worry about the mechanics or finer points of grammar. Be concerned only with getting ideas from your head onto paper. Don't think that you have to get all of your ideas onto paper at one sitting. If you repeat this brainstorming several times, you'll continue to discover new ideas and hidden angles about your subject.

Do not judge your ideas. The goal here is quantity: You'll produce at least 10 times more ideas if you write freely than if you stop and evaluate each idea. Too much self-criticism will stifle your productivity and creativity.

Write answers to the questions, "What does the reader want to know about this?" and "What is important about this?" These questions should be answered in your paper because they are questions that the readers will ask as they read your paper.

Remember that there is no one "right way" to write or start writing. Accept whatever comes to mind. Although an English teacher may not love what you write, don't worry about it; you'll revise it later. For now, just get the flow started.

Some of your ideas will come out of your head ready for battle, like Athena from the head of Zeus. However, more often your ideas will resemble weak and tottering infants needing nourishment before they can stand alone. This doesn't mean that your idea is poor. Rather, it merely indicates that you've not yet discovered all of the parts of the idea and their relationships to other ideas. Don't be too critical of yourself. Our brains can handle only a few ideas at one time. It's analogous to learning to juggle: You can learn to juggle three or four balls, but only after much practice. Even then, you'll still be limited as to how many balls you can juggle at one time.

Many people try to generate ideas by making outlines. Such outlines restrict and frustrate us because they do not function like our brain does. An outline merely lists words, while our brain works with concepts, patterns, and relationships. Unlike an outline, our brain lets ideas go in their own direction, the way branches go from the trunk of a tree. As each branch grows, we see a pattern that maps the landscape of our thoughts. Outlines are useful only *after* you begin to see this landscape. Imposing an outline on your thoughts before you see relationships and patterns restricts your thoughts, thereby explaining why most people dislike outlines and why they usually fail to help us see the interrelationships of our ideas. But if outlines don't work, how can you discover the parts of a subject and their relationships?

One way to see the relationships of your ideas is to use *clustering*, a simple and natural way to write about and discover your thoughts. Unlike an outline, clustering imposes no structure on your thoughts. Here's how to do it:

Write a phrase summarizing the subject of your writing. Draw a circle around that phrase. I'll use clustering as an example:

Figure 2–1
The first step in clustering is to write a phrase summarizing your subject. Draw a circle around that phrase.

Then write your ideas that relate to the subject. Draw a circle around each idea and connect it with a line to related ideas. This expands our example to this:

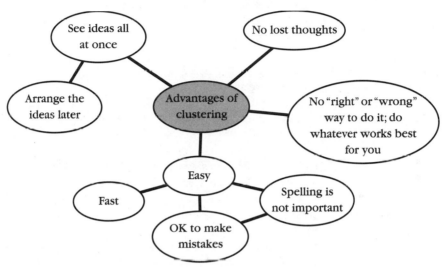

Figure 2–2
Jot down your ideas that relate to the subject. Connect each idea to related ideas.

Heed John Steinbeck's advice: "Write freely and as rapidly as possible and throw the whole thing on paper. Never correct or rewrite until the whole thing is down. Rewrite in process is usually found to be an excuse for not going on."

Brainstorm until you feel that you're repeating yourself and have thus reached a point of diminishing returns. Confront conflicts and write about everything that comes to your mind, including the obvious. Ignore the messiness and don't worry about grammar, punctuation, or style—your list of ideas is for your eyes only. Be embarrassed only when mistakes appear in the final draft.

Organizing Your Ideas

> There is in writing the constant joy of
> sudden discovery, of happy accident.
> — H.L. Mencken

> Order and simplification are the first
> steps toward the mastery of a subject.
> — Thomas Mann

How do I know what I think, until I see
what I say?
— E.M. Forster

Writing is all a matter of choice and
arrangement.
— Graham Greene

The key to effective writing is organization. To write effectively, you must understand the decisions that organize your ideas and words.

Readers are seldom forced to read your papers. It's up to you to organize and write the paper so that it interests readers. They need only enough information to know what your paper is about and why they should read it. Readers will probably study your article for its content, not for the joy of reading your prose.

The ability to efficiently organize one's ideas, and therefore one's writing, is what separates poor writers from effective writers. It is also the most hidden part of the writing process. Most students are not taught organizational skills, and therefore their decisions about how to organize material often seem random or odd. This, combined with a poor understanding of other aspects of writing, causes students to waste much time struggling with organization. Experienced writers know that writing and organization are often difficult, frustrating, unpredictable, and exciting. However, they are not alarmed by their task because they understand the decision-making process used to write effectively.

The first and most basic decisions you make when writing are identifying your subject and audience. If you misjudge your subject or audience, all that follows will be extremely difficult.

Define the Subject and Purpose of Your Writing

The discipline of the writer is to learn to
be still and listen to what his subject has
to tell him.
— Rachel Carson

Most of the knowledge and much of the
genius of the research worker lie behind
his selection of what is worth observing.
It is a crucial choice, often determining
the brilliant discoverer from the . . . plodder.
— Alan Gregg

Defining the subject and purpose of your writing is the most important decision you'll make in writing, for it will influence how you write your paper. If you start in the wrong direction, you'll waste much time and effort trying to change your course. Effective writers define their subject clearly so that their readers have no trouble understanding it. In poor writing, the subject is absent, obscured, or so broad that it is unmanageable.

Brainstorming about your experiment or research will help you define the subject of many papers that you write. Other times you'll be assigned the subject for your writing assignments. Teachers usually don't make such assignments just to keep students busy. Rather, teachers make these assignments because they know that students learn when they write. When choosing your subject, remember that it is much easier to write about something that interests you than about something that bores you. You cannot write effectively about a subject if you do not understand it.

Know Your Audience

> The first and most important step in writing is to know who you're writing to—
> who they are and what they want.
> — David G. Lyon

Who you write for won't change your subject, but it should affect your approach to writing. The audiences for your writing will vary greatly. For example, the audience for a term paper or laboratory report will be a professor or teaching assistant, while that for a grant proposal or publication will be scientists who know something about your research.

To identify your audience, ask yourself these questions:

Who is my audience? Whom am I writing for?

What do they already know?

What do they need to know?

When answering these questions, remember that much of what you write is for your audience's benefit and that a major goal of your writing is to enhance or expand the readers' comprehension. Therefore, start at a point from which your intended readers can follow and don't tell them what they already know. This is especially important for scientists, because science is complex and often contains many unusual terms. These complexities increase the concentration of the writing, and therefore are amplified by poor writing. Just as archers cannot expect a target to swell to meet their poorly aimed arrows, writers cannot expect their readers to understand poorly written papers. Tiny errors made when aiming an arrow produce shots that widely miss the mark, as does sloppy writing that fails to communicate. Poor aim and complacency always cause you to miss the target, whether it be a bull's-eye or a reader's comprehension.

The responsibility for presenting understandable information rests with the writer. Therefore, aim carefully with your writing, and write to suit your audience, not yourself. Do this by using language that your readers will understand, by writing at the proper level of difficulty, and by giving readers the information they need. Start with what readers are familiar with and progress in a way that follows the readers' interests and knowledge. Knowing your audience will help you communicate better because you won't waste readers' time by telling them what they already know.

The initial attention of your audience will be automatic if you are writing about a subject that interests them or if they must read what you have written. For example, most people who study the genetics of *Drosophila* will at least start to read a paper entitled "The Genetics of *Drosophila*." To *keep* readers' interest you must present information concisely and logically, lead readers with evidence and explanations, and not force them to reread a sentence or consult a dictionary. Such writing is both a courtesy and an insurance policy, even with a captive audience, because it ensures that your paper will be read. If you write well about an important subject, you'll have no trouble keeping readers' interest.

Gathering Information

> Trouble in writing clearly . . . reflects
> troubled thinking, usually an incomplete
> grasp of the facts or their meaning.
>   — Barbara Tuchman

> Knowledge is of two kinds. We know a
> subject ourselves, or we know where
> we can find information upon it.
>   — Samuel Johnson

> Men give me credit for genius; but all
> the genius I have lies in this: when I
> have a subject on hand, I study it pro-
> foundly.
>   — Alexander Hamilton

> If I have seen further . . . it is by stand-
> ing upon the shoulders of giants.
>   — Isaac Newton[1]

---

[1]Scientists often repeat this famous metaphor that Newton used to express his intellectual debt to Descartes and others who preceded him. As if to prove its own point, the metaphor was not Newton's invention. Robert Burton, an English clergyman who died before Newton was born, had written, "A dwarf standing on the shoulders of giants may see farther than a giant himself." Similarly, Burton acknowledged that 1600 years earlier the poet Lucan wrote, "Pigmies placed on the shoulders of giants see more than the giants themselves."

> The next best thing to knowing some-
> thing is knowing where to find it.
>   — Anonymous

This stage of writing involves gathering data, notes, books, reprints, and other materials relevant to your subject. Gathering this information is critical to writing because, as in cooking, you can't produce a good product unless you start with the right ingredients. Fortunately, obtaining information for a science paper is more a physical than a mental chore. All that's required is tracking down details that relate to your subject. This may involve finding references in the library, doing or repeating an experiment, documenting a method, or doing a statistical test. If you're writing about your research, much of your information will come from your lab or field notebook (see "Laboratory and Field Notebooks" p. 271).

One of the most valuable places to gather information, and one of a scientist's most valuable laboratories, is a library. Others' ideas often clarify your thoughts, as they did for Charles Darwin:

> In October 1838 . . . I happened to read for amusement Malthus on popula-
> tion, and being well prepared to appreciate the struggle for existence which
> everywhere goes on from long continued observation of the habits of animals
> and plants, it at once struck me that under these circumstances favorable vari-
> ations would tend to be preserved and unfavorable ones to be destroyed. The
> result of this would be the formation of a new species.

Unfortunately, the sheer magnitude of information in most libraries often makes it difficult to find what you want. Don't be intimidated by a library—you need to know only a few things to start using it efficiently.

Card Catalog   The card catalog is a catalog of all of the library's books. These books are listed by title, author, and subject, so you need not know all of the details about a particular book to locate it. Most libraries have a computerized card catalog in which each book has a unique call number.

Many public libraries use the Dewey Decimal System, in which books are catalogued from 000 to 999 according to subject. For example, the 500 code denotes science, and 600 denotes technology and applied sciences.

Most large libraries, including those at most universities, use the Library of Congress System to catalogue books. This system has more subdivisions and does not use numbers such as 570.3988, as does the Dewey Decimal System. The classifications of primary interest to scientists include Q (science), R (medicine), and S (agriculture). Each of these categories is further divided into specific categories. For example, books about molecular biology are grouped in QH 506.

Circulation Desk   The circulation desk is where you check books into and out of the library. This is usually the "main desk" of a library. If you want to check out a book, take the book to the circulation desk.

Reserve Desk   The reserve desk is where you can check out materials placed "on reserve" at the library. These materials typically include books, papers, and old tests. You'll probably not be able to remove these materials from the library for more than a few hours. If your teacher tells you that he or she has placed some-thing on reserve at the library, get information about it at the reserve desk.

Reference Desk The reference desk is where to get help in locating reference materials. Here are some of the publications used often by scientists:

Scientific Abstracts publishes two issues per month and two volumes per year, and adds about 300,000 citations per year from about 9,000 journals. Papers are numbered. To find the abstract, look up the paper's reference number in the accompanying volume. Papers are referenced according to author, subject, taxonomic group, and genus or genus–species names.

Suppose you're considering writing a paper about how roots of plants perceive and respond to gravity (root gravitropism). You would start by look-ing up roots in the subject index. Figure 2–3 shows how that section appears in the November 1, 1990 (Vol. 90, No. 9) issue of *Scientific Abstracts*:

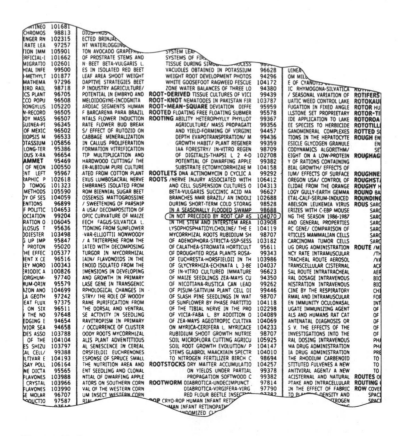

Figure 2–3
Part of a page from the November 1, 1990 (Vol. 90, No. 9) issue of
Biological Abstracts. Note that the entry identified by *in not preceded by root cap as* is reference number 104070. Reprinted with permission.

Now suppose you're interested in reading the paper identified by the phrase *"in not preceded by root cap as,"* which is listed as reference number 104070. To find the abstract and bibliographic information for that paper, look up that reference in *Scientific Abstracts.* (Figure 2–4)

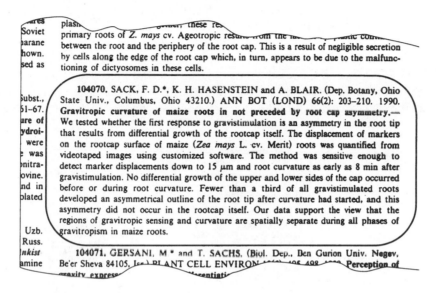

Figure 2–4
The abstract of reference number 104070. This article appears on pages 203–210 of volume 66 of *Annals of Botany,* 1990. Used with permission.

If you want to read the entire paper, look it up in *Annals of Botany,* Vol. 66, No. 2, pp. 203–210, 1990. Citations in abstracting services such as *Chemical Abstracts* and *Scientific Abstracts* usually lag several months behind the publications.

General Science Index includes nontechnical articles that will give you an overview of the subject and prepare you for technical articles.

Science Digest summarizes articles selected from about 200 journals.

Index Medicus is a monthly survey of more than 2,800 journals received by the National Library of Medicine. *Index Medicus* is a major source of literature in medicine and biomedicine.

Current Contents is a weekly index that reproduces the table of contents of recent journals in many fields of science. *Current Contents* provides a quick way to scan the titles of articles published in many journals, and also lists the addresses of authors so you can write for a reprint of an article.

Environment Abstracts gives an excellent list of sources classified by topics.

Science Citation Index is published every two months. It lets a user look for a specific author and learn who has cited that author in their publications. Full bibliographic information about each citation is included in the *Source Index*, an accompanying volume; use the *Source Index* alone to find new papers. *Science Citation Index* is a way to identify authors studying a particular subject and is useful for tracing all references on a topic or by an author. Citations lag less than a year behind publications.

Unlike many other sources, *Science Citation Index* allows a user to go forward from a particular citation. Suppose you're interested in knowing what other scientists are studying topics related to Fred Sack's work on root gravitropism. To learn this, look up SACK FD in the *Citation Index*. Figure 2–5 shows what you'll find in the November–December (No. 6C) 1990 issue.

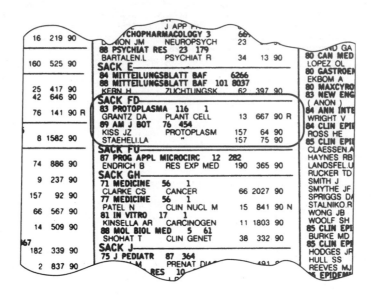

Figure 2–5
Part of a page from November–December (No. 6C) 1990 issue of *Citation Index.* Used with permission.

Sack's paper in *Protoplasma* was cited by D.A. Grantz in *Plant Cell,* Vol. 13, p. 667, 1990 (the "R" indicates that Grantz's paper is a review article). Similarly, Sack's paper in *American Journal of Botany* was cited by two authors in *Protoplasma*: J.Z. Kiss and L.A. Staeheli. These papers are probably relevant to your interest in root gravitropism. You can get full descriptions of these articles by looking them up in the journals in which they're published or in the *Source Index* section of the *Science Citation Index.*

Other useful references for writing to learn science include the *World Wide Web, Chemical Abstracts, Environment Index, Pollution Abstracts,* and *Zoological Record.* If you have trouble finding reference information, ask a reference librarian for help.

Tools of the Trade

Writing, like any skill, requires tools. In this book, you'll learn about many of these tools, including how to discover, organize, and refine your ideas. However, you'll also need these:

- *Access to a computer with a word-processing package*: Word processing enables you to easily record, revise, and format your writing. Moreover, most word-processing programs have spelling-checkers that identify words that could be misspelled. Don't worry if you cannot afford to buy a computer; virtually all universities have computer labs at which you can use a computer. To learn more about word processing, see "Writing with a Word Processor," pp. 36–37.

- *Dictionary*: Writing without a dictionary is like playing basketball without a backboard: You are at a tremendous disadvantage. Moreover, without a dictionary you'll probably fail the test of this ancient Chinese proverb: "The beginning of wisdom is to call things by their right names." To help ensure that you best express your ideas, buy and use a good dictionary such as *The American Heritage Dictionary* or *Merriam Webster's Collegiate Dictionary*. You'll find these dictionaries at any good bookstore. Remember, however, that a dictionary is descriptive, not prescriptive. Thus, although a dictionary will include the word *ain't*, you shouldn't use *ain't* in your papers.

- *Thesaurus*: A thesaurus is organized like a dictionary; you merely look up a word and the thesaurus supplies you with a list of related words and phrases. The most commonly used thesauri include *The New Roget's Thesaurus* and *Webster's Collegiate Thesaurus*. When you want to express your ideas completely, you'll find a thesaurus priceless, essential, invaluable, . . .

- *Reference books*: Your textbook is an excellent reference book. However, collect more. A personal library is a great asset to any scientist.

A writer's tools include anything that makes writing easier. Agatha Christie needed only "a steady table and a typewriter," whereas John Steinbeck said that he needed "pencils that are round. A hexagonal pencil cuts my fingers after a long day." The tools listed above are used by most scientists; consider using them yourself. If they help you, great. If not, try a round pencil.

Periodicals Section   Included among periodicals are recent issues of journals and magazines. Although old copies of journals and magazines are usually bound and filed on shelves of the library, the most recent issues of journals are usually kept in a separate room for reading. Visit this room regularly so you can stay up-to-date on what's new in science.

On-line Services   Many libraries have computerized "on-line" services that can greatly speed your search for information. For example, *Bibliographic Retrieval Services* lists research papers relevant to a particular subject, whereas services such as *InfoTrac* list recent publications from popular journals and magazines. Depending on the service you use, you can request a reference list, abstracts of articles you want, or complete copies of the articles. These searches can be done in a few minutes, and all you must provide is a list of key words. While on-line services may not identify every paper that you're interested in, they will give you a good start. A computer search usually won't identify relevant literature published before the 1960s.

To efficiently gather information in the library, you must record information systematically and completely. Here are some tips to help you efficiently gather information in a library:

Record all of the bibliographic information about the reference so that you won't have to return to the library to check on a page number or volume of a journal. If you're using an on-line service, get a print-out of the reference. Figure 2–6 shows an example of a complete listing of a reference (for more information about citing references, see Chapter 7):

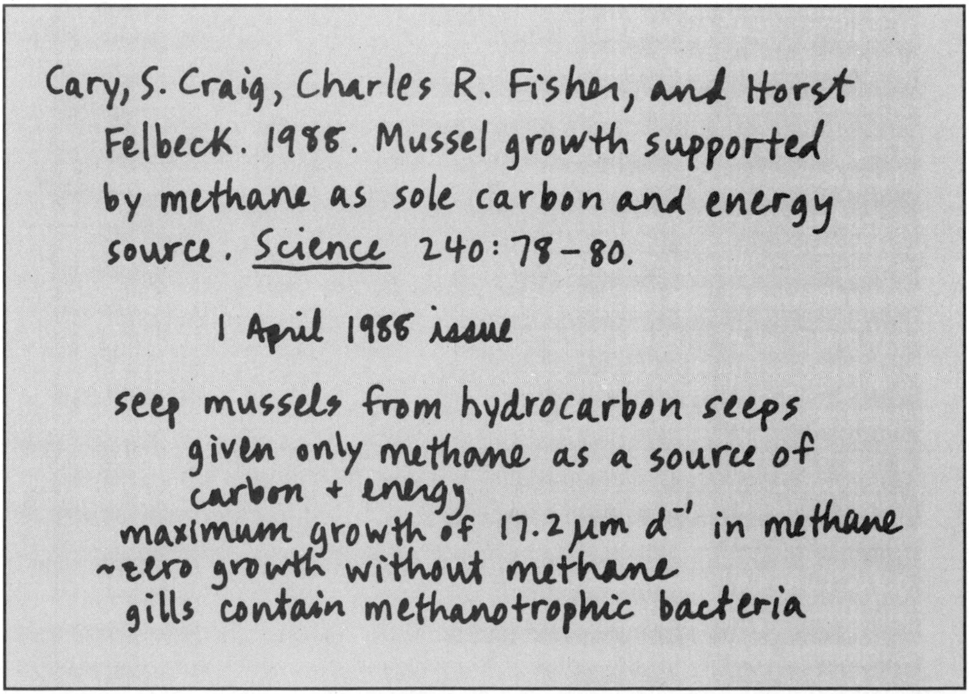

Cary, S. Craig, Charles R. Fisher, and Horst Felbeck. 1988. Mussel growth supported by methane as sole carbon and energy source. Science 240: 78–80.

1 April 1988 issue

seep mussels from hydrocarbon seeps
     given only methane as a source of
     carbon + energy
maximum growth of 17.2 μm d⁻¹ in methane
~zero growth without methane
     gills contain methanotrophic bacteria

Figure 2–6
Reference card containing complete bibliographic information and notes about an article.

Use index cards, and write one idea per card. You can sort these cards later. Photocopy papers that look especially useful. If you prefer to take notes on full-size pages, begin each topic on a separate page. On each card write the usefulness of and important information contained in the reference. These cards, containing bibliographic details and information about the article or book, comprise an annotated bibliography. Such a bibliography will help you record and coordinate resources and information.

If a book or paper is unavailable in your library, ask the reference librarian if he or she can help you get the book or paper via an interlibrary loan. Note on the index card where you obtained the book or paper.

Unless you want to quote a paper or book, take notes in your own words. This will help you get away from being awed by other people's words and will help you become more comfortable with your own ideas and writing. This, along with taking notes in incomplete sentences, will also help you avoid accidental plagiarism. Moreover, it will force you to understand the material. If you don't understand the material, you will be unable to take good notes in your own words.

Make sure your notes are complete so that you know the context of the author's ideas. Don't use too much shorthand—you might forget what your abbreviations mean.

Nothing discredits a writer more than a mistake. When readers discover such mistakes, they conclude that all of the writer's work may be riddled with mistakes. Fortunately, accurate writing requires more effort than brilliance. Check the accuracy of all of your facts.

If you gather information from scientific journals, you'll realize that a relatively small number of journals contain much of the important information about your topic. For example, papers of the 300 most-cited scientists for 1961-76 appeared in only 86 journals. Five of these journals accounted for 10% of the citations, and 10 of the journals accounted for half of the citations. Similarly, in a bibliography on schistosomiasis (snail fever, caused by an infection by blood flukes) covering 110 years and containing 10,286 references, almost half of the references were from only 50 of the 1,738 journals that were cited.

Although gathering information is relatively easy, don't underestimate its importance. Nothing wastes more of a writer's time than the paralysis that accompanies not having details to document what you want to say.

Starting to Write

> "Writing is easy. All you have to do is
> stare at a blank sheet of paper until
> drops of blood form on your forehead."
> — Gene Fowler

No matter how much you've gathered information, organized your thoughts, or researched a topic, there comes a time when you must face a blank page or computer screen. Many people have much trouble starting a writing assignment. As soon as they sit down to write, they instead find themselves in front of a vending machine, walking down the hall, or deciding that they can't start writing until they rearrange their papers, wash their car, or sharpen all of their pencils. This problem is referred to as "writer's block" and usually results from writers either not knowing what they're writing about or from trying to get everything onto paper at one sitting. No one, not even the world's best writers, can write things in final form at their first attempt. Rather, they write effectively by using a process involving organizing your ideas, creating a tentative outline, producing a first draft, and, finally, revising their work.

Organizing Your Ideas

> The card-player begins by arranging his
> hand for maximum sense. Scientists do
> the same thing with the facts they gather.
> — Isaac Asimov

> Science consists of grouping facts so that
> general laws or conclusions may be
> drawn from them.
> — Charles Darwin

Science is organized knowledge, and writing is the means to that organization. Organizing your ideas is an exercise in labelling your ideas and reviewing the products of your brainstorming. Begin by reading your notes. Many of your ideas will be irrelevant to your subject or audience and should be eliminated, no matter how interesting. Also eliminate ideas that are relevant but insignificant. If you include those ideas, you'll imply that all of your ideas have the same importance. Doing this will lose your readers before you can spring the punch line.

Try to discover a natural, common-sense sequence among the topics. You may find your ideas easier to shuffle and rearrange if they are listed on index cards, one idea or example per card. You can then divide and subdivide your subject by moving the cards around until they fall into a logical pattern. Although this kind of "writing architecture" seems simple, it is critical because it commits you to a specific course of action. You can later overcome mistakes in execution, but you can't afford mistakes in architecture. If you fail to label and organize your thoughts, you won't be satisfied with your paper no matter how well you build it. Similarly, if you haven't taken the time to organize your thoughts, you'll probably never find time to revise your paper once it's written. But that's exactly what you'll have to do if you string ideas together randomly or without forethought.

Effectively labelling your thoughts also simplifies the next step of the writing process: creating a tentative outline of your paper.

Creating a Tentative Outline

> The first rule of style is to have something to say. The second rule of style is to control yourself when, by chance, you have two things to say; say first one, then the other, not both at the same time.
> — George Polya

Labelling your thoughts produces parts of an outline that almost puts itself together. This is much easier than trying to create an outline as the first step in writing, which often limits your perspective. When making your outline, don't be handcuffed by a formal structure of roman numerals or numbered subtopics. Focus instead on the *function* of the outline, which is to help you organize and manipulate subjects so you can identify your ideas and communicate clearly. Remember: *the only real requirement of any outline is that it be useful.*

Preparing an outline is a powerful way to decide what elements are essential to your idea and in fixing their logical relationships. It is also an efficient way to determine the key points of your writing, for it is a skeleton of the first draft of your paper; it will tell you what you've done and suggest the outcome of your paper. It is simple to create: It's merely the logical listing of the ideas that you've already labelled. This forms a tentative outline for your paper.

Your outline should highlight the logic of your ideas. Ask yourself these questions to determine if your outline is in good shape:

Are the topics balanced? Have you devoted too much or too little space to any topics?

Have you abruptly jumped from one subject to another?

Are there gaps in the logic? If the gaps are Grand Canyons, rethink your plan.

Did you gather enough information?

Have you discovered all of your ideas?

Continue to brainstorm as you examine your outline. Is your outline complete? If not, add other topics. Similarly, delete all irrelevant topics. Your outline will tell you if you're on the right track. If you're on the wrong track, the outline will tell you that also. Your outline will also help you divide your paper into sections and subsections that alert the reader to what's coming.

Take the time now to rethink, clarify, and reorganize your ideas; it will pay dividends later.

Going from Design to Construction: Creating a First Draft

> Science consists in grouping facts so that
> general laws or conclusions may be
> drawn from them.
> — Charles Darwin

> If a man can group his ideas, he is a
> good writer.
> — Robert Lewis Stevenson

> People who cannot put strings of sen-
> tences in good order cannot think.
> — Richard Mitchell

You're now ready to piece together a first draft of your paper. To do this, all you need to do is expand and build on each section of the outline, one section at a time. Again, don't worry about grammar, details, or style. Parts will be vague, while others will seem disjointed and thin. Accept that the first draft of your paper will need more work—just get your ideas into a logically arranged text. At this stage of writing, as in previous ones, writing continues to be a means of thinking, thereby helping you create more knowledge than you've been given.

If you have trouble writing the first draft of your paper, don't accept the self-delusion of excuses such as, "I know this but just can't find the right words." Instead, carefully review what you've done. Have you gathered enough information? Have you organized the information logically? Don't accept the excuse of "writer's block"—if you've gathered the information or completed your experiments, your ideas are already there. Any trouble you have in writing about your ideas probably results from "science block" rather than from "writer's block."

Reviewing Your First Draft

Read your first draft and give yourself a mental pat-on-the-back. Although what you've written looks rough and still needs work, you've accomplished the most critical part of the work. You're halfway home.

As you reread your paper you'll probably be tempted to add to your notes, thus beginning the next stage of the writing process: revising. However, for now just concentrate on the big picture and make sure that you've gotten the main idea, content, and organization. Then ask yourself these questions:

Do you really have something to say? This question, which will focus your thinking, is also the question that your readers will ask as they read your paper. No one will care how well or badly you express yourself if you have nothing to say. Moreover, it is impossible to write well about anything if you have nothing to say.

Clear writing requires clear thinking. However, if your ideas are a bit muddled when you start to write, the act of writing, itself, may help clarify your thinking and thereby clarify your writing. Most of your ideas will not become clear until you've written them down.

Have you told readers everything they need to know? It's easy for writers, when immersed in a subject, to lose sight of how much background the reader needs to grasp their ideas. Too much detail bores readers, while too little confuses them. Be sure that you've written your paper for your audience.

Did you stray from your outline or make changes that detract from an orderly presentation? If so, ask yourself why you made those changes. They'll probably need to be corrected, but may indicate new ideas that should stay in the paper.

If you're writing a paper for publication, ask yourself, "Have I answered an important scientific question?" We make little progress with "potboiler" publications containing randomly gathered facts that writers wishfully claim may "shed some light" on a nonexistent problem.

If the answers to these questions are no, rethink your plans. No amount of writing can disguise poorly designed experiments, an author having nothing to say, or an author who doesn't understand what he or she is writing about. For example, here's what science students wrote when they tried to write about a subject they didn't quite understand:

A fossil is an extinct animal. The older it is, the more extinct it is.

Artificial insemination is when the farmer does it to the cow and not the bull.

When you breathe, you inspire. When you do not breathe, you expire.

Think about what you've written and check for simple, but subtle, mistakes. For example, one student wrote a note to me stating that "I have a temperature." I wrote her a reply saying that I, too, had a temperature, as did everything else in the universe. Similarly, a scientist wrote that, "The solvents were evaporated *in vacuo* at 40°C under a stream of nitrogen." Although including the phrase *in vacuo* makes this sentence sound sophisticated, it also helps hide the absurdity of a vacuum made of nitrogen. Finally, a safety-conscious student wrote in his laboratory report that "Goggles are required to do this experiment." I wonder if those goggles were also required to write the lab report. If so, those are some *very* talented goggles.

Put Your Paper Aside

Regardless of what you think of your first draft, put it aside for a couple of days. If you try to revise your paper too soon, you'll read what you think you wrote, not what you actually wrote. Setting your paper aside will also help you be objective about what you wrote. As Maria von Ebner-Eschenbach wrote, "The manuscript in the drawer either rots or ripens."

If you have told readers what they need to know, if you have presented information in a logical way, and if you have answered a significant scientific question, then you're ready for the next stage of the writing process: revising and rethinking what you've written so that you'll communicate effectively with your readers.

Richard Feynman

Physics: 1920 to Today

Richard Feynman (1918–1988) was a renowned physicist. He worked on the early development of the atomic bomb and developed the Feynman diagram, a system of notation used to describe and calculate subatomic reactions. Feynman did his most important work when he was only 30 years old. Along with Julia Schwinger, Feynman developed a new theory for quantum electrodynamics more powerful than the 1927 Heisenberg theory. For that work, Feynman won the Nobel Prize in physics in 1965.

Feynman's *Lectures on Physics*, from which this article comes, are widely used as a textbook and were originally given to students at California Institute of Technology.

It is a little difficult to begin at once with the present view, so we shall first see how things looked in about 1920 and then take a few things out of that picture. Before 1920, our world picture was something like this: The "stage" on which the universe goes is the three-dimensional *space* of geometry, as described by Euclid, and things change in a medium called *time*. The elements on the stage are *particles*, for example the atoms, which have some *properties*. First, the property of inertia: if a particle is moving it keeps on going in the same direction unless *forces* act upon it. The second element, then, is *forces*, which were then thought to be of two varieties: First, an enormously complicated, detailed kind of interaction force which held the various atoms in different combinations in a complicated way, which determined whether salt would dissolve faster or slower when we raise the temperature. The other force that was known was a long-range interaction—a smooth and quiet attraction—which varied inversely as the square of the distance, and was called *gravitation*. This law was known and was very simple. Why things remain in motion when they are moving, or why there is a law of gravitation was, of course, not known.

A description of nature is what we are concerned with here. From this point of view, then, a gas, and indeed *all* matter, is a myriad of moving

Physics: 1920 to Today by Richard Feynman. Feynman, Leighton, Sands. *The Feynman Lectures on Physics*, Vol. 2. Copyright © 1974 by Addison-Wesley.

particles. Thus many of the things we saw while standing at the seashore can immediately be connected. First the pressure: this comes from the collisions of the atoms with the walls or whatever; the drift of the atoms, if they are all moving in one direction on the average, is wind; the *random* internal motions are the *heat*. There are waves of excess density, where too many particles have collected, and so as they rush off they push up piles of particles farther out, and so on. This wave of excess density is *sound*. It is a tremendous achievement to be able to understand so much. . . .

What *kinds* of particles are there? There were considered to be 92 at that time: 92 different kinds of atoms were ultimately discovered. They had different names associated with their chemical properties.

The next part of the problem was, *what are the short-range forces?* Why does carbon attract one oxygen or perhaps two oxygens, but not three oxygens? What is the machinery of interaction between atoms? Is it gravitation? The answer is no. Gravity is entirely too weak. But imagine a force analogous to gravity, varying inversely with the square of the distance, but enormously more powerful and having one difference. In gravity everything attracts everything else, but now imagine that there are *two kinds* of "things," and that this new force (which is the electrical force, of course) has the property that likes *repel* but unlikes *attract*. The "thing" that carries this strong interaction is called *charge*.

Then what do we have? Suppose that we have two unlikes that attract each other, a plus and a minus, and that they stick very close together. Suppose we have another charge some distance away. Would it feel any attraction? It would feel *practically none*, because if the first two are equal in size, the attraction for the one and the repulsion for the other balance out. Therefore there is very little force at any appreciable distance. On the other hand, if we get *very close* with the extra charge, *attraction* arises, because the repulsion of likes and attraction of unlikes will tend to bring unlikes closer together and push likes farther apart. Then the repulsion will be *less* than the attraction. This is the reason why the atoms, which are constituted out of plus and minus electric charges, feel very little force when they are separated by appreciable distance (aside from gravity). When they come close together, they can "see inside" each other and rearrange their charges, with the result that they have a very strong interaction. The ultimate basis of an interaction between the atoms is *electrical*. Since this force is so enormous, all the plusses and all minuses will normally come together in as intimate a combination as they can. All things, even ourselves, are made of fine-grained, enormously strongly interacting plus and minus parts, all neatly balanced out. Once in a while, by accident, we may rub off a few minuses or a few plusses (usually it is easier to rub off minuses), and in those circumstances we find the force of electricity *unbalanced*, and we can then see the effects of these electrical attractions.

To give an idea of how much stronger electricity is than gravitation, consider two grains of sand, a millimeter across, thirty meters apart. If the force between them were not balanced, if everything attracted everything else instead of likes repelling, so that there were no cancellation, how much force would there be? There would be a force of *three million tons* between the two! You see, there is very, *very* little excess or deficit of the number of negative or positive charges necessary to produce appreciable electrical effects. That is, of course, the reason why you cannot see the difference between an electrically charged or uncharged thing—so few particles are involved that they hardly make a difference in the weight or size of an object.

With this picture the atoms were easier to understand. They were thought to have a "nucleus" at the center, which is positively electrically charged and very massive, and the nucleus is surrounded by a certain number of "electrons" which are very light and negatively charged. Now we go a little ahead in our story to remark that in the nucleus itself there were found two kinds of particles, protons and neutrons, almost of the same weight and very heavy. The protons are electrically charged and the neutrons are neutral. If we have an atom with six protons inside its nucleus, and this is surrounded by six electrons (the negative particles in the ordinary world of matter are all electrons, and these are very light compared with the protons and neutrons which make nuclei), this would be atom number six in the chemical table, and it is called carbon. Atom number eight is called oxygen, etc., because the chemical properties depend upon the electrons on the *outside*, and in fact only upon *how many* electrons there are. So the *chemical* properties of a substance depend only on a number, the number of electrons. (The whole list of elements of the chemists really could have been called 1, 2, 3, 4, 5, etc. Instead of saying "carbon," we could say "element six," meaning six electrons, but of course, when the elements were first discovered, it was not known that they could be numbered that way, and secondly, it would make everything look rather complicated. It is better to have names and symbols for these things, rather than to call everything by number.)

More was discovered about the electrical force. The natural interpretation of electrical interaction is that two objects simply attract each other: plus against minus. However, this was discovered to be an inadequate idea to represent it. A more adequate representation of the situation is to say that the existence of the positive charge, in some sense, distorts, or creates a "condition" in space, so that when we put the negative charge in, it feels a force. This potentiality for producing a force is called an *electric field*. When we put an electron in an electric field, we say it is "pulled." We then have two rules: (a) charges make a field, and (b) charges in fields have forces on them and move. The reason for this will become clear when we discuss the following phenomena: If we were to charge a body, say a comb, electrically, and then place a charged piece of paper at a distance and move the comb back and forth, the paper will respond by always pointing to the comb. If we shake it faster, it will be discovered that the paper is a little behind, *there is a delay* in the action. (At the first stage, when we move the comb rather slowly, we find a complication which is *magnetism*. Magnetic influences have to do with *charges in relative motion*, so magnetic forces and electric forces can really be attributed to one field, as two different aspects of exactly the same thing. A changing electric field cannot exist without magnetism.) If we move the charged paper farther out, the delay is greater. Then an interesting thing is observed. Although the forces between two charged objects should go inversely as the *square* of the distance, it is found, when we shake a charge, that the influence extends *very much farther out* than we would guess at first sight. That is, the effect falls off more slowly than the inverse square.

Here is an analogy: If we are in a pool of water and there is a floating cork very close by, we can move it "directly" by pushing the water with another cork. If you looked only at the two *corks*, all you would see would be that one moved immediately in response to the motion of the other— there is some kind of "*interaction*" between them. Of course, what we really do is to disturb the *water*, the *water* then disturbs the other cork. We could make up a "law" that if you pushed the water a little bit, an object close by in the water would move. If it were farther away, of course, the second cork would scarcely move, for we move the water *locally*. On the other hand, if we jiggle the cork a new phenomenon is involved, in which the motion of the water moves the water there, etc., and *waves* travel away, so that by jiggling, there is an influence *very much farther out*, an oscillatory influence, that cannot be understood from the direct interaction. Therefore the idea of direct interaction must be replaced with the existence of the water, or in the electrical case, with what we call the *electromagnetic field*.

The electromagnetic field can carry waves; some of these waves are *light*, others are used in *radio broadcasts*, but the general name is *electro-magnetic waves*. These oscillatory waves can have various *frequencies*. The only thing that is really different from one wave to another is the *frequency of oscillation*. If we shake a charge back and forth more and more rapidly, and look at the effects, we get a whole series of different kinds of effects, which are all unified by specifying but one number, the number of oscillations per second. The usual "pickup" that we get from electric currents in the circuits in the walls of a building have a frequency of about one hundred cycles per second. If we increase the frequency to 500 or 1000 kilocycles (1 kilocycle = 1000 cycles) per second, we are "on the air," for this is the frequency range which is used for radio broadcasts. (Of course it has nothing to do with the *air!* We can have radio broadcasts without any air.) If we again increase the frequency, we come into the range that is used for FM and TV. Going still further, we use certain short waves, for example for *radar*. Still higher, and we do not need an instrument to "see" the stuff, we can see it with the human eye. In the range of frequency from 5×10^{14} to 5×10^{15} cycles per second our eyes would see the oscillation of the charged comb, if we could shake it that fast, as red, blue, or violet light, depending on the frequency. Frequencies below this range are called infrared, and above it, ultraviolet. The fact that we can see in a particular frequency range makes that part of the electromagnetic spectrum no more impressive than the other parts from a physicist's standpoint, but from a human standpoint, of course, it *is* more interesting. If we go up even higher in frequency, we get x-rays. X-rays are nothing but very high-frequency light. If we go still higher, we get gamma rays. These two terms, x-rays and gamma rays, are used almost synonymously. Usually electromagnetic rays coming from nuclei are called gamma rays, while those of high energy from atoms are called x-rays, but at the same frequency they are indistinguishable physically, no matter what their source. If we go to still higher frequencies, say to 10^{24} cycles per second, we find that we can make those waves artificially, for example with the synchrotron here at Caltech. We can find electromagnetic waves with stupendously high frequencies—with even a thousand times more rapid oscillation—in the waves found in *cosmic rays*. These waves cannot be controlled by us.

Quantum Physics

Having described the idea of the electromagnetic field, and that this field can carry waves, we soon learn that these waves actually behave in a strange way which seems very unwavelike. At higher frequencies they behave much more like *particles!* It is *quantum mechanics*, discovered just after 1920, which explains this strange behavior. In the years before 1920, the picture of space as a three-dimensional space, and of time as a separate thing, was changed by Einstein, first into a combination which we call space-time, and then still further into a *curved* space-time to represent gravitation. So the "stage" is changed into space-time, and gravitation is presum-

Table 1. The Electromagnetic Spectrum.

Frequency in oscillations/sec	Name	Rough behavior
10^2	Electical disturbance	Field
5×10^5–10^6	Radio broadcast	Waves
10^8	FM–TV	
10^{10}	Radar	
5×10^{14}–10^{15}	Light	
10^{18}	X-rays	Particle
10^{21}	γ-rays, nuclear	
10^{24}	γ-rays, "artificial"	
10^{27}	γ-rays, in cosmic rays	

ably a modification of space-time. Then it was also found that the rules for the motions of particles were incorrect. The mechanical rules of "inertia" and "forces" are *wrong*—Newton's laws are *wrong*—in the world of atoms. Instead, it was discovered that things on a small scale behave *nothing like* things on a large scale. That is what makes physics difficult—and very interesting. It is hard because the way things behave on a small scale is so "unnatural"; we have no direct experience with it. Here things behave like nothing we know of, so that it is impossible to describe this behavior in any other than analytic ways. It is difficult, and takes a lot of imagination.

Quantum mechanics has many aspects. In the first place, the idea that a particle has a definite location and a definite speed is no longer allowed; that is wrong. To give an example of how wrong classical physics is, there is a rule in quantum mechanics that says that one cannot know both where something is and how fast it is moving? The uncertainty of the momentum and the uncertainty of the position are complementary, and the product of the two is constant. We can write the law like this: $\Delta x \, \Delta p \geq h/2\pi$, but we shall explain it in more detail later. This rule is the explanation of a very mysterious paradox: if the atoms are made out of plus and minus charges, why don't the minus charges simply sit on top of the plus charges (they attract each other) and get so close as to completely cancel them out? *Why are atoms so big?* Why is the nucleus at the center with the electrons around it? It was first thought that this was because the nucleus was so big; but no, the nucleus is *very small*. An atom has a diameter of about 10^{-8} cm. The nucleus has a diameter of about 10^{-13} cm. If we had an atom and wished to see the nucleus, we would have to magnify it until the whole atom was the size of a large room, and then the nucleus would be a bare speck which you could just about make out with the eye, but very nearly *all the weight* of the atom is in that infinitesimal *nucleus*. What keeps the electrons from simply falling in? This principle: If they were in the nucleus, we would know their position precisely, and the uncertainty principle would then require that they have a very *large* (but uncertain) momentum, i.e., a very large *kinetic energy*. With this energy they would break away from the nucleus. They make a compromise: they leave themselves a little room for this uncertainty and then jiggle with a certain amount of minimum motion in accordance with this rule. (Remember that when a crystal is cooled to absolute zero, . . . the atoms do not stop moving, they still jiggle. Why? If they stopped moving, we would know where they were and that they had zero motion, and that is against the uncertainty principle. We cannot know where they are and how fast they are moving, so they must be continually wiggling in there!)

Another most interesting change in the ideas and philosophy of science brought about by quantum mechanics is this: it is not possible to predict *exactly* what will happen in any circumstance. For example, it is possible to arrange an atom which is ready to emit light, and we can measure when it has emitted light by picking up a photon particle, which we shall describe shortly. We cannot, however, predict *when* it is going to emit the light or, with several atoms, *which one* is going to. You may say that this is because there are some internal "wheels" which we have not looked at closely enough. No, there *are* no internal wheels; nature, as we understand it today, behaves in such a way that it is *fundamentally impossible* to make a precise prediction of *exactly what will happen* in a given experiment. This is a horrible thing; in fact, philosophers have said before that one of the fundamental requisites of science is that whenever you set up the same conditions, the same thing must happen. This is simply *not true*, it is *not* a fundamental condition of science. The fact is that the same thing does not happen, that we can find only an average, statistically, as to what happens. Nevertheless, science has not completely collapsed. Philosophers, incidentally, say a great deal about what is *absolutely necessary* for science, and it is always, so far as one can see, rather naive, and probably wrong. For example, some philosopher or other said it is fundamental to the scientific effort that if an experiment is performed in, say, Stockholm, and then the same experiment is done in, say, Quito, the *same results* must occur. That is quite false. It is not necessary that science do that; it may be a *fact of experience*, but it is not necessary. For example, if one of the experiments is to look out at the sky and see the aurora borealis in Stockholm, you do not see it in Quito; that is a different phenomenon. "But," you say, "that is something that has to do with the outside; can you close yourself up in a box in Stockholm and pull down the shade and get any difference?" Surely. If we take a pendulum on a universal joint, and pull it out and let go, then the pendulum will swing almost in a plane, but not quite. Slowly the plane keeps changing in Stockholm, but not in Quito. The blinds are down, too. The fact that this happened does not bring on the destruction of science. What is the fundamental hypothesis of science, the fundamental philosophy? We stated it in the first chapter: *the sole test of the validity of any idea is experiment*. If it turns out that most experiments work out the same in Quito as they do in Stockholm, then those "most experiments" will be used to formulate some general law, and those experiments which do not come out the same we will say were a result of the environment near Stockholm. We will invent some way to summarize the results of the experiment, and we do not have to be told ahead of time what this way will look like. If we are told that the same experiment will always produce the same result, that is all very well, but if when we try it, it does *not*, then it does *not*. We just have to take what we see, and then formulate all the rest of our ideas in terms of our actual experience.

Returning again to quantum mechanics and fundamental physics, we cannot go into details of the quantum-mechanical principles at this time, of course, because these are rather difficult to understand. We shall assume that they are there, and go on to describe what some of the consequences are. One of the consequences is that things which we used to consider as waves also behave like particles, and particles behave like waves; in fact everything behaves the same way. There is no distinction between a wave and a particle. So quantum mechanics *unifies* the idea of the field and its waves, and the particles, all into one. Now it is true that when the frequency is low, the field aspect of the phenomenon is more evident, or more useful as an approximate description in terms of everyday experiences. But as the frequency increases, the particle aspects of the phenomenon become more evident with the equipment with which we usually make the measurements. In fact, although we mentioned many frequencies, no phenomenon directly involving a frequency has yet been detected above approximately 10^{12} cycles per second. We only *deduce* the higher frequencies from the energy of the particles, by a rule which assumes that the particle-wave idea of quantum mechanics is valid.

Thus we have a new view of electromagnetic interaction. We have a new kind of *particle* to add to the electron, the proton, and the neutron. That new particle is called a *photon*. The new view of the interaction of electrons and protons that is electromagnetic theory, but with everything quantum-mechanically correct, is called *quantum electrodynamics*. This fundamental theory of the interaction of light and matter, or electric field and charges, is our greatest success so far in physics. In this one theory we have the basic rules for all ordinary phenomena except for gravitation and nuclear processes. For example, out of quantum electrodynamics come all known electrical, mechanical, and chemical laws: the laws for the collision of billiard balls, the motions of wires in magnetic fields, the specific heat of carbon monoxide, the color of neon signs, the density of salt, and the reactions of hydrogen and oxygen to make water are all consequences of this one law. All these details can be worked out if the situation is simple enough for us to make an approximation, which is almost never, but often we can understand more or less what is happening. At the present time no exceptions are found to the quantum-electrodynamic laws outside the nucleus, and there we do not know whether there is an exception because we simply do not know what is going on in the nucleus.

In principle, then, quantum electrodynamics is the theory of all chemistry, and of life, if life is ultimately reduced to chemistry and therefore just to physics because chemistry is already reduced (the part of physics which is involved in chemistry being already known). Furthermore, the same quantum electrodynamics, this great thing, predicts a lot of new things. In the first place, it tells the properties of very high-energy photons, gamma rays, etc. It predicted another very remarkable thing: besides the electron, there should be another particle of the same mass, but of opposite charge, called a *positron*, and these two, coming together, could annihilate each other with the emission of light or gamma rays. (After all, light and gamma rays are all the same, they are just different points on a frequency scale.) The generalization of this, that for each particle there is an antiparticle, turns out to be true. In the case of electrons, the antiparticle has another name—it is called a positron, but for most other particles, it is called antiso-and-so, like antiproton or antineutron. In quantum electrodynamics, *two numbers* are put in and most of the other numbers in the world are supposed to come out. The two numbers that are put in are called the mass of the electron and the charge of the electron. Actually, that is not quite true, for we have a whole set of numbers for chemistry which tells how heavy the nuclei are. That leads us to the next part.

Nuclei and Particles

What are the nuclei made of, and how are they held together? It is found that the nuclei are held together by enormous forces. When these are released, the energy released is tremendous compared with chemical energy, in the same ratio as the atomic bomb explosion is to a TNT explosion, because, of course, the atomic bomb has to do with changes inside the nucleus, while the explosion of TNT has to do with the changes of the electrons on the outside of the atoms. The question is, what are the forces which hold the protons and neutrons together in the nucleus? Just as the electrical interaction can be connected to a particle, a photon, Yukawa suggested that the forces between neutrons and protons also have a yield of some kind, and that when this field jiggles it behaves like a particle. Thus there could be some other particles in the world besides protons and neutrons, and he was able to deduce the properties of these particles from the already known characteristics of nuclear forces. For example, he predicted they should have a mass of two or three hundred times that of an electron; and lo and behold, in cosmic rays there was discovered a particle of the right mass! But it later turned out to be the wrong particle. It was called a μ-meson, or muon.

However, a little while later, in 1947 or 1948, another particle was found, the π-meson, or pion, which satisfied Yukawa's criterion. Besides the proton and the neutron, then, in order to get nuclear forces we must add the pion. Now, you say, "Oh, great!, with this theory we make quantum nucleodynamics using the pions just like Yukawa wanted to do, and see if it works, and everything will be explained." Bad luck. It turns out that the calculations that are involved in this theory are so difficult that no one has ever been able to figure out what the consequences of the theory are, or to check it against experiment, and this has been going on now for almost twenty years!

So we are stuck with a theory, and we do not know whether it is right or wrong, but we do know that it is a *little* wrong, or at least incomplete. While we have been dawdling around theoretically, trying to calculate the consequences of this theory, the experimentalists have been discovering some things. For example, they had already discovered this μ-meson or muon, and we do not yet know where it fits. Also, in cosmic rays, a large number of other "extra" particles were found. It turns out that today we have approximately thirty particles, and it is very difficult to understand the relationships of all these particles, and what nature wants them for, or what the connections are from one to another. We do not today understand these various particles as different aspects of the same thing, and the fact that we have so many unconnected particles is a representation of the fact that we have so much unconnected information without a good theory. After the great successes of quantum electrodynamics, there is a certain amount of knowledge of nuclear physics which is rough knowledge, sort of half experience and half theory, assuming a type of force between protons and neutrons and seeing what will happen, but not really understanding where the force comes from. Aside from that, we have made very little progress. We have collected an enormous number of chemical elements. In the chemical case, there suddenly appeared a relationship among these elements which was unexpected, and which is embodied in the periodic table of Mendeléev. For example, sodium and potassium are about the same in their chemical properties and are found in the same column in the Mendeléev chart. We have been seeking a Mendeléev-type charge for the new particles. One such chart of the new particles was made independently by Gell-Mann in the U.S.A. and Nishijima in Japan. The basis of their classification is a new number, like the electric charge, which can be assigned to each particle, called its "strangeness," S. This number is conserved, like the electric charge, in reactions which take place by nuclear forces.

In table 2 are listed all the particles. We cannot discuss them much at this stage, but the table will at least show you how much we do not know. Underneath each particle its mass is given in a certain unit, called the Mev. One Mev is equal to 1.782×10^{-27} gram. The reason this unit was chosen is historical, and we shall not go into it now. More massive particles are put higher up on the chart; we see that a neutron and a proton have almost the same mass. In vertical columns we have put the particles with the same electrical charge, all neutral objects in one column, all positively charged ones to the right of this one, and all negatively charged objects to the left.

Particles are shown with a solid line and "resonances" with a dashed one. Several particles have been omitted from the table. These include the important zero-mass, zero-charge particles, the photon and the graviton, which do not fall into the baryon-meson-lepton classification scheme, and also some of the newer resonances (K^*, ψ, η). The antiparticles of the mesons are listed in the table, but the antiparticles of the leptons and baryons would have to be listed in another table which would look exactly like this one reflected on the zero-charge column. Although all of the particles except the electron, neutrino, photon, graviton, and proton are unstable, decay products have been shown only for the resonances. Strangeness assignments are not applicable for leptons, since they do not interact strongly with nuclei.

All particles which act together with the neutrons and protons are called *baryons*, and the following ones exist: There is a "lambda," with a mass of

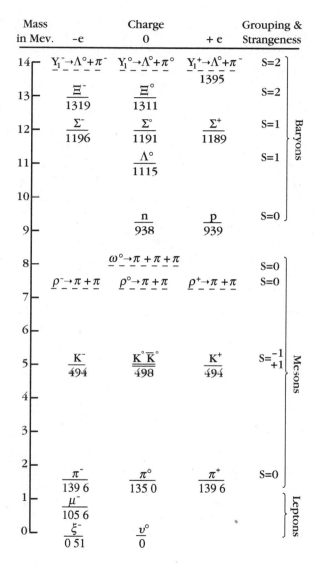

Mass in Mev.	−e	Charge 0	+ e	Grouping & Strangeness	
14	$Y_1^- \to \Lambda^\circ + \pi^-$	$Y_1^\circ \to \Lambda^\circ + \pi^\circ$	$Y_1^+ \to \Lambda^\circ + \pi^-$ 1395	S=2	
13	Ξ^- 1319	Ξ° 1311		S=2	Baryons
12	Σ^- 1196	Σ° 1191	Σ^+ 1189	S=1	
11		Λ° 1115		S=1	
10					
9		n 938	p 939	S=0	
8		$\omega^\circ \to \pi + \pi + \pi$		S=0	
7	$\rho^- \to \pi + \pi$	$\rho^\circ \to \pi + \pi$	$\rho^+ \to \pi + \pi$	S=0	
6					
5	K^- 494	$K^\circ \overline{K}^\circ$ 498	K^+ 494	S=$\begin{smallmatrix}-1\\+1\end{smallmatrix}$	Mesons
4					
3					
2	π^- 139 6	π° 135 0	π^+ 139 6	S=0	
1	μ^- 105 6				Leptons
0	ξ^- 0 51	v° 0			

Table 2
Elementary Particles.

1154 Mev, and three others, called sigmas, minus, neutral, and plus, with several masses almost the same. There are groups or multiplets with almost the same mass, within one or two percent. Each particle in a multiplet has the same strangeness. The first multiplet is the proton-neutron doublet, and then there is a singlet (the lambda) then the sigma triplet, and finally the xi doublet. Very recently, in 1961, even a few more particles were found. Or *are* they particles? They live so short a time, they disintegrate almost instantaneously, as soon as they are formed, that we do not know whether they should be considered as new particles, or some kind of "resonance" interaction of a certain definite energy between the A and π products into which they disintegrate.

In addition to the baryons the other particles which are involved in the nuclear interaction are called *mesons*. There are first the pions, which come in three varieties, positive, negative, and neutral; they form another multiplet. We have also found some new things called *K*-mesons, and they occur as a doublet, K^+ and K°. Also, every particle has its antiparticle, unless a particle *is its own* antiparticle. For example, the π^- and the π^+ are antiparticles, but the π° is its own antiparticle. The K^- and K^+ are antiparticles, and the K° and K° are their own antiparticles. In addition, in 1961 we also found some more mesons or *maybe* mesons which disintegrate almost immediately. A thing called ω which goes into three pions has a mass 780 on this scale, and somewhat less certain is an object which disintegrates into two pions. These particles, called mesons and baryons, and the antiparticles of the mesons are on the same chart, but the antiparticles of the baryons must be put on another chart, "reflected" through the charge-zero column.

Just as Mendeléev's chart was very good, except for the fact that there were a number of rare earth elements which were hanging out loose from it, so we have a number of things hanging out loose from this chart—particles which do not interact strongly in nuclei, have nothing to do with a nuclear interaction, and do not have a strong interaction (I mean the powerful kind of interaction of nuclear energy). These are called leptons, and they are the following: there is the electron, which has a very small mass on this scale, only 0.510 Mev. Then there is that other, the μ-meson, the muon, which has a mass much higher, 206 times as heavy as an electron. So far as we can tell, by all experiments so far, the difference between the electron and the muon is nothing but the mass. Everything works exactly the same for the muon as for the electron, except that one is heavier than the other. Why is there another one heavier; what is the use for it? We do not know. In addition, there is a lepton which is neutral, called a neutrino, and this particle has zero mass. In fact, it is now known that there are *two*

Table 3. Elementary Interactions.

Coupling	Strength*	Law
Photon to charge particles	$\sim 10^{-2}$	Law known
Gravity to all energy	$\sim 10^{-40}$	Law known
Weak decays	$\sim 10^{-5}$	Law partly known
Mesons to baryons	~ 1	Law unknown (some rules known)

*The strength is a dimensionless measure of the coupling constant involved in each interaction ($\sim$ means "approximately").

different kinds of neutrinos, one related to electrons and the other related to muons.

Finally, we have two other particles which do not interact strongly with the nuclear ones: one is a photon, and perhaps, if the field of gravity also has a quantum-mechanical analog (a quantum theory of gravitation has not yet been worked out), then there will be a particle, a graviton, which will have zero mass.

What is this "zero mass"? The masses given here are the masses of the particles *at rest*. The fact that a particle has zero mass means, in a way, that it cannot *be* at *rest*. A photon is never at rest, it is always moving at 186,000 miles a second. We will understand more what mass means when we understand the theory of relativity, which will come in due time.

Thus we are confronted with a large number of particles, which together seem to be the fundamental constituents of matter. Fortunately, these particles are not *all* different in their *interactions* with one another. In fact, there seem to be just *four kinds* of interaction between particles which, in the order of decreasing strength, are the nuclear force, electrical interactions, the beta-decay interaction, and gravity. The photon is coupled to all charged particles and the strength of the interaction is measured by some number, which is 1/137. The detailed law of this coupling is known, that is quantum electrodynamics. Gravity is coupled to all *energy*, but its coupling is extremely weak, much weaker than that of electricity. This law is also known. Then there are the so-called weak decays—beta decay, which causes the neutron to disintegrate into proton, electron, and neutrino, relatively slowly. This law is only partly known. The so-called strong interaction, the meson-baryon interaction, has a strength of 1 in this scale, and the law is completely unknown, although there are a number of known rules, such as that the number of baryons does not change in any reaction.

This then, is the horrible condition of our physics today. To summarize it, I would say this: outside the nucleus, we seem to know all; inside it, quantum mechanics is valid—the principles of quantum mechanics that have not been found to fail. The stage on which we put all of our knowledge, we would say, is relativistic space-time; perhaps gravity is involved in space-time. We do not know how the universe got started, and we have never made experiments which check our ideas of space and time accurately, below some tiny distance, so we only know that our ideas work above that distance. We should also add that the rules of the game are the quantum-mechanical principles, and those principles apply, so far as we can tell, to the new particles as well as to the old. The origin of the forces in nuclei leads us to new particles, but unfortunately they appear in great profusion and we lack a complete understanding of their interrelationship, although we already know that there are some very surprising relationships among them. We seem gradually to be groping toward an understanding of the world of sub-atomic particles, but we really do not know how far we have yet to go in this task.

Understanding What You've Read

What does Feynman convey about himself and his view of science? Cite examples to make your point.

Like many scientists, Feynman used the lecture hall to present his ideas. How does this affect the style and content of his essay?

Why did Feynman believe physics is in "horrible condition"?

Feynman's style has been described as "both casual and rushed." Do you agree? Why or why not?

What is the "uncertainty principle"? Write a short essay describing the principle and its importance.

Charles Darwin

Recapitulation and Conclusion

No book has transformed our perceptions of ourselves and our place in nature more than Charles Darwin's *On the Origin of Species.* Darwin (1809–1882) was part of a scientifically inclined family: his father was a rich physician, and his grandfather, Erasmus Darwin, was a physician-poet and amateur naturalist. When he was young, Darwin abandoned his study of medicine (and greatly disappointed his family) because he was horrified by the idea of performing operations on children without anesthesia. However, Darwin did remain interested in natural history. After graduating from Cambridge at the age of 22, he set sail aboard the H.M.S. *Beagle* because he thought it would be an opportunity for "collecting, observing and noting anything worthy to be noted in natural history."[1]

The cruise was planned to last only two years, but eventually stretched to almost five. While sailing along the coast of South America, Darwin saw small variations in species. The most striking of these variations was among finches scattered among the Galapagos Islands. These and other observations made Darwin question the immutability of species and formed the basis for his ideas about evolution. Twenty years after returning to England, he published *On the Origin of Species* (1859), starting a furor over whether evolution applied to humans. Darwin stood his ground and, in 1871, published

---

[1]Contrary to myth, Darwin was not the naturalist aboard the H.M.S. *Beagle* when it set sail. The ship's surgeon, Robert McKormick, originally held the official position of naturalist (Darwin disliked McKormick's approach to science). When McKormick was sent home to England in April of 1832, Darwin became the official naturalist aboard the ship. For more about this interesting story, see Chapter 2 of *Ever Since Darwin: Reflections in Natural History* by Stephen Jay Gould (New York: W.W. Norton & Co., 1977).

The Descent of Man.[2]
    Here is a chapter entitled "Recapitulation and Conclusion" from the second edition of *On the Origin of Species.*[3]

I have now recapitulated the facts and considerations which have thoroughly convinced me that species have been modified, during a long course of descent. This has been effected chiefly through the natural selection of numerous successive, slight, favourable variations; aided in an important manner by the inherited effects of the use and disuse of parts; and in an unimportant manner, that is in relation to adaptive structures, whether past or present, by the direct action of external conditions, and by variations which seem to us in our ignorance to arise spontaneously. It appears that I formerly underrated the frequency and value of these latter forms of variation, as leading to permanent modifications of structure independently of natural selection. But as my conclusions have lately been much misrepresented, and it has been stated that I attribute the modification of species exclusively to natural selection, I may be permitted to remark that in the first edition of this work, and subsequently, I placed in a most conspicuous position—namely, at the close of the Introduction—the following words: "I am convinced that natural selection has been the main but not the exclusive means of modification." This has been of no avail. Great is the power of steady misrepresentation; but the history of science shows that fortunately this power does not long endure.

It can hardly be supposed that a false theory would explain, in so satisfactory a manner as does the theory of natural selection, the several large classes of facts above specified. It has recently been objected that this is an unsafe method of arguing; but it is a method used in judging of the common events of life, and has often been used by the greatest natural philosophers. The undulatory theory of light has thus been arrived at; and the belief in the revolution of the earth on its own axis was until lately supported by hardly any direct evidence. It is no valid objection that science as yet throws no light on the far higher problem of the essence or origin of life. Who can explain what is the essence of the attraction of gravity? No one now objects to following out the results consequent on this unknown ele-

[2]Although parts of Darwin's writing are often criticized as stodgy Victorian prose—obscure, convoluted, and ambiguous—his *On the Origin of Species* is a masterpiece of organization, technical description, reporting, induction, and documentation. It was also a popular success: its first printing sold out on its first day. Darwin's logic forced scientists to accept his theory of natural selection despite all of the opposition that dogma, ignorance, and prejudice could muster. For a more lively and direct writing about the subject for a general audience, read the works of Thomas Huxley, a friend and advocate of Darwin.

[3]The phrase "by the Creator," in the final sentence of this chapter, did not appear in the first edition of *On the Origin of Species*. It was added to the second edition to conciliate angry clerics. Darwin later wrote, "I have long since regretted that I truckled to public opinion and used the Pentateuchal term of creation, by which I really meant 'appeared' by some wholly unknown process."

ment of attraction; notwithstanding that Leibnitz formerly accused Newton of introducing "occult qualities and miracles into philosophy."

I see no good reason why the views given in this volume should shock the religious feelings of any one. It is satisfactory, as showing how transient such impressions are, to remember that the greatest discovery ever made by man, namely, the law of the attraction of gravity, was also attacked by Leibnitz, "as subversive of natural, and inferentially of revealed, religion." A celebrated author and divine has written to me that "he has gradually learnt to see that it is just as noble a conception of the Deity to believe that He created a few original forms capable of self-development into other and needful forms, as to believe that He required a fresh act of creation to supply the voids caused by the action of His laws."

Why, it may be asked, until recently did nearly all the most eminent living naturalists and geologists disbelieve in the mutability of species? It cannot be asserted that organic beings in a state of nature are subject to no variation; it cannot be proved that the amount of variation in the course of long ages is a limited quantity; no clear distinction has been, or can be, drawn between species and well-marked varieties. It cannot be maintained that species when intercrossed are invariably sterile, and varieties invariably fertile; or that sterility is a special endowment and sign of creation. The belief that species were immutable productions was almost unavoidable as long as the history of the world was thought to be of short duration; and now that we have acquired some idea of the lapse of time, we are too apt to assume, without proof, that the geological record is so perfect that it would have afforded us plain evidence of the mutation of species, if they had undergone mutation.

But the chief cause of our natural unwillingness to admit that one species has given birth to other and distinct species, is that we are always slow in admitting great changes of which we do not see the steps. The difficulty is the same as that felt by so many geologists, when Lyell first insisted that long lines of inland cliffs had been formed, and great valleys excavated, by the agencies which we see still at work. The mind cannot possibly grasp the full meaning of the term of even a million years; it cannot add up and perceive the full effects of many slight variations, accumulated during an almost infinite number of generations.

Although I am fully convinced of the truth of the views given in this volume under the form of an abstract, I by no means expect to convince experienced naturalists whose minds are stocked with a multitude of facts all viewed, during a long course of years, from a point of view directly opposite to mine. It is so easy to hide our ignorance under such expressions as the "plan of creation," "unity of design," &c., and to think that we give an explanation when we only re-state a fact. Any one whose disposition leads him to attach more weight to unexplained difficulties than to the explanation of a certain number of facts will certainly reject the theory. A few naturalists, endowed with much flexibility of mind, and who have already begun to doubt the immutability of species, may be influenced by this volume; but I look with confidence to the future,—to young and rising naturalists, who will be able to view both sides of the question with impartiality. Whoever is led to believe that species are mutable will do good service by conscientiously expressing his conviction; for thus only can the load of prejudice by which this subject is overwhelmed be removed.

Several eminent naturalists have of late published their belief that a multitude of reputed species in each genus are not real species; but that other species are real, that is, have been independently created. This seems to me a strange conclusion to arrive at. They admit that a multitude of forms, which till lately they themselves thought were special creations, and which are still thus looked at by the majority of naturalists, and which consequently have all the external characteristic features of true species,—they admit that these have been produced by variation, but they refuse to extend the same view to other and slightly different forms. Nevertheless they do not pretend that they can define, or even conjecture, which are the created forms of life, and which are those produced by secondary laws. They admit variation as a vera causa in one case, they arbitrarily reject it in another, without assigning any distinction in the two cases. The day will come when this will be given as a curious illustration of the blindness of preconceived opinion. These authors seem no more startled at a miraculous act of creation than at an ordinary birth. But do they really believe that at innumerable periods in the earth's history certain elemental atoms have been commanded suddenly to flash into living tissues? Do they believe that at each supposed act of creation one individual or many were produced? Were all the infinitely numerous kinds of animals and plants created as eggs or seed, or as full grown? And in the case of mammals, were they created bearing the false marks of nourishment from the mother's womb? Undoubtedly some of these same questions cannot be answered by those who believe in the appearance or creation of only a few forms of life, or of some one form alone. It has been maintained by several authors that it is as easy to believe in the creation of a million beings as of one; but Maupertuis' philosophical axiom "of least action" leads the mind more willingly to admit the smaller number; and certainly we ought not to believe that innumerable beings within each great class have been created with plain, but deceptive, marks of descent from a single parent.

As a record of a former state of things, I have retained in the foregoing paragraphs, and elsewhere, several sentences which imply that naturalists believe in the separate creation of each species; and I have been much censured for having thus expressed myself. But undoubtedly this was the general belief when the first edition of the present work appeared. I formerly spoke to very many naturalists on the subject of evolution, and never once met with any sympathetic agreement. It is probable that some did then believe in evolution, but they were either silent, or expressed themselves so ambiguously that it was not easy to understand their meaning. Now things are wholly changed, and almost every naturalist admits the great principle of evolution. There are, however, some who still think that species have suddenly given birth, through quite unexplained means, to new and totally different forms: but, as I have attempted to show, weighty evidence can be opposed to the admission of great and abrupt modifications. Under a scientific point of view, and as leading to further investigation, but little advantage is gained by believing that new forms are suddenly developed in an inexplicable manner from old and widely different forms, over the old belief in the creation of species from the dust of the earth.

It may be asked how far I extend the doctrine of the modification of species. The question is difficult to answer, because the more distinct the forms are which we consider, by so much the arguments in favour of community of descent become fewer in number and less in force. But some arguments of the greatest weight extend very far. All the members of whole classes are connected together by a chain of affinities, and all can be classed on the same principle, in groups subordinate to groups. Fossil remains sometimes tend to fill up very wide intervals between existing orders.

Organs in a rudimentary condition plainly show that an early progenitor had the organ in a fully developed condition; and this in some cases implies an enormous amount of modification in the descendants. Throughout whole classes various structures are formed on the same pattern, and at a very early age the embryos closely resemble each other. Therefore I cannot doubt that the theory of descent with modification embraces all the members of the same great class or kingdom. I believe that animals are descended from at most only four or five progenitors, and plants from an equal or lesser number.

Analogy would lead me one step farther, namely, to the belief that all animals and plants are descended from some one prototype. But analogy may be a deceitful guide. Nevertheless all living things have much in common, in their chemical composition, their cellular structure, their laws of growth, and their liability to injurious influences. We see this even in so trifling a fact as that the same poison often similarly affects plants and animals; or that the poison secreted by the gall-fly produces monstrous growth on the wild rose or oak-tree. With all organic beings, excepting perhaps some of the very lowest, sexual reproduction seems to be essentially similar. With all, as far as is at present known, the germinal vesicle is the same; so that all organisms start from a common origin. If we look even to the two main divisions—namely, to the animal and vegetable kingdoms—certain low forms are so far intermediate in character that naturalists have disputed to which kingdom they should be referred. As Professor Asa Gray has remarked, "the spores and other reproductive bodies of many of the lower algae may claim to have first a characteristically animal, and then an unequivocally vegetable existence." Therefore, on the principle of natural selection with divergence of character, it does not seem incredible that, from some such low and intermediate form, both animals and plants may have been developed; and, if we admit this, we must likewise admit that all the organic beings which have ever lived on this earth may be descended from some one primordial form. But this inference is chiefly grounded on analogy, and it is immaterial whether or not it be accepted. No doubt it is possible, as Mr. G.H. Lewes has urged, that at the first commencement of life many different forms were evolved; but if so, we may conclude that only a very few have left modified descendants. For, as I have recently remarked in regard to the members of each great kingdom, such as the Vertebrata, Articulata, &c., we have distinct evidence in their embryological, homologous, and rudimentary structures, that within each kingdom all the members are descended from a single progenitor.

When the views advanced by me in this volume, and by Mr. Wallace, or when analogous views on the origin of species are generally admitted, we can dimly foresee that there will be a considerable revolution in natural history. Systematists will be able to pursue their labours as at present; but they will not be incessantly haunted by the shadowy doubt whether this or that form be a true species. This, I feel sure and I speak after experience, will be no slight relief. The endless disputes whether or not some fifty species of British brambles are good species will cease. Systematists will have only to decide (not that this will be easy) whether any form be sufficiently constant and distinct from other forms, to be capable of definition; and if definable, whether the differences be sufficiently important to deserve a specific name. This latter point will become a far more essential consideration than it is at present; for differences, however slight, between any two forms, if not blended by intermediate gradations, are looked at by most naturalists as sufficient to raise both forms to the rank of species.

Hereafter we shall be compelled to acknowledge that the only distinction between species and well-marked varieties is, that the latter are known, or believed to be connected at the present day by intermediate gradations whereas species were formerly thus connected. Hence, without rejecting the consideration of the present existence of intermediate gradations between any two forms, we shall be led to weigh more carefully and to value higher the actual amount of difference between them. It is quite possible that forms now generally acknowledged to be merely varieties may hereafter be thought worthy of specific names; and in this case scientific and common language will come into accordance. In short, we shall have to treat species in the same manner as those naturalists treat genera, who admit that genera are merely artificial combinations made for convenience. This may not be a cheering prospect; but we shall at least be freed from the vain search for the undiscovered and undiscoverable essence of the term species.

The other and more general departments of natural history will rise greatly in interest. The terms used by naturalists, of affinity, relationship, community of type, paternity, morphology, adaptive characters, rudimentary and aborted organs, & c., will cease to be metaphorical, and will have a plain signification. When we no longer look at an organic being as a savage looks at a ship, as something wholly beyond his comprehension; when we regard every production of nature as one which has had a long history; when we contemplate every complex structure and instinct as the summing up of many contrivances, each useful to the possessor, in the same way as any great mechanical invention is the summing up of the labour, the experience, the reason, and even the blunders of numerous workmen; when we thus view each organic being, how far more interesting—I speak from experience—does the study of natural history become!

A grand and almost untrodden field of inquiry will be opened, on the causes and laws of variation, on correlation, on the effects of use and disuse, on the direct action of external conditions, and so forth. The study of domestic productions will rise immensely in value. A new variety raised by man will be a more important and interesting subject for study than one more species added to the infinitude of already recorded species. Our classifications will come to be, as far as they can be so made, genealogies; and will then truly give what may be called the plan of creation. The rules for classifying will no doubt become simpler when we have a definite object in view. We possess no pedigrees or armorial bearings; and we have to discover and trace the many diverging lines of descent in our natural genealogies, by characters of any kind which have long been inherited.
Rudimentary organs will speak infallibly with respect to the nature of long lost structures. Species and groups of species which are called aberrant, and which may fancifully be called living fossils, will aid us in forming a picture of the ancient forms of life. Embryology will often reveal to us the structure, in some degree obscured, of the prototypes of each great class.

When we can feel assured that all the individuals of the same species, and all the closely allied species of most genera, have within a not very remote period descended from one parent, and have migrated from some one birth-place; and when we better know the many means of migration, then, by the light which geology now throws, and will continue to throw, on former changes of climate and of the level of the land, we shall surely be enabled to trace in an admirable manner the former migrations of the inhabitants of the whole world. Even at present, by comparing the differences between the inhabitants of the sea on the opposite sides of a continent, and the nature of the various inhabitants on that continent in relation to their apparent means of immigration, some light can be thrown on ancient geography.

The noble science of Geology loses glory from the extreme imperfection of the record. The crust of the earth with its imbedded remains must not be looked at as a well-filled museum, but as a poor collection made at hazard and at rare intervals. The accumulation of each great fossiliferous formation will be recognised as having depended on an unusual concurrence of favourable circumstances, and the blank intervals between the successive stages as having been of vast duration. But we shall be able to gauge with some security the duration of these intervals by a comparison of the preceding and succeeding organic forms. We must be cautious in attempting to correlate as strictly contemporaneous two formations, which do not include many identical species, by the general succession of the forms of life. As species are produced and exterminated by slowly acting and still existing causes, and not by miraculous acts of creation; and as the most important of all causes of organic change is one which is almost independent of altered and perhaps suddenly altered physical conditions, namely, the mutual relation of organism to organism,—the improvement of one organism entailing the improvement or the extermination of others; it follows, that the amount of organic change in the fossils of consecutive formations probably serves as a fair measure of the relative, though not actual lapse of time. A number of species, however, keeping in a body might remain for a long period unchanged, whilst within the same period several of these species by migrating into new countries and coming into competition with foreign associates, might become modified; so that we must not overrate the accuracy of organic changes as a measure of time.

In the future I see open fields for far more important researches. Psychology will be securely based on the foundation already well laid by Mr. Herbert Spencer, that of the necessary acquirement of each mental power and capacity by gradation. Much light will be thrown on the origin of man and his history.

Authors of the highest eminence seem to be fully satisfied with the view that each species has been independently created. To my mind it accords better with what we know of the laws impressed on matter by the Creator, that the production and extinction of the past and present inhabitants of the world should have been due to secondary causes, like those determining the birth and death of the individual. When I view all beings not as special creations, but as the lineal descendants of some few beings which lived long before the first bed of the Cambrian system was deposited, they seem to me to become ennobled. Judging from the past, we may safely infer that not one living species will transmit its unaltered likeness to a distant futurity. And of the species now living very few will transmit progeny of any kind to a far distant futurity; for the manner in which all organic beings are grouped, shows that the greater number of species in each genus, and all the species in many genera, have left no descendants, but have become utterly extinct. We can so far take a prophetic glance into futurity as to foretell that it will be the common and widely-spread species, belonging to the larger and dominant groups within each class, which will ultimately prevail and procreate new and dominant species. As all the living forms of life are the lineal descendants of those which lived long before the Cambrian epoch, we may feel certain that the ordinary succession by generation has never once been broken, and that no cataclysm has desolated the whole world. Hence we may look with some confidence to a secure future of great length. And as natural selection works solely by and for the good of each being, all corporeal and mental endowments will tend to progress towards perfection.

   It is interesting to contemplate a tangled bank, clothed with many plants of many kinds, with birds singing on the bushes, with various insects flitting about, and with worms crawling through the damp earth, and to reflect that these elaborately constructed forms, so different from each other, and dependent upon each other in so complex a manner, have all been produced by laws acting around us. These laws, taken in the largest sense, being Growth with Reproduction; Inheritance which is almost implied by reproduction: Variability from the indirect and direct action of the conditions of life, and from use and disuse: a Ratio of Increase so high as to lead to a Struggle for Life, and as a consequence to Natural Selection, entailing Divergence of Character and the Extinction of less-improved forms. Thus, from the war of nature, from famine and death, the most exalted object which we are capable of conceiving, namely, the production of the higher animals, directly follows. There is grandeur in this view of life, with its several powers, having been originally breathed by the Creator into a few forms or into one; and that, whilst this planet has gone cycling on according to the fixed law of gravity, from so simple a beginning endless forms most beautiful and most wonderful have been, and are being evolved.

Understanding What You've Read

Darwin devoted only one paragraph of *On the Origin of Species* to humans, yet that paragraph was a focal point of controversies that followed:

> In the future I see open fields for far more important researches. Psychology will be securely based on the foundation already well laid by Mr. Herbert Spencer, that of the necessary acquirement of each mental power and capacity by gradation. Much light will be thrown on the origin of man and his history.

The last sentence of this paragraph has been referred to as "the understatement of the nineteenth century." Why was this controversial? Do you agree with Darwin? Why or why not?

Darwin presented an overwhelming amount of evidence to support what he called "one long argument" for natural selection. Why are keen observations and detailed notes important to a scientist?

Does natural selection act on human behavior? Are certain traits "better" than others? Why or why not?

Darwin is regarded as an excellent writer as well as a first-class scientist. What about Darwin's writing makes it effective?

Darwin ended the first paragraph of this chapter by discussing the misrepresentation of his ideas. How have Darwin's ideas been misrepresented?

Recall when you first learned about "evolution." In what context did you learn about the topic? How have your ideas about evolution changed since then?

What about natural selection is most logical? Does any aspect seem illogical?

Should a school board or group of politicians be allowed to ban a biology book that discusses evolutionary theory from use in public schools? Write a short essay explaining your answer.

Although Darwin acknowledged that "analogy may be a deceitful guide," he often used analogy to strengthen his argument for natural selection. What analogy did Darwin use in *On the Origin of Species* to introduce his ideas?

Exercises

1. Use clustering and brainstorming to gather your ideas about the following quotations about science:

    > Destroying species is like tearing pages out of an unread book, written in a language humans hardly know how to read, about the place where they live.
    > — Robert Holmes III

    > [Science] is the least self-centered, the least narcissistic of the sciences—the one that, by taking us out of ourselves, leads us to re-establish the link with nature and to shake ourselves free from our spiritual isolation. — Jean Rostand

    > The role of science today, like the role of every other science, is simply to describe, and when it explains it does not mean that it arrives at finality; it only means that some descriptions are so charged with significance that they expose the relationship of cause and effect. — Donald Culross Peattie

    > We prefer economic growth to clean air. — Charles Barden

2. Go to the library and examine an issue of *Current Contents*. How could you use *Currents Contents* to write about science?

3. Locate the reference desk in the library. Use information there to locate five articles about a particular scientific topic of your choice. Make a reference card for each of the articles.

4. Choose any topic of science and define it for an audience of your classmates. Write the following about the subject:

    Five statements

    Three yes/no questions

    A thesis statement based on one of these questions

    Two facts supporting this statement

    A paragraph summarizing the subject

5. Choose a subject, create an audience, and brainstorm for 20 minutes. Then organize the ideas and create a tentative outline for the subject.

6. The human population is growing extremely fast. Consider these data:

Year	Population (millions)
8000 BC	5
4000 BC	86
1 AD	133
1650	545
1750	728
1800	906
1850	1130
1900	1610
1950	2400
1960	2998
1970	3659
1980	4551
1990	5300
2000	6500+ (projected)

Write a short essay describing the importance of these data. Do not repeat the data—discuss what they mean.

7. Go to the library and browse through recent issues of at least four different science journals. How do they differ? Do they have different audiences? If so, how can you tell? In which of the journals would you like to publish an article?

Journal	Features	Audience
1.		
2.		
3.		
4.		

8. Choose a paper from one of the journals you listed in the previous question. What hypotheses were the scientists testing? What were their assumptions? Do you agree with their conclusion? If given a chance, how would you rewrite the paper?

9. Go to the library and examine a science-related thesis, research paper, review article, book, and newspsper. How does the writing differ? How is it similar?

10. Discuss, support, or refute the ideas contained in these quotations:

Biology

Nothing in biology makes sense except in the light of evolution. — Theodosius Dobzhansky, 1900–1975

Evolution is the most powerful and the most comprehensive idea that has ever arisen on Earth. — Julian Huxley, 1887–1975

Eventually, we'll realize that if we destroy the ecosystem, we destroy ourselves. — Jonas Salk, 1914–1995

Today, the theory of evolution is an accepted fact for everyone but a fundamentalist minority, whose objections are based not on reasoning but on doctrinaire adherence [sic] to religious principles. — James D. Watson, 1928–

Chemistry

Neither cookery nor chemistry [has] been able to make milk out of grass. — William Paley (1743–1805)

A chemical compound once formed would persist for ever, if no alteration took place in the surrounding conditions. — Thomas Henry Huxley (1825–1895)

Geology

Geology gives us a key to the patience of God. — Josiah Gilbert Holland (1819–1881)

What clearer evidence could we have had of the different formation of these rocks, and of the long interval which separated their formation, had we actually seen them emerging from the bosom of the deep? . . . The mind seemed to grow giddy by looking so far into the abyss of time. — James Hutton (1726–1797)

Physics

Heat cannot be lessened or absorbed without the production of living force, or its equivalent attraction through space. — James Prescott Joule (1818–1889)

We shall never get people whose time is money to take much interest in atoms. — Samuel Butler (1835–1902)

Entropy is time's arrow. — Sir Arthur Stanley Eddington (1882–1944)

A theory is more impressive the greater is the simplicity of its premises, the more different are the kinds of things it relates, and the more extended its range of applicability. Therefore, the deep impression which classical thermodynamics made on me. It is the only physical theory of universal content which I am convinced, that within framework of applicability of its basic concepts will never be overthrown. — Albert Einstein (1879–1955)

CHAPTER THREE

Revising is Rethinking:

Writing to Communicate

Writing is largely a matter of application and hard work, of writing and rewriting endlessly until you are satisfied that you have said what you want to say as clearly and as simply as possible. For me that usually means many, many revisions.
— Rachel Carson

. . . [T]o eliminate the vice of wordiness is to ensure the virtue of emphasis, which depends more on conciseness than on any other factor. Wherever we can make twenty-five words do the work of fifty, we halve the area in which looseness and disorganization can flourish, and by reducing the span of attention required we can increase the force of thought. To make our words count for as much as possible is surely the simplest as well as the hardest secret of style.
— Wilson Follett, *Modern American Usage*

I have rewritten—often several times—every word I

have ever published. My pencils outlast their erasers.
— Vladimir Nabokov

As your experience grows you will find that revising is pleasurable, even though its purpose is the discovery of your own failings.
— Jacques Barzun

Don't be afraid to seize whatever you have written and cut it to ribbons. It can always be restored to its original condition in the morning.
— E.B. White

I'm probably the world's worst writer. But I'm the world's best rewriter.
— James Michener

The difficulty is not to write but to write what you mean, not to affect your reader but to affect him precisely as you wish.
— Robert Louis Stevenson

Only the hand that erases can write the true thing.
— Meister Eckhart

Having imagination, it takes you an hour to write a paragraph that, if you were unimaginative, would take you only a minute.
— Franklin P. Adams

Meaning is not what you start with, but what you end up with.
— Peter Elbow

When a thought is too weak to be expressed simply, it should be rejected.
— Vauven Argues

Revise, revise, revise.
— William Least Heat-Moon

The objective of writing a first draft of your paper was to identify and organize your ideas—to identify what you knew and, in the process, discover something that you didn't know. Thus, you wrote the first draft primarily for yourself. Now it's time to rethink your ideas and communicate with your audience. To do this you must revise your work.

Learning how to revise your writing is the key to becoming a good writer. Revision is often hard work because it traps you in a double bind: You can't find the right words until you know exactly what you want to say, but you can't know exactly what you are saying until you find just the right words. Although revising what you've written is difficult, it is the essence of good writing. It involves rewriting *and rethinking* what you've already written (then rewriting what you've rewritten) while constantly and carefully studying every word, sentence, and paragraph to determine if they say *exactly* what you mean. Revising is essential because, while preparing a first draft, you often must write what you'll delete before discovering what you want to keep. Similarly, to know what you're trying to say you often must look at what you've written and ask if you've said it. Good writing is revision, not magic in producing a first draft. As many writers like to say, "There is no good writing, only good revision."[1]

Revising improves your writing and thinking by increasing the clarity and forcefulness of your message. You'll go over and over what you've written. All great writers do. For example, James Thurber's wife rated a first draft of one of Thurber's manuscripts as "high school stuff," to which Thurber replied, "Wait until the seventh draft."

People unwilling to revise and rethink their writing miss an important way to learn about their subject and improve communication with their audience. William Faulkner rewrote *The Sound and the Fury* five times, and Hemingway wrote the last page of *A Farewell to Arms* 39 times—until, as he later said, he got "the words right." Revising your writing is where the hard work begins. To again quote Hemingway, "Wearing down seven #2 pencils is a good day's work." You'll wear down most of your pencils when you start revising your work because, contrary to what some scientists believe, you cannot think of everything at once as you write. What you write will constantly change and grow as your writing teaches you more about your subject. Therefore, think of your paper as a flexible framework rather than as a rigid box. This framework flexes most when you revise your paper.

---

[1]Many writers balk at having to revise a paper. Nevertheless, such revisions are essential to learning about your subject and to producing an effective paper or essay. Moreover, the revision process mimics science. Scientists routinely repeat and revise their experiments to best attack hypotheses they're testing. For example, in 1941 George Beadle and E.L. Tatum used X-rays to form 1,000 mutated spore-cultures of the fungus *Neurospora crassa*, an orange bread-mold. They fed each of these cultures special diets, noting which diets promoted growth and which did not. In the 299th culture, vitamin B_6 (pyridoxine) restored proper growth. Beadle and Tatum concluded that radiation had damaged a gene responsible for making an enzyme needed for synthesis of vitamin B_6. From this work, Beadle and Tatum announced the single-gene single-enzyme hypothesis, for which they shared the 1958 Nobel Prize in physiology and medicine. Their work, which helped put modern genetics on a chemical basis, remains important for chemically controlling genetic diseases.

Rewriting Is Rethinking

> What makes me happy is rewriting. In
> the first draft you get your ideas and
> your theme clear, if you are using some
> kind of metaphor you get that estab-
> lished, and certainly you have to know
> where you're coming out. But the next
> time through it's like cleaning house,
> getting rid of all the junk, getting things
> in the right order, tightening things up.
> I like the process of making writing neat.
> — Ellen Goodman

Many students view revising as a trivial, perfunctory clean-up involving little more than moving commas, while others view revising as punishment for not writing well the first time. Both of these attitudes inhibit your effectiveness because they ignore the power of writing as a learning tool. Revision improves your writing because *rewriting is rethinking*. No matter how mechanical, rewriting is critical to learning, understanding, and communicating.

You'll find many mistakes in the first draft of your paper. Some of these mistakes will be glaring: incomplete sentences, periods substituted for commas, and garbled paragraphs. Don't be alarmed by these mistakes; they're part of getting ideas out of your head and onto paper. Having done that with the first draft of your paper, what you must now do is correct the mistakes and produce a coherent document, both of which involve revising your work.

Revising your work is important for several reasons:

Revision improves learning. When done properly, revising your work reveals gaps and other problems that were disguised in your original paper by poor writing. Filling these gaps requires that you write new material. Moreover, it often takes you to a point from which you can't proceed without either connecting ideas or eliminating material. Choosing between these options requires that you analyze the structure of your writing, which is synonymous with examining the structure of your scientific argument. This leads to reinvestigating your ideas, which extends beyond discovering and recording data to interpretation. When you revise, you rethink.

Revision helps you communicate better with your audience. When you wrote the first draft of your paper, you wrote so that you could identify your ideas and discover new ideas. If you do not revise your work, all you'll have is a running account of your thinking. This may be interesting and valuable to you, but probably will not impress others.

Revision helps you scrutinize what you've written. You'll be able to polish, sharpen, correct mistakes and misquotes, and focus your writing. More importantly, you'll be able to make your writing more accurate, precise, consistent, concise, and fair—all of the things that will help you learn more about your subject and help you communicate better with readers.

Although revising your writing is hard work, it is also rewarding. It is like furnishing a house that you built: You'll be able to see the outline and general direction of the paper, while molding your paper into a comfortable, readable form. Just as the remodeled house eventually becomes livable, so too will your paper communicate well with readers.

The principles described in the upcoming chapters will tell you how to revise your papers. You'll be unable to apply all of the principles at once, because removing one layer of errors often makes another layer appear. Thus, you should expect to revise your paper several times, until you've incorporated all of the principles discussed in the upcoming chapters. You'll apply many principles rather than make one or two major changes in the paper. Although any one of the changes that you make may seem insignificant, their cumulative effect determines whether you write well or write poorly. In writing, the victory goes to those who recognize and correct the many small problems that impede learning and communication.

To write well, you must learn to effectively revise your work. The first step is to delete clutter from your first draft. When you delete clutter, clear thinking becomes clear writing.

Deleting Clutter: Simplicity and Conciseness

If men would only say what they have
to say in plain terms, how much more
eloquent they would be.
— Samuel Coleridge, *On Style*

Brevity in writing is the best insurance
for its perusal.
— Rudolph Virchow

Everything should be made as simple as
possible, but not simpler.
— Albert Einstein

Very simple ideas are within the reach of
only very complicated minds.
— Remy De Gourmont

> Beauty of style and harmony and grace
> and good rhythm depend on simplicity.
> — Plato

> Don't try to save junk just because it
> took you a long time to write it.
> — David Eddings

Linus Pauling said, "Science is the search for truth." If you hide the truth with vague, pretentious clutter, you've not advanced science. To advance science and to do yourself justice, you must delete the clutter from your writing.

Many books about scientific writing urge scientists to "be brief." However, this rule is useless because effective writing is not synonymous with brief writing. Rather, effective writing is *concise* writing that takes the shortest time to be understood. Calvin Coolidge's wife learned the difference between brief and concise when she asked her husband what the preacher had talked about at church. "Silent Cal" replied, "Sin." When his wife then asked "Well, what about it?" Calvin replied, "He was against it." Coolidge's responses were certainly brief, yet frustratingly incomplete and not concise.

To write clearly and concisely you must choose your words for minimum clutter and maximum strength. This first involves ruthlessly stripping your writing to its essentials by removing unnecessary words, pompous frills, and irrelevant details. All of this clutter is fuzz that annoys readers by obscuring and lessening the importance of your writing. Take Ernest Hemingway's crude but valuable advice: "The most essential gift for a good writer is a built-in shock-proof shit-detector."

Clutter appears in all kinds of writing and is especially common when people try to overstate simple ideas. Educators are notorious for this. Consider this example:

> Realization has grown that the curriculum or the experiences of learners
> change and improve only as those who are most directly involved examine
> their goals, improve their understandings and increase their skills in perform-
> ing the tasks necessary to reach newly defined goals. This places the focus
> upon the teacher, lay citizen and learner as partners in curricular improve-
> ment and as the individuals who must change, if there is to be curriculum
> change.

When you run this through your clutter decoder, you get something like this:

> Teachers, parents, and students must help if we are to change the curricu-
> lum.

Notice that the fog caused by the original version was not due to technical words —it was due merely to overstating a simple idea.

Tightening your writing into fewer and shorter words also saves space and readers' time while making your message more obvious and powerful. However, the trick is not merely to delete words—anyone can do that—but to know what words to delete and why.

Delete words that contribute nothing to what you're trying to say. Just as removing dead limbs and pruning can help a tree grow taller, so too can deleting unnecessary words strengthen a paper. However, just as too much pruning kills a tree, so too can too much deleting weaken your writing. Therefore, strive not for mere brevity, but rather for *conciseness*—brevity *and* completeness. To do this, get rid of the fuzz in your writing. A compulsion to keep everything means not that you write well, but merely that you lack good scientific judgment. So how long should a paper be? Only as long as it is interesting.

Although writing can be too compact and terse, wordiness is a more common problem. This is because it is harder to write a concise paper than a wordy one. As American humorist Ambrose Bierce once scribbled at the end of a letter to a friend, "Forgive me for writing such a long letter. I didn't have time to write a short one."

Whenever Possible, Use Simple Words and Sentences

> Genius is the ability to reduce the complicated to the simple.
> — C.W. Ceram

> A man of true science . . . uses but few hard words, and those only when none other will answer his purpose; whereas the smatterer in science . . . thinks that by mouthing hard words, he proves that he understands hard things.
> — Herman Melville

> Never fear big words. Long words name little things. All big things have little names, such as Life and death, Peace and war, or dawn, day, night, live, home. Learn to use little words in a big way—it is hard to do. But they say what you mean. When you don't know what you mean, use big words: They often fool little people.
> — SSC BOOKNEWS, July 1981

> Writing is easy. All you have to do is cross out the wrong words.
> — Mark Twain

> A poorly built sentence of twenty words is about four times as hard to attend to

and to understand as an equally poor
sentence of ten words.
— Walter Pitkin

Small words are the best, and the old
small words are best of all.
— Winston Churchill

Alice had not the slightest idea what
Latitude was, or Longitude either, but
she thought they were nice grand
words to say.
— Lewis Carroll, *Alice's*
Adventures in Wonderland

Many scientists are like Alice—they insist on using as many nice, grand words as possible, even when those words are unnecessary and when the writer is unsure of the words' meanings. Consequently, these writers often write tedious, incomprehensible papers and leave themselves open to the reprimand that Alice received: "Speak English!" said the Eaglet. "I don't know the meaning of half those long words, and, what's more, I don't believe you do either."

The Unavoidable Big Words

Bertrand Russell, when asked about the relationship of simple words to effective writing, said that, "Big men write little words, little men write big words." Although this is usually true in popular writing, it's not always true when writing about science. For example, there are no short, simple words to replace big words such as *endoplasmic reticulum, photosynthesis,* or *chromatography*. Other scientific words are even bigger. For example, the scientific name of the Dahlemense strain of tobacco mosaic virus has 1,185 letters, that of bovine glutamate dehydrogenase has 3,641 letters, and the name of tryptophane synthetase A (a listing of 267 amino acids) has 1,913 letters. Similarly, the longest word in the *Oxford English Dictionary Supplement* is *pneumonoultramicroscopicsilicovolcanoconiosis,* a 45-letter monstrosity that refers to a lung disease gotten by some miners. Coming in at a close second is *hepaticocholangiocholecystenterostomies,* a medical term referring to the surgical creation of new communications between gall bladders and hepatic ducts and between intestines and gall bladders. However, even these giants seem like runts when compared to the systematic name for DNA of the human mitochondria that contains 16,569 nucleotide sequences (published in key form in the April 9, 1981 issue of *Nature*); that word has about 207,000 letters.

Proofreader's Marks

You'll speed your editing by using proofreader's marks, a set of symbols that save you from having to repeat your editing instructions.

Problem	Symbol	Example
Omitted word	(caret)	study describes *the* ∧ effects
Omitted letter	(caret)	that b∘ok
Transpose letters	∼	f∼rom the ocean
Transpose words	∼	was only exposed
Capitalize word	☰ (three short underlines)	these data
Word should be lowercase	/ (slash)	These data
Italicize	— (underline once)	Homo sapiens
Separate words	| (draw vertical line between)	read|carefully
Delete word	—— (draw line through)	the ~~nice~~ data
Delete letter	/ (draw line through)	the nic∤e data
Do not leave a space	⌒ (sideways parentheses)	the e͡nd
Wrong letter	/ (draw line through and add correct letter above)	*f* ∤emale
Wrong word	—— (draw line through and add correct word above)	*These* ~~This~~ data
Start a new paragraph	¶ (paragraph symbol)	female. ¶ In contrast
Let it stand	(stet)	The energy ~~needs~~ *stet*
Insert comma	∧ (caret)	red, small, and poisonous

Text continued on page 92.

Let's apply these marks to this essay about anabolic steroids.

anabolic sterods function marvelously to develope maintaain ma le sed characteristics. We definitely know that they are synthetic derivatives of Testosterone because we have made them. They not only effect one phyiscally, but also psychologically and mentally. Some reviews of the subject suggest that anabolic steroids may even have an addictive potential similar to the many, many other drugs that are abused everyday by people. Anabolic steroids are psychoactive compounds, as evidenced by their well-documented affects on behavior. Neuronal anabolic steroid receptors have been identified in the brain, suggesting the neurochemical basis for their psychoactive effects. Most addicts suffer symptoms withdrawal and headaches was unsuccessful in his efforts to cut down on use, and continued to use the steroids despite their knowledge that he was having emotional and martial problems related to their use. These were just a few of examples to emphasize the possibility of psychoactive substance dependence. In conclusion, their patients developed a dependence on anabolicsteroids that was strikingly similar to dependencies seen with other drugs.

Here's how this essay looks after revision:

anabolic sterods ; are synthetic derivatives of ~~testosterone~~, a hormone that ~~function marvelously~~ to develop~~,~~ and maintains ~~male sex~~ in characteristics. ~~We definitely know that they are synthetic derivatives of Testosterone because we have made them. They not only effect one~~ Anabolic steroids affect people physically, ~~but also psychologically~~ and may be as and mentally. ~~Some reviews of the subject suggest that anabolic steroids may even have an~~ addictive ~~potential similar to the~~ as drugs such as cocaine. Indeed, our brains have receptors for ~~many, many other drugs that are abused everyday by people.~~ Anabolic steroids ~~are psychoactive compounds, as evidenced by their well-documented affects~~ ¶ Addiction to anabolic steroids resembles that to other ~~on behavior. Neuronal anabolic steroid receptors have been identified in the~~ drugs. For example, when deprived of anabolic steroids, ~~brain, suggesting the neurochemical basis for their psychoactive effects.~~ ~~Most~~ addicts suffer |symptoms| withdrawal |~~and were~~ headaches. ~~was unsuccessful in his~~ such as However, most addicts do not reduce their use of the ~~efforts to cut down on use, and continued to use the steroids despite their~~ drug, despite knowing the harm it causes. ~~knowledge that he was having emotional and martial problems related to their use. These were just a few of examples to emphasize the possibility of psychoactive substance dependence. In conclusion, their patients developed a dependence on anabolic steroids that was strikingly similar to dependencies seen with other drugs.~~

Here's the product:

> Anabolic steroids are synthetic derivatives of testosterone, a hormone that develops and maintains sex characteristics in males. Anabolic steroids affect people physically and mentally, and may be as addictive as drugs such as cocaine. Indeed, our brains have receptors for anabolic steroids.
>
> Addiction to anabolic steroids resembles that to other drugs. For example, when deprived of anabolic steroids, most addicts suffer withdrawal symptoms such as headaches. However, most addicts do not reduce their use of the drug, despite knowing the harm it causes.

Simple writing communicates well and indicates clear thinking. Consequently, the most effective writing uses short, simple words and phrases. For example, Lincoln's 701-word second inaugural address contained 505 one-syllable words and 122 two-syllable words. Similarly, some of the greatest scientists of all time, including Louis Pasteur, Albert Einstein, and Marie Curie, explained their complex ideas with simple words that other scientists and the public could understand and appreciate. These and other scientists knew that simple writing communicates more effectively than complex writing. They also knew that their clear, simple writing contributed as much to their stature and fame as did their brilliant ideas and discoveries.

If you want to communicate with your readers, use simple words and phrases. This produces concise writing, which can carry a tremendous punch. For example, the most powerful sentence in the New Testament is also its shortest: "Jesus wept." Similarly, the Lord's Prayer has only 56 words, the Twenty-third Psalm 118 words, the Gettysburg Address 258 words (196 of which are one-syllable), Newton's First Law of Motion 29 words, the First Amendment to the U.S. Constitution 45 words, the Ten Commandments only 75 words, and the Golden Rule only 11 words. For comparison, the U.S. Department of Agriculture's directive for pricing cabbage contains 15,629 words. You can guess which of these documents is hardest to read.

Just as simple words such as *life*, *death*, *love*, and *earth* describe profound concepts, simple sentences and paragraphs often convey much wisdom. Prefer the simple, understandable word or sentence to one that is stilted and long-winded. For example, writing that "The biota in question in this particular study exhibited a

100% mortality rate" wastes readers' time and shows that the writer is either unwilling or incapable of thinking clearly. Instead, write, "All of the rats died." Aim at functional beauty, not superficial grace, and constantly ask what can be deleted, shortened, or simplified in your writing. Simplicity is important in science, especially when stating conclusions for the first time. Also, as you'll learn in Chapter 5, simplicity has an added benefit: It simplifies punctuation.

I can't list all of the ways that scientists clutter their writing. However, here are the most common causes and solutions.

Delete Unnecessary Words

> The most valuable of all talents is that of never using two words when one will do.
> — Thomas Jefferson

> Use as many words as you need, and not one that you can get by without.
> — Robert Jordan

Each unnecessary word or phrase, no matter how small, tires readers and lessens their receptivity to your ideas. Therefore, delete windy phrases and "bankrupt" words that, rather than add information, only delay readers from grasping your ideas. For example, consider deleting these words and phrases:

As you know (if the readers know what you're about to say, don't waste their time by re-telling them the information)

Needless to say (if it's needless to say, then don't say it)

I might add that (just add it)

It is important to state here that (this suggests that other material that you've included may be unimportant)

It is worth noting that (this suggests that other things you've said are not worth noting)

When time permits (This may sound poetic, but it's inaccurate. Time cannot permit anything.)

As already stated (if you've already said it, don't say it again)

It goes without saying that (if it goes without saying, let it go without saying)

From this point of view, it is relevant to mention that

It is of interest to note that

In order to keep the problem in perspective, we would like to emphasize that

It is considered, in this connection, that

It appears that

As a matter of fact

As such

For the sake of

In the case of

In some instances

Take this opportunity

The fact that

The reason is

A distance of

A period of

As of this date

In a very real sense

In my opinion

Surely

Doubtless

Quite

As a matter of fact,

It stands to reason that

It has been shown that

It has long been known that

It has been demonstrated that

It must be remembered that

It may seem that

It is worthy to note that

It is clear that

It may be mentioned that

It is worth pointing out that

It is interesting to note that

It is significant to note that

It is relevant to mention here that

It is well known that

It should be kept in mind in this particular connection that

It is clearly obvious from data in Fig. 2 that

It has come to our attention that

It may be said that

For all intents and purposes

Generally

Virtually

Actually

Really

Kind of

Definitely

A lot

Perhaps

Sort of

Very

Practically

Certain

Given

Redundant words and phrases are also unnecessary. Lawyers are notorious for the poor writing that accompanies their use of redundant phrases such as *null and void, aid and abet, sum and substance, irrelevant and immaterial, cease and desist,* and *give and convey*—all of which are often buried in a blizzard of whereases and hereinafters. Although redundancy is essential for safety, it is awful in scientific writing. In writing, redundancies usually appear as tautologies, which are phrases that say the same thing twice using different words. Like stuttering in a speech, redundancy in writing detracts from communication. Don't confuse restatement with explanation and don't cover the same ground twice. Say something once, and say it well.

Delete redundant words and phrases such as:

~~viable~~ alternative

~~ultimate~~ outcome

~~past~~ medical history

~~young~~ juvenile

~~authentic~~ replica

period ~~of time~~

ask ~~the question~~

~~total of~~ 41 people

~~previously~~ found

~~advance~~ planning

by ~~means of~~

hurry ~~up~~

~~end~~ product

combine ~~together~~

consensus ~~of opinion~~

each ~~and every~~

~~entirely~~ eliminated

few ~~in number~~

~~personal~~ opinion

reduce ~~down~~

separate ~~out~~

disappear ~~from sight~~

~~come to an~~ end

for ~~the purpose of~~

~~duly~~ noted

~~mutual~~ cooperation

red ~~in color~~

~~already~~ existing

~~currently~~ underway

two ~~different~~ methods

~~absolutely~~ essential

9 a.m. ~~in the morning~~

~~advance~~ notice

~~general~~ consensus

~~absolutely~~ complete

assembled ~~together~~

~~exactly~~ the same

~~in close~~ proximity

after the ~~conclusion of~~

due to ~~the fact that~~

~~in order~~ to

near ~~the place of~~

~~completely~~ unanimous

cooperate ~~together~~

endorse ~~on the back~~

~~final~~ outcome

~~personal~~ friend

recur ~~again~~

resume ~~again~~

~~in connection~~ with

spherical ~~in shape~~

~~science of~~ physics

plan ~~ahead for the future~~

~~wish to~~ thank

~~most~~ unique

introduce ~~a new~~

~~basic~~ fundamentals

~~completely~~ eliminate

never ~~before~~

~~continue to~~ remain

~~first~~ began

mix ~~together~~

~~private~~ industry

~~the question as~~ to whether

basic ~~and fundamental~~

~~various~~ differences

~~each~~ individual

if ~~at all~~ possible

~~perform a~~ study

join ~~together~~

slow ~~up~~

hectares ~~of land~~

~~all~~ throughout

large ~~in size~~

~~exactly~~ identical

~~underlying~~ purpose

~~viable~~ solution

~~past~~ experience

~~joint~~ partnership

while ~~at the same time~~

~~at the time~~ when

~~and so on~~ and so forth

~~completely~~ finish

each ~~individual~~

~~initial~~ preparation

~~absolute~~ necessity

~~advance~~ reservation

~~close~~ proximity

~~separate~~ entities

~~currently~~ being

had done ~~previously~~

none ~~at all~~

~~joint~~ cooperation

any ~~and all~~

~~completely~~ finish

~~future~~ plans

~~unexpected~~ surprise

bisect ~~into two parts~~

~~most~~ unique

~~one and~~ the same

blame ~~it on~~

~~actual~~ facts

~~basic~~ essentials

~~definite~~ decision

few ~~in number~~

~~usual~~ custom

subject ~~matter~~

~~equally~~ as effective

~~two equal~~ halves

~~have~~ need ~~for~~

~~any and~~ all

~~full and~~ complete

~~various~~ different

near ~~the vicinity of~~

unusual ~~in nature~~

~~active~~ consideration

~~baffling~~ enigma

~~conclusive~~ proof

~~basic~~ fundamentals

brief ~~in duration~~

merge ~~together~~

repeat ~~the same~~

until ~~such time as~~

the ~~actual~~ number

~~conclusive~~ proof

stunted ~~in growth~~

~~hard~~ evidence

assemble ~~together~~

during ~~the course of~~

revert ~~back~~

~~advance~~ plan

~~current~~ status

~~honest~~ truth

~~overall~~ plan

repeat ~~again~~

this ~~particular~~ instance

whether ~~or not~~

by ~~means of~~

balance ~~against one another~~

range ~~all the way~~ from

add ~~an additional~~

during ~~the course of~~

~~excess~~ verbiage

~~mutual~~ cooperation

refer ~~back~~

file ~~away~~

protrude ~~out~~

write ~~up~~

because ~~of the fact that~~

few ~~in number~~

nominated for ~~the position of~~

~~make a~~ study ~~of~~

~~absolutely~~ essential

~~completely~~ surround

~~deliberately~~ chosen

~~quite~~ impossible

~~wholly~~ new

~~customary~~ practice

rarely ~~ever~~

any ~~and all~~

combine ~~into one~~

~~final~~ outcome

one ~~and the same~~

first ~~and foremost~~

refer ~~back~~ to

~~different~~ species

all ~~of~~

~~uniformly~~ consistent

cancel ~~out~~

for ~~the purpose of~~

continue ~~on~~

~~early~~ beginnings

~~necessary~~ requisite

debate ~~about~~

repeat ~~again~~

~~in~~ between

~~still~~ remain

circulate ~~around~~

~~close~~ scrutiny   ~~completely~~ full

~~consequent~~ results   ~~equal~~ halves

~~definite~~ proof   consensus ~~of opinion~~

enclosed ~~herewith~~   ~~end~~ result

~~entirely~~ eliminate   fewer ~~in number~~

~~in conjunction~~ with   if ~~it is assumed that~~

by ~~means of~~   ~~at a~~ later ~~date~~

~~serious~~ crisis   smaller ~~in size~~

subject ~~matter~~   ~~new~~ initiatives

join ~~together~~   mix ~~together~~

blend ~~together~~   is ~~defined as~~

Avoid placing adjectives before absolute words such as *dead, extinct, unique,* and *final* that resist modifiers. Also avoid throat-clearing words such as *basically* and *ideally*. These words add nothing to a sentence's meaning. Similarly, avoid tagging nouns with words that identify those nouns. For example:

Our geological survey was done 100 km south of ~~the city of~~ Chicago, Illinois during ~~the month of~~ April. The scientists included ~~a group of~~ physicists.

Also watch for redundant nouns such as *size* (large ~~in size~~), *color* (green ~~in color~~), *weight* (heavy ~~in weight~~), *process* (~~the process of~~ cellular respiration), *concept* (~~the concept of~~ science), and *number* (few ~~in number~~). Write that a rock formation is brown, not that it is brown *in color*. Be a word-miser, but don't hesitate to use a longer word, phrase, or sentence when it makes your writing more clear, more precise, or more accurate. Also watch for words such as *phenomenon, virtually, element, objective, primary,* and *constitute*; some scientists use these words to dress up a simple statement or to give an air of impartiality to a biased judgment.

Rid your writing of pretentious, pseudointellectual words such as *utilize* and *facilitate*—just say *use* and *make*. If you insist on using pretentious, bloated language, your writing will resemble a failing dieter who can't resist second helpings and desserts, neither of which help them achieve their goal of losing weight. Writers who can't resist inserting unnecessary phrases such as, "*It is interesting to note that,*" produce fat, lifeless papers that few scientists will want to read and even fewer editors will want to publish.

Many scientists insist on using Latin or Greek derivations of Anglo-Saxon words rather than the simpler Anglo-Saxon words themselves. These folks write sentences such as, "I prefer an abbreviated phraseology, distinguished for its lucidity," when all they mean is, "I like short, clear words." Prefer the short, simple word. *Avoid hippopotomonstrosesquipedalian construction of written prose.* To write effectively, avoid ~~excess~~ verbiage. For example,

POMPOUS: Prices were impacted adversely by the season's aridity, which was deleterious to agriculture.

IMPROVED: The dry summer hurt the crops and increased prices.

As George Orwell said, "If it is possible to cut a word out, always cut it out."

Replace Large Words and Phrases With Simple Words

A word may be a fine-sounding word, of an unusual length, and very imposing from its learning and novelty, and yet in the connection in which it is introduced, may be quite pointless and irrelevant.
— William Hazlitt

If you don't understand something, use longer words. Share the ignorance with your readers.
— George Orwell

The pseudo prestige of long and difficult words transcends the useful scientific term and diffuses widely through our papers. Simple things are made complicated, and the complex is made incomprehensible. Chaos reigns. The so-called medical literature is stuffed to bursting with junk, written in a hopscotch style characterized by a Brownian movement of uncontrolled parts of speech which seethe in restless unintelligibility.
— William B. Bean

Most of the fundamental ideas of science are essentially simple, and may, as a rule, be expressed in a language comprehensible to everyone.
— Albert Einstein

Many scientists use wordy and excessively formal language either to show others how smart they are or to compensate for their professional shortcomings. These writers write to impress rather than to express, and therefore ignore their responsibility of communicating with readers.

You'll communicate most effectively, and therefore impress your readers most favorably, if you choose simple, direct words over long-winded, pretentious words. Furthermore, expressing your ideas in simple terms is a sure way to determine if your reasoning is logical. If it is, your logic will be obvious when you simplify your writing. If it isn't, that too will be clear, and you'll probably have a tendency to leave the clutter in your writing as a way of covering your lack of thought and logic. Instead of relying on cluttered, poor writing to deceive readers, rethink and rewrite what you've written. If you write poorly, you'll only divert attention from your ideas.[2]

Here are some long-winded words and phrases (in the right column) that can be replaced by simpler words (in the left column):

about	the order of; as regards; with regard to; concerning the matter of; in regard to; approximately; in the neighborhood of; in the approximate amount of; in reference to
adjust	make an adjustment
after	subsequent to; at the conclusion of; following
agree	are found to be in agreement; are of the same opinion
allow	afford an opportunity to
also	additionally
although	despite the fact that; notwithstanding the fact that
always	in all cases
analyze	perform an analysis of
apparently	it is apparent that
are	have been shown to be
as	in view of the fact that
ask	inquire; request
aware	cognizant of
because	due to the fact that; as a consequence of; as a result of; in light of the fact that; owing to the fact that; because of the fact that; on account of the fact that; inasmuch as; on the grounds that; prior to this point in time; accounted for by the fact
before	in advance of; prior to

[2]Many authors divert readers' attention by using unnecessarily complex writing. Siegfried (Siegfried, John J. 1970. A first lesson in econometrics. *Journal of Political Economy* 78: 1378–1379) humorously attacked this strategy with a "rigorous" mathematical demonstration that 1 + 1 = 2.

begin	initiate; commence
build	construct
can	has the ability to; is able to; is in a position to; has the opportunity to; has the capacity for
cannot	not in a position to
cause	give rise to
change	modification
claim	allege
consider	take into consideration
copy	duplicate
cut	incision
decide	arrive at a decision
describe	give an account of
despite	in spite of the fact that
do	perform; accomplish; achieve; implement; carry out
during	in the course of
early	ahead of schedule
ease	facilitate
education	educational process
end	terminate; conclude; finalize
enough	sufficient
examine	make an examination of
happen	eventuate
exceeds	in excess of
except	with the exception of; with the only difference being
expect	anticipate
explain	elucidate
extra	superfluous; additional
face	confront
few	small number of

find	ascertain the location of; locate
finish	bring to a conclusion
first	initial; the thing to do before anything else
for	in the amount of; on behalf of
for example	an example of this is the fact that
free	disengage
frequently	is often the case that
full	replete
fully	to the fullest possible extent
get	procure; obtain
give	donate; contribute
go	proceed
goal	objective
happen	eventuate
have	am in possession of
help	facilitate; assist; render assistance to
judge	adjudicate
idea	concept
if	in the event that; assuming that; under conditions in which; in case
important	of great importance
in	during the month of
inaccurate	not a high order of accuracy
inhibited	produced an inhibitory efffect
instead	in lieu of
investigate	conduct an investigation of
is	has been shown to be; is widely acknowledged to be
know	be cognizant of
later	subsequently
like	along the lines of; similar in character to

list	enumerate
long	lengthy
lost weigh	experienced a weight loss
later	subsequently
many	a considerable number of; a large number of
may	the chance that; is possible that; could happen that
measure	quantify
meet	come in contact with; make the acquaintance of; encounter
methods	methodology
most	the majority of; the predominant number of
much	a considerable amount of; a great deal of; quite a large quantity of
must	is necessary that
named	by the name of
near	in the vicinity of; in close proximity to
next	subsequent
never	in no case; on no occasion
notice	take cognizance of
now	at the present time; at this point in time; presently; in this day and age
often	in many cases
opposite	antithesis
part	component
perhaps	it may well be that
please	I would appreciate it if you would; I would like to ask that
propose	make a proposal
previously	at an earlier date
purify	achieve purification
rapidly	at a rapid rate
rarely	in rare cases; in only a very small number of cases
regularly	on a regular basis

remind	call attention to the fact that
replaces	supersedes
require	involve the necessity of
result	resultant effect
risk	jeopardize
satisfactorily	in a satisfactory manner
say	assert
saw	was witness to
several	a number of
show	demonstrate
some	a number of
sometimes	in some cases
soon	in the not-too-distant future
start	initiate
stress	place a major emphasis on
study	undertake an examination of
surgery	surgical procedure
then	after this has been done
tiny	miniscule
try	endeavor
twice	on two separate occasions
use	utilize
were	proved to be
while	during the time that
without	in the absence of
yearly	on an annual basis

Do not use imaginary words such as *analyzation, interpretate, irregardless,* and *unequivocably.* Replace these non-existent words with *analysis, interpret, regardless,* and *unequivocally.* Finally, remember that *inflammable* and *flammable* both mean *combustible.* If you want to say that something will not burn, just say that it's *nonflammable* or *noncombustible.*

Don't let your writing sound like Dr. Ray Stantz (played by Dan Akroyd) of *Ghostbusters*. Ray is bright, but always babbles about the latest scientific gizmos: "If the ionization rate is constant for all ectoplasmic entities, we could really bust some heads—in a spiritual sense." Conversely, Dr. Peter Venkman (Bill Murray) is some- one like the rest of us and has no idea what phrases such as "total protonic rever- sal" or "PKE valences" mean. Peter tells Ray, ". . . just tell me what the hell is going on."

Take Peter's advice. Just tell readers what is going on. You'll do that best by simplifying your writing.

Simplify Your Writing

> Genius is the ability to reduce the com- plicated to the simple.
> — C.W. Ceram

> It is the essence of genius to make use of the simplest ideas.
> — Charles Peguy

> On the whole, I think the pains which my father took over the literary part of the work was very remarkable. He often laughed or grumbled at himself for the difficulty which he found in writing English, saying, for instance, that if a bad arrangement of a sentence was pos- sible, he would be sure to adopt it . . . When a sentence got hopelessly involved, he would ask himself "now what do you want to say?" and his answer written down, would often dis- entangle the confusion.
> — Francis Darwin, *The Life of Charles Dar win*

Just as unneeded pounds slow the movement of an overweight person, so too do excess words rob your writing of vitality by obscuring your ideas. Verbiage inhibits communication because the more a reader has to concentrate on, the less attention he or she can give to an idea. However, simple writing can also be flat and dry. The product resembles unsalted meat and potatoes—edible, but hardly memorable. To produce memorable writing that communicates effectively, you must make your writing clear and concise.

Write as simply and concisely as you can. Although wordiness is not synonymous with redundancy, it is a first cousin. Moreover, it usually accompanies shallow thought or poor understanding. Therefore, don't be a windbag—don't use 20 words to say what you can with 15. Remember that there's no direct relationship between the length of what you write and its significance. For example, Watson's and Crick's paper in the April, 1953 issue of *Nature* describing the structure of DNA was only one page long and reported no experimental data, yet it revolutionized science (see discussion on pages 4–6).

If you find yourself using long words and elaborate phrases, remember that short phrases of short words usually communicate more effectively than do long phrases of long words. This is why Erasmus Darwin (Charles's grandfather) left the long words out of his medical encyclopedia that he wrote in 1794; he did so because, "a short periphrasis is easier to be understood, and less burdensome to the memory." If you need to look up what *periphrasis* means, you'll know what he meant.

Simplify your writing by asking yourself what can be eliminated from your writing. Use the fewest number of words to say exactly what you mean in a way that cannot be misunderstood.

Rachel Carson

The Obligation to Endure

Few books have had a greater impact than Rachel Carson's (1907–1964) *Silent Spring*, one of the most important books of the twentieth century. Carson wrote *Silent Spring* in response to a letter that she received in January of 1958 from Olga Huckins describing how a small part of the world had been made lifeless by pesticides. Carson's response was that, "There would be no peace for me if I kept silent."

Carson, who had spent most of her professional life as a writer and marine biologist with the U.S. Bureau of Fisheries (now the U.S. Fish and Wildlife Service), began to study the use and effects of pesticides such as DDT. In 1962, she published *Silent Spring*. Most magazines, fearing lost advertising income, refused to publish excerpts of the book, despite Carson's fame for writing *The Sea Around Us* several years earlier.[1] Chemical companies spent hundreds of thousands of dollars to block publication, and one manufacturer of baby food even claimed that her article would cause, "unwarranted fear" to mothers who used their product. The book was attacked as scientifically unsound and emotionally motivated, and Carson was maligned as "ignorant," "hysterical," and "not a real scientist." Nevertheless, Carson's powerful message, meticulous research, and brilliant writing made *Silent Spring* a best seller. Moreover, *Silent Spring* prompted President Kennedy to create a special panel of

[1]The only exception to this was *The New Yorker*, which serialized *Silent Spring* before the book's publication. *The Sea Around Us*, published in 1951, won a National Book Award and went into eleven printings in its first year.

his Science Advisory Committee to study the use of pesticides. That panel's report vindicated Carson and stimulated the government to act against pollution. *Silent Spring* helped launch the movement that made "ecology" a household word.

Here is "The Obligation to Endure" from *Silent Spring*. This chapter uses imaginative fiction and scientific exposition to persuade readers.

The history of life on earth has been a history of interaction between living things and their surroundings. To a large extent, the physical form and the habits of the earth's vegetation and its animal life have been molded by the environment. Considering the whole span of earthly time, the opposite effect, in which life actually modifies its surroundings, has been relatively slight. Only within the moment of time represented by the present century has one species—man—acquired significant power to alter the nature of his world.

During the past quarter century this power has not only increased to one of disturbing magnitude but it has changed in character. The most alarming of all man's assaults upon the environment is the contamination of air, earth, rivers, and sea with dangerous and even lethal materials. This pollution is for the most part irrecoverable; the chain of evil it initiates not only in the world that must support life but in living tissues is for the most part irreversible. In this now universal contamination of the environment, chemicals are the sinister and little-recognized partners of radiation in changing the very nature of the world—the very nature of its life. Strontium 90, released through nuclear explosions into the air, comes to earth in rain or drifts down as fallout, lodges in soil, enters into the grass or corn or wheat grown there, and in time takes up its abode in the bones of a human being, there to remain until his death. Similarly, chemicals sprayed on croplands or forests or gardens lie long in soil, entering into living organisms, passing from one to another in a chain of poisoning and death. Or they pass mysteriously by underground streams until they emerge and, through the alchemy of air and sunlight, combine into new forms that kill vegetation, sicken cattle, and work unknown harm on those who drink from once-pure wells. As Albert Schweitzer has said, "Man can hardly even recognize the devils of his own creation."

It took hundreds of millions of years to produce the life that now inhabits the earth—eons of time in which that developing and evolving and diversifying life reached a state of adjustment and balance with its surroundings. The environment, rigorously shaping and directing the life it supported, contained elements that were hostile as well as supporting. Certain rocks gave out dangerous radiation; even within the light of the sun, from which all life draws its energy, there were short-wave radiations with power to injure. Given time—time not in years but in millennia—life adjusts, and a balance has been reached. For time is the essential ingredient; but in the modern world there is no time.

The rapidity of change and the speed with which new situations are created follow the impetuous and heedless pace of man rather than the deliberate pace of nature. Radiation is no longer merely the background radiation of rocks, the bombardment of cosmic rays, the ultraviolet of the sun that have existed before there was any life on earth; radiation is now the unnatural creation of man's tampering with the atom. The chemicals to which life is asked to make its adjustment are no longer merely the calcium and silica and copper and all the rest of the minerals washed out of the rocks and carried in rivers to the sea; they are the synthetic creations of man's inventive mind, brewed in his laboratories, and having no counterparts in nature.

To adjust to these chemicals would require time on the scale that is nature's; it would require not merely the years of a man's life but the life of generations. And even this, were it by some miracle possible, would be futile, for the new chemicals come from our laboratories in an endless stream; almost five hundred annually find their way into actual use in the United States alone. The figure is staggering and its implications are not easily grasped—500 new chemicals to which the bodies of men and animals are required somehow to adapt each year, chemicals totally outside the limits of biologic experience.

Among them are many that are used in man's war against nature. Since the mid-1940's over 200 basic chemicals have been created for use in killing insects, weeds, rodents, and other organisms described in the modern vernacular as "pests"; and they are sold under several thousand different brand names.

These sprays, dusts, and aerosols are now applied almost universally to farms, gardens, forests, and homes—nonselective chemicals that have the power to kill every insect, the "good" and the "bad," to still the song of birds and the leaping of fish in the streams, to coat the leaves with a deadly film, and to linger on in soil—all this though the intended target may be only a few weeds or insects. Can anyone believe it is possible to lay down such a barrage of poisons on the surface of the earth without making it unfit for all life? They should not be called "insecticides," but "biocides."

The whole process of spraying seems caught up in an endless spiral. Since DDT was released for civilian use, a process of escalation has been going on in which ever more toxic materials must be found. This has happened because insects, in a triumphant vindication of Darwin's principle of the survival of the fittest, have evolved super races immune to the particular insecticide used, hence a deadlier one has always to be developed—and then a deadlier one than that. It has happened also because, for reasons to be described later, destructive insects often undergo a "flareback," or resurgence, after spraying, in numbers greater than before. Thus the chemical war is never won, and all life is caught in its violent crossfire.

Along with the possibility of the extinction of mankind by nuclear war, the central problem of our age has therefore become the contamination of man's total environment with such substances of incredible potential for harm—substances that accumulate in the tissues of plants and animals and even penetrate the germ cells to shatter or alter the very material of heredity upon which the shape of the future depends.

Some would-be architects of our future look toward a time when it will be possible to alter the human germ plasm by design. But we may easily be doing so now by inadvertence, for many chemicals, like radiation, bring about gene mutations. It is ironic to think that man might determine his own future by something so seemingly trivial as the choice of an insect spray.

All this has been risked—for what? Future historians may well be amazed by our distorted sense of proportion. How could intelligent beings seek to control a few unwanted species by a method that contaminated the entire environment and brought the threat of disease and death even to their own kind? Yet this is precisely what we have done. We have done it, moreover, for reasons that collapse the moment we examine them. We are told that the enormous and expanding use of pesticides is necessary to maintain farm production. Yet is our real problem not one of *overproduction?* Our farms, despite measures to remove acreages from production and to pay farmers *not* to produce, have yielded such a staggering excess of crops that the American taxpayer in 1962 is paying out more than one billion dollars a year as the total carrying cost of the surplus-food storage program. And is the situation helped when one branch of the Agriculture Department tries to reduce production while another states, as it did in 1958, "It is believed generally that reduction of crop acreages under provisions of the Soil Bank will stimulate interest in use of chemicals to obtain maximum production on the land retained in crops."

All this is not to say there is no insect problem and no need of control. I am saying, rather, that control must be geared to realities, not to mythical situations, and that the methods employed must be such that they do not destroy us along with the insects.

The problem whose attempted solution has brought such a train of disaster in its wake is an accompaniment of our modern way of life. Long before the age of man, insects inhabited the earth—a group of extraordinarily varied and adaptable beings. Over the course of time since man's advent, a small percentage of the more than half a million species of insects have come into conflict with human welfare in two principal ways: as competitors for the food supply and as carriers of human disease.

Disease-carrying insects become important where human beings are crowded together, especially under conditions where sanitation is poor, as in time of natural disaster or war or in situations of extreme poverty and deprivation. Then control of some sort becomes necessary. It is a sobering fact, however, as we shall presently see, that the method of massive chemical control has had only limited success, and also threatens to worsen the very conditions it is intended to curb.

Under primitive agricultural conditions the farmer had few insect problems. These arose with the intensification of agriculture—the devotion of immense acreages to a single crop. Such a system set the stage for explosive increases in specific insect populations. Single-crop farming does not take advantage of the principles by which nature works; it is agriculture as an engineer might conceive it to be. Nature has introduced great variety into the landscape, but man has displayed a passion for simplifying it. Thus he undoes the built-in checks and balances by which nature holds the species within bounds. One important natural check is a limit on the amount of suitable habitat for each species. Obviously then, an insect that lives on wheat can build up its population to much higher levels on a farm devoted to wheat than on one in which wheat is intermingled with other crops to which the insect is not adapted.

The same thing happens in other situations. A generation or more ago, the towns of large areas of the United States lined their streets with the noble elm tree. Now the beauty they hopefully created is threatened with complete destruction as disease sweeps through the elms, carried by a beetle that would have only limited chance to build up large populations and to spread from tree to tree if the elms were only occasional trees in a richly diversified planting.

Another factor in the modern insect problem is one that must be viewed against a background of geologic and human history: the spreading of thousands of different kinds of organisms from their native homes to invade new territories. This worldwide migration has been studied and graphically described by the British ecologist Charles Elton in his recent book *The Ecology of Invasions*. During the Cretaceous Period, some hundred million years ago, flooding seas cut many land bridges between continents and living things found themselves confined in what Elton calls "colossal separate nature reserves." There, isolated from others of their kind, they developed many new species. When some of the land masses were joined again, about 15 million years ago, these species began to move out into new territories—a movement that is not only still in progress but is now receiving considerable assistance from man.

The importation of plants is the primary agent in the modern spread of species, for animals have almost invariably gone along with the plants, quarantine being a comparatively recent and not completely effective innovation. The United States Office of Plant Introduction alone has introduced almost 200,000 species and varieties of plants from all over the world. Nearly half of the 180 or so major insect enemies of plants in the United States are accidental imports from abroad, and most of them have come as hitchhikers on plants.

In new territory, out of reach of the restraining hand of the natural enemies that kept down its numbers in its native land, an invading plant or
animal is able to become enormously abundant. Thus it is no accident that
our most troublesome insects are introduced species.

These invasions, both the naturally occurring and those dependent on
human assistance, are likely to continue indefinitely. Quarantine and massive chemical campaigns are only extremely expensive ways of buying time.
We are faced, according to Dr. Elton, "with a life-and-death need not just to
find new technological means of suppressing this plant or that animal";
instead we need the basic knowledge of animal populations and their relations to their surroundings that will "promote an even balance and damp
down the explosive power of outbreaks and new invasions."

Much of the necessary knowledge is now available but we do not use it.
We train ecologists in our universities and even employ them in our governmental agencies but we seldom take their advice. We allow the chemical
death rain to fall as though there were no alternative, whereas in fact there
are many, and our ingenuity could soon discover many more if given
opportunity.

Have we fallen into a mesmerized state that makes us accept as
inevitable that which is inferior or detrimental, as though having lost the
will or the vision to demand that which is good? Such thinking, in the
words of the ecologist Paul Shepard, "idealizes life with only its head out of
water, inches above the limits of toleration of the corruption of its own
environment. . . . Why should we tolerate a diet of weak poisons, a home
in insipid surroundings, a circle of acquaintances who are not quite our
enemies, the noise of motors with just enough relief to prevent insanity?
Who would want to live in a world which is just not quite fatal?"

Yet such a world is pressed upon us. The crusade to create a chemically
sterile, insect-free world seems to have engendered a fanatic zeal on the
part of many specialists and most of the so-called control agencies. On
every hand there is evidence that those engaged in spraying operations
exercise a ruthless power. "The regulatory entomologists . . . function as
prosecutor, judge and jury, tax assessor and collector and sheriff to enforce
their own orders," said Connecticut entomologist Neely Turner. The most
flagrant abuses go unchecked in both state and federal agencies.

It is not my contention that chemical insecticides must never be used. I
do not contend that we have put poisonous and biologically potent chemicals indiscriminately into the hands of persons largely or wholly ignorant of
their potentials for harm. We have subjected enormous numbers of people
to contact with these poisons, without their consent and often without their
knowledge. If the Bill of Rights contains no guarantee that a citizen shall be
secure against lethal poisons distributed either by private individuals or by
public officials, it is surely only because our forefathers, despite their considerable wisdom and foresight, could conceive of no such problem.

I contend, furthermore, that we have allowed these chemicals to be used with little or no advance investigation of their effect on soil, water, wildlife, and man himself. Future generations are unlikely to condone our lack of prudent concern for the integrity of the natural world that supports all life.

There is still very limited awareness of the nature of the threat. This is an era of specialists, each of whom sees his own problem and is unaware of or intolerant of the larger frame into which it fits. It is also an era dominated by industry, in which the right to make a dollar at whatever cost is seldom challenged. When the public protests, confronted with some obvious evidence of damaging results of pesticide applications, it is fed little tranquilizing pills of half truth. We urgently need an end to these false assurances, to the sugar coating of unpalatable facts. It is the public that is being asked to assume the risks that the insect controllers calculate. The public must decide whether it wishes to continue on the present road, and it can do so only when in full possession of the facts. In the words of Jean Rostand, "The obligation to endure gives us the right to know."

Understanding What You've Read

Where and why does Carson use value-laden as compared to neutral language?

Write a paragraph describing why we should or shouldn't (1) use pesticides to increase crop-yields, and (2) sacrifice forests to land developers. Try to persuade your audience to act.

Rachel Carson said that, "As cruel a weapon as the cave man's club, the chemical barrage has been hurled against the fabric of life." Do you agree with this statement? Write a short essay describing your answer.

By effectively arguing against indiscriminate dumping of chemicals into the environment, *Silent Spring* became as important to stirring the American conscience as Tom Paine's *Common Sense*, Upton Sinclair's *The Jungle*, and Harriet Beecher Stowe's *Uncle Tom's Cabin*. What other science books have influenced society? Why did they have such a strong impact?

Isaac Asimov

Organic Synthesis

The late Isaac Asimov (1920–1992) is the same scientist that wrote the controversial essay entitled "The
Next Frontiers for Science" in Chapter One. After
several years on the faculty of Boston University's
School of Medicine, Asimov left to pursue his con-
suming passion: writing about science and science
fiction.
    This selection, from his *An Intelligent Man's
Guide to Science*, shows Asimov's practical style of
writing about science.

After Wöhler had produced urea from ammonium cyanate and chemists
had formed various other organic molecules by trial and error, in the
1850's there came a chemist who went systematically and methodically
about the business of synthesizing organic substances in the laboratory. He
was the Frenchman Pierre Eugène Marcellin Berthelot. He prepared a number of simple organic compounds from still simpler inorganic compounds
such as carbon monoxide. Berthelot built his simple organic compounds up
through increasing complexity until he finally had ethyl alcohol, among
other things. It was "synthetic ethyl alcohol," to be sure, but absolutely
indistinguishable from the "real thing," because it *was* the real thing.

    Ethyl alcohol is an organic compound familiar to all and highly valued
by most. No doubt the thought that the chemist could make ethyl alcohol
from coal, air, and water (coal to supply the carbon, air the oxygen, and
water the hydrogen), without the necessity of fruits or grain as a starting
point, must have created enticing visions and endowed the chemist with a
new kind of reputation as a miracle worker. At any rate, it put organic synthesis on the map.

    For chemists, however, Berthelot did something even more significant.
He began to form products that did not exist in nature. He took "glycerol,"
a component obtained from the breakdown of the fats of living organisms,
and combined it with acids not known to occur naturally in fats (although
they occurred naturally elsewhere). In this way he obtained fatty substances
which were not quite like those that occurred in organisms.

    Thus Berthelot laid the groundwork for a new kind of organic chem-
istry—the synthesis of molecules that nature could not supply. This meant

the possible formation of a kind of "synthetic" which might be a substitute—perhaps an inferior substitute—for some natural compound that was hard or impossible to get in the needed quantity. But it also meant the possibility of "synthetics" which were improvements on anything in nature.

This notion of improving on nature in one fashion or another, rather than merely supplementing it, has grown to colossal proportions since Berthelot showed the way. The first fruits of the new outlook were in the field of dyes.

The beginnings of organic chemistry were in Germany. Wöhler and Liebig were both German, and other men of great ability followed them. Before the middle of the nineteenth century, there were no organic chemists in England even remotely comparable to those in Germany. In fact, English schools had so low an opinion of chemistry that they taught the subject only during the lunch recess, not expecting (or even perhaps desiring) many students to be interested. It is odd, therefore, that the first feat of synthesis with world-wide repercussions was actually carried through in England.

It came about in this way. In 1845, when the Royal College of Science in London finally decided to give a good course in chemistry, it imported a young German to do the teaching. He was August Wilhelm von Hofmann, only 27 at the time, and he was hired at the suggestion of Queen Victoria's husband, the Prince Consort Albert (who was himself of German birth).

Hofmann was interested in a number of things, among them coal tar, which he had worked with on the occasion of his first research project under Liebig. Coal tar is a black, gummy material given off by coal when it is heated strongly in the absence of air. The tar is not an attractive material, but it is a valuable source of organic chemicals. In the 1840's, for instance, it served as a source of large quantities of reasonably pure benzene and of a nitrogen-containing compound called "aniline" which is related to benzene.

About ten years after he arrived in England, Hofmann came across a 17-year-old boy studying chemistry at the college. His name was William Henry Perkin. Hofmann had a keen eye for talent and knew enthusiasm when he saw it. He took on the youngster as an assistant and set him to work on coal-tar compounds. Perkin's enthusiasm was tireless. He set up a laboratory in his home and worked there as well as at school.

Hofmann, who was also interested in medical applications of chemistry, mused aloud one day in 1856 on the possibility of synthesizing quinine, a natural substance used in the treatment of malaria. Now those were the days before structural formulas had come into their own. The only thing known about quinine was its composition, and no one at the time had any idea of just how complicated its structure was.

Blissfully ignorant of its complexity, Perkin, at the age of 18, tackled the problem of synthesizing quinine. He began with allyltoluidine, one of his coal-tar compounds. This molecule seemed to have about half the numbers of the various types of atoms that quinine had in its molecule. If he put two of these molecules together and added some missing oxygen atoms (say by mixing in some potassium dichromate, known to add oxygen atoms to chemicals with which it was mixed), Perkin thought he might get a molecule of quinine.

Naturally this approach got Perkin nowhere. He ended with a dirty, red-brown goo. Then he tried aniline in place of allyltoluidine and got a black-ish goo. This time, though, it seemed to him that he caught a purplish glint in it. He added alcohol to the mess, and the colorless liquid turned a beautiful purple. At once Perkin thought of the possibility that he had discovered something that might be useful as a dye.

Dyes had always been greatly admired, and expensive, substances. There were only a handful of good dyes—dyes that stained fabric permanently and brilliantly and did not fade or wash out. There was dark blue indigo, from the indigo plant; there was "Tyrian purple," from a snail (so-called because ancient Tyre grew rich on its manufacture—in the later Roman Empire the royal children were born in a room with hangings dyed with Tyrian purple, whence the phrase "born to the purple"); and there was reddish alizarin, from the madder plant ("alizarin" came from Arabic words meaning "the juice"). To these inheritances from ancient and medieval times later dyers had added a few tropical dyes and inorganic pigments (today used chiefly in paints).

This explains Perkin's excitement about the possibility that his purple substance might be a dye. At the suggestion of a friend, he sent a sample to a firm in Scotland which was interested in dyes, and quickly the answer came back that the purple compound had good properties. Could it be supplied cheaply? Perkin proceeded to patent the dye (there was considerable argument as to whether an 18-year-old could obtain a patent, but eventually he obtained it), to quit school, and to go into business.

His project wasn't easy. Perkin had to start from scratch, preparing his own starting materials from coal tar with equipment of his own design. Within six months, however, he was producing what he named "Aniline Purple"—a compound not found in nature and superior to any natural dye in its color range.

French dyers, who took to the new dye more quickly than did the more conservative English, named the color "mauve," from the mallow (Latin name "malva"), and the dye itself came to be known as "mauveine." Quickly it became the rage (the period being sometimes referred to as the "Mauve Decade"), and Perkin grew rich. At the age of 23 he was the world authority on dyes.

The dam had broken. A number of organic chemists, inspired by Perkin's astonishing success, went to work synthesizing dyes, and many succeeded. Hofmann himself turned to this new field, and in 1858 he synthesized a red-purple dye which was later given the name "magenta" by the

French dyers (then, as now, arbiters of the world's fashions). The dye was named for the Italian city where the French defeated the Austrians in a battle in 1859.

Hofmann returned to Germany in 1865, carrying his new interest in dyes with him. He discovered a group of violet dyes still known as "Hofmann's violets."

Chemists also synthesized the natural dyestuffs in the laboratory. Karl Graebe of Germany and Perkin both synthesized alizarin in 1869 (Graebe applying for the patent one day sooner than Perkin), and in 1880 the German chemist Adolf von Baeyer worked out a method of synthesizing indigo. (For his work on dyes von Baeyer received the Nobel Prize in chemistry in 1905.)

Perkin retired from business in 1874, at the age of 35, and returned to his first love, research. By 1875 he had managed to synthesize coumarin (a naturally-occurring substance which has the pleasant odor of new-mown hay); this served as the beginning of the synthetic perfume industry.

Perkin alone could not maintain British supremacy against the great development of German organic chemistry, and by the turn of the century "synthetics" had become virtually a German monopoly. But during World War I, Great Britain and the United States, shut off from the products of the German chemical laboratories, were forced to develop chemical industries of their own.

Achievements in synthetic organic chemistry could not have proceeded at anything better than a stumbling pace if chemists had had to depend upon fortunate accidents such as the one that had been seized upon by Perkin. Fortunately the structural formulas of Kekulé, presented three years after Perkin's discovery, made it possible to prepare blueprints, so to speak, of the organic molecule. No longer did chemists have to try to prepare quinine by sheer guesswork and hope; they had methods for attempting to scale the structural heights of the molecule step by step, with advance knowledge of where they were headed and what they might expect.

Chemists learned how to alter one group of atoms to another; to open up rings of atoms and to form rings from open chains; to split groups of atoms in two, and to add carbon atoms one by one to a chain. The specific method of doing a particular architectural task within the organic molecule is still often referred to by the name of the chemist who first described the details. For instance, Perkin discovered a method of adding a two-carbon atom group by heating certain substances with chemicals named acetic anhydride and sodium acetate. This is still called the "Perkin Reaction." Perkin's teacher, Hofmann, discovered that a ring of atoms which included a nitrogen could be treated with a substance called methyl iodide in the presence of silver compound in such a way that the ring was eventually broken and the nitrogen atom removed. This is the "Hofmann Degradation." In 1877 the French chemist Charles Friedel, working with the American chemist James Mason Crafts, discovered a way of attaching a short carbon chain to a benzene ring by the use of heat and aluminum chloride. This is now known as the "Friedel-Crafts Reaction."

In 1900 the French chemist Victor Grignard discovered that magnesium metal, properly used, could bring about a rather large variety of different joinings of carbon chains. For the development of these "Grignard Reactions" he shared in the Nobel Prize in chemistry in 1912. The French chemist Paul Sabatier, who shared it with him, had discovered (with J. B. Senderens) a method using finely divided nickel to bring about the addition of hydrogen atoms in those places where a carbon chain possessed a double bond. This is the "Sabatier-Senderens Reduction."

In other words, by noting the changes in the structural formulas of substances subjected to a variety of chemicals and conditions, organic chemists worked out a slowly growing set of ground rules on how to change one compound into another at will. It wasn't easy. Every compound and every change had its own peculiarities and difficulties. But the main paths were blazed, and the skilled organic chemist found them clear signs toward progress in what had formerly seemed a jungle.

Knowledge of the manner in which particular groups of atoms behaved could also be used to work out the structure of unknown compounds. For instance, when simple alcohols react with metallic sodium and liberate hydrogen, only the hydrogen linked to an oxygen atom is released, not the hydrogens linked to carbon atoms. On the other hand, some organic compounds will take on hydrogen atoms under appropriate conditions while others will not. It turns out that compounds that add hydrogen generally possess double or triple bonds and add the hydrogen at those bonds. From such information a whole new type of chemical analysis of organic compounds arose; the nature of the atom-groupings was determined, rather than just the numbers and kinds of various atoms present. The liberation of hydrogen by the addition of sodium signified the presence of an oxygen-bound hydrogen atom in the compound; the acceptance of hydrogen meant the presence of double or triple bonds. If the molecule was too complicated for analysis as a whole, it could be broken down into simpler portions by well-defined methods; the structures of the simpler portions could be worked out and the original molecule deduced from those.

Using the structural formula as a tool and guide, chemists could work out the structure of some useful naturally occurring organic compound (analysis) and then set about duplicating it or something like it in the laboratory (synthesis). One result was that something which was rare, expensive or difficult to obtain in nature might become cheaply available in quantity in the laboratory. Or, as in the case of the coal-tar dyes, the laboratory might create something that fulfilled a need better than did similar substances found in nature.

One startling case of a deliberate improvement on nature involves the drug cocaine. Cocaine is found in the leaves of the coca plant, which is native to Bolivia and Peru (but is now grown chiefly in Java). The South American Indians would chew coca leaves, finding it an antidote to fatigue and a source of happiness-sensation. The Scottish physician Sir Robert Christison introduced the plant to Europe, and eventually cocaine was isolated as the active principle. In 1884 the American physician Carl Koller discovered that cocaine could be used as a local anesthetic when added to the mucous membranes around the eye. Eye operations could then be performed without pain. Cocaine could also be used in dentistry, allowing teeth to be extracted without pain.

Anesthetics had come into general use about 40 years before that. The American surgeon Crawford Williamson Long in 1842 had used ether to put a patient to sleep during tooth extractions. In 1846 the American dentist William Thomas Green Morton conducted a surgical operation under ether at the Massachusetts General Hospital. Morton usually gets the credit for the discovery, because Long did not describe his feat in the medical journals until after Morton's public demonstration. In any case, doctors were quite aware that anesthesia had finally converted surgery from torture-chamber butchery to something that was at least humane and, with the addition of antiseptic conditions, even lifesaving. For that reason any further advance in anesthesia was seized upon with great interest, and this included cocaine.

There were several drawbacks to cocaine. In the first place, it induced troublesome side-effects and could even kill patients sensitive to it. Secondly, it could bring about addiction and had to be used skimpily and with caution. (Cocaine is one of the dangerous "dopes." Up to 20 tons of it are produced illegally each year and sold with tremendous profits to a few and tremendous misery to many, despite world-wide efforts to stop the traffic.) Thirdly, the molecule is fragile, and heating cocaine to sterilize it of any bacteria leads to changes in the molecule that interfere with its anesthetic effects.

The structure of the cocaine molecule is rather complicated:

The double ring on the left is the fragile portion, and that is the difficult one to synthesize. (The synthesis of cocaine wasn't achieved until 1923, when the German chemist Richard Willstätter managed it.) However, it occurred to chemists that they might synthesize similar compounds in which the double ring was not closed. This would make the compound both easier to form and more stable. The synthetic substance might possess the anesthetic properties of cocaine, perhaps without the undesirable side-effects.

For some 20 years German chemists tackled the problem, turning out dozens of compounds, some of which were pretty good. The most successful modification was obtained in 1909, when a compound with the following formula was prepared:

Compare this with the formula for cocaine and you will see the similarity, and also the important fact that the double ring no longer exists. This simpler molecule—stable, easy to synthesize, with good anesthetic properties and very little in the way of side-effects—does not exist in nature. It is a "synthetic substitute" far better than the real thing. It is called "procaine," but is better known to the public by the trade-name Novocaine.

A series of other anesthetics have been synthesized in the half-century since, and, thanks to chemistry, doctors and dentists have an assortment of effective and safe pain-killers at hand.

Man now has at his disposal all sorts of synthetics of great potential use and misuse: explosives, poison gases, insecticides, weed-killers, antiseptics, disinfectants, detergents, drugs—almost no end of them, really. But synthesis is not merely the handmaiden of consumer needs. It can also be placed at the service of pure chemical research.

It often happens that a complex compound, produced either by living tissue or by the apparatus of the organic chemist, can only be assigned a tentative molecular structure, after all possible deductions have been drawn from the nature of the reactions it undergoes. In that case, a way out is to synthesize a compound by means of reactions designed to yield a molecular structure like the one that has been deduced. If the properties of the resulting compound are identical with the compound being investigated in the first place, the assigned structure becomes something in which a chemist can place his confidence.

An impressive case in point involves hemoglobin, the main component of the red blood cells and the pigment that gives the blood its red color. In 1831 the French chemist L. R. LeCanu split hemoglobin into two parts, of which the smaller portion, called "heme," made up 4 per cent of the mass of hemoglobin. Heme was found to have the empirical formula $C_{34}H_{32}O_4N_4Fe$. Compounds like heme were known to occur in other vitally important substances, both in the plant and animal kingdoms, and so the structure of the molecule was a matter of great moment to biochemists. For nearly a century after LeCanu's isolation of heme, however, all that could be done was to break it down into smaller molecules. The iron atom (Fe) was easily removed, and what was left then broke up into pieces roughly a quarter the size of the original molecule. These fragments were found to be "pyrroles"—molecules built on rings of five atoms, of which four are carbon and one nitrogen. Pyrrole itself has the following structure:

$$
\begin{array}{ccc}
& CH \text{---} CH & \\
/\!\!/ & & \backslash\!\backslash \\
CH & & CH \\
\backslash & & / \\
& NH &
\end{array}
$$

The pyrroles actually obtained from heme possessed small groups of atoms containing one or two carbon atoms attached to the ring in place of one or more of the hydrogen atoms.

In the 1920's the German chemist Hans Fischer tackled the problem further. Since the pyrroles were one quarter the size of the original heme, he decided to try to combine four pyrroles and see what he got. What he finally succeeded in getting was a four-ring compound which he called "porphin" (from a Greek word meaning "purple," because of its purple color). Porphin would look like this:

However, the pyrroles obtained from heme in the first place contained small "side-chains" attached to the ring. These remained in place when the pyrroles were joined to form porphin. The porphin with various side-chains attached make up a family of compounds called the "porphyrins." It was obvious to Fischer upon comparing the properties of heme with those of the porphyrins he had synthesized that heme (minus its iron atom) was a porphyrin. But which one? No fewer than 15 different compounds could be formed from the various pyrroles obtained from heme, according to Fischer's reasoning, and any one of those 15 might be heme itself.

A straightforward answer could be obtained by synthesizing all 15 and testing the properties of each one. Fischer put his students to work preparing, by painstaking chemical reactions that allowed only a particular structure to be built up, each of the 15 possibilities. As each different porphyrin was formed, he compared its properties with those of the natural porphyrin of heme.

In 1928 he discovered that the porphyrin numbered nine in his series was the one he was after. The natural variety of porphyrin is therefore called "porphyrin IX" to this day. It was a simple procedure to convert porphyrin IX to heme by adding iron. Chemists at last felt confident that they knew the structure of that important compound. Here is the structure of heme, as worked out by Fischer:

For this achievement Fischer was awarded the Nobel Prize in chemistry in 1930.

As a postscript to the story of organic chemical synthesis, let me say that in 1945 two young chemists at Harvard, Robert B. Woodward and William E. von Doering, succeeded in synthesizing quinine, the hopeless objective of Perkin that had started it all. And here, if you are curious, is the structural formula of quinine:

$$
\begin{array}{c}
\text{OH} \\
| \\
\text{CH} \longrightarrow \text{CH} - \text{N} - \text{CH}_2 \\
\end{array}
$$

Understanding What You've Read

How would you describe Asimov's style of writing? Does he explain things well? Does he oversimplify or leave readers with unanswered questions?

Cite examples of how Asimov conveys concepts in a historical context, all while mixing in biographical stories and industrial history to maintain readers' interest.

Asimov's style has been described as "seductive." Do you agree?

What does Asimov tell you about his view of science?

Werner Heisenberg

The Uncertainty Principle

To appreciate the importance of Werner Karl Heisenberg's (1901–1976) ideas, one must consider the small world of atoms. In 1911, Ernest Rutherford presented a complex model of the atom, claiming that it consisted of a positively charged nucleus and negatively charged and orbiting electrons. Even after the nucleus was found to include protons and chargeless neutrons, Rutherford's model was still wrong. The problem was that the laws of classical physics did not apply to atoms; electron orbits, unlike planetary orbits, did not decay. Thus, physicists had to devise a new theory that would include new ideas about electrons. Rather than being a charged, solid particle, electrons were considered a cross between pure energy and matter—that is, a quantum of radiation. This meant that radiation could be treated as both a wave and a particle. This differed significantly from the view of classical mechanics, which treated radiation as pure waves.

Werner Heisenberg made his contribution to quantum mechanics in 1927 when he realized that the position and velocity of a given electron—which classical physics would have described with two independent variables—could not be predicted simultaneously. That is, the very act of measuring an electron's velocity would affect its position and vice versa; the best that could be done would be to establish a *probability* for one variable or another. This came to be known as the *Heisenberg Uncertainty Principle*, and was the basis for Heisenberg winning the Nobel Prize for physics in 1932. The implications of Heisenberg's ideas are important: Indeterminacy is not the result of human limitations, but instead exists in the very nature of matter.

In this essay, Heisenberg describes his ideas.

"The Uncertainty Principle" by Werner Heisenberg from *Physics and Philosophy*, pp. 89–91; Volume 19 in *World Perspectives*, planned and edited by Ruth Nanda Anshen. Copyright © 1958 by Werner Heisenberg. Reprinted by permission of HarperCollins Publishers, Inc.

Kant says that whenever we observe an event we assume that there is a foregoing event from which the other event must follow according to some rule. This is, as Kant states, the basis of all scientific work. In this discussion it is not important whether or not we can always find the foregoing event from which the other one followed. Actually we can find it in many cases. But even if we cannot, nothing can prevent us from asking what this foregoing event might have been and to look for it. Therefore, the law of causality is reduced to the method of scientific research; it is the condition which makes science possible. Since we actually apply this method, the law of causality is "a priori" and is not derived from experience.

Is this true in atomic physics? Let us consider a radium atom, which can emit an α-particle. The time for the emission of the α-particle cannot be predicted. We can only say that in the average the emission will take place in about two thousand years. Therefore, when we observe the emission we do not actually look for a foregoing event from which the emission must according to a rule follow. Logically it would be quite possible to look for such a foregoing event, and we need not be discouraged by the fact that hitherto none has been found. But why has the scientific method actually changed in this very fundamental question since Kant?

Two possible answers can be given to that question. The one is: We have been convinced by experience that the laws of quantum theory are correct and, if they are, we know that a foregoing event as cause for the emission at a given time cannot be found. The other answer is: We know the foregoing event, but not quite accurately. We know the forces in the atomic nucleus that are responsible for the emission of the α-particle. But this knowledge contains the uncertainty which is brought about by the interaction between the nucleus and the rest of the world. If we wanted to know why the α-particle was emitted at that particular time we would have to know the microscopic structure of the whole world including ourselves, and that is impossible. Therefore, Kant's arguments for the a priori character of the law of causality no longer apply.

A similar discussion could be given on the a priori character of space and time as forms of intuition. The result would be the same. The a priori concepts which Kant considered an undisputable truth are no longer contained in the scientific system of modern physics.

Still they form an essential part of this system in a somewhat different sense. In the discussion of the Copenhagen interpretation of quantum theory it has been emphasized that we use the classical concepts in describing our experimental equipment and more generally in describing that part of the world which does not belong to the object of the experiment. The use of these concepts, including space, time and causality, is in fact the condition for observing atomic events and is, in this sense of the word *a priori*. What Kant had not foreseen was that these a priori concepts can be the conditions for science and at the same time can have only a limited range of applicability. When we make an experiment we have to assume a causal chain of events that leads from the atomic event through the apparatus finally to the eye of the observer; if this causal chain was not assumed, nothing could be known about the atomic event. Still we must keep in mind that classical physics and causality have only a limited range of applicability. It was the fundamental paradox of quantum theory that could not be foreseen by Kant. Modern physics has changed Kant's statement about the possibility of synthetic judgments a priori from a metaphysical one into a practical one. The synthetic judgments a priori thereby have the character of a relative truth.

Understanding What You've Read

Why is it impossible to describe precisely the forces that cause the emission of an alpha particle?

Heisenberg claims that the fundamental concern of science is no longer to find a cause to explain an observation. Why?

In this essay, Heisenberg refers to Kant. Who was Kant? Why was it necessary for Heisenberg to refer to Kant's notion of a priori to explain the basis of the uncertainty principle?

Is Kant's notion of causality still applicable in science? Why or why not?

Heisenberg wrote that, "[We should] abandon all attempts to construct perpetual models of atomic processes." Discuss what he meant.

Exercises

1.   Discuss, support, or refute the ideas contained in these quotations:

Writing About Science

In science the important thing is to modify and change one's ideas as science advances. — Herbert Spencer

Science is the father of knowledge, but opinion breeds ignorance. — Hippocrates

If you cannot—in the long run—tell everyone what you have been doing, your doing has been worthless. — Erwin Schrödinger

There are passages in every novel whose first writing is the last. But it's the joint and cement between these passages that take a great deal of rewriting . . . Each sentence is a skeleton accompanied by enormous activity of rejection. — Thornton Wilder

Science has given to this generation the means of unlimited disaster or of unlimited progress. There will remain the greater task of directing knowledge lastingly towards the purpose of peace and human good. — Winston Churchill

Biology

Every individual alive today, the highest as well as the lowest, is derived in an unbroken line from the first and lowest forms. — August Frederick Leopold Weismann

One hundred trout are needed to support one man for a year. The trout, in turn, must consume 90,000 frogs, that must consume 27 million grasshoppers that live off of 1,000 tons of grass. — G. Tyler Miller, Jr.

It's humbling to think that all animals, including human beings, are parasites of the plant world. — Isaac Asimov

Chemistry

I mean by elements certain primitive and simple, or perfectly unmingled bodies which not being made of other bodies . . . are the ingredients of which all those called perfectly mixed bodies are immediately compounded, and into which they are ultimately resolved. — Robert Boyle (1627-1691)

I am convinced that the future progress of chemistry as an exact science depends very much upon the alliance with mathematics. — A. Frankland (1825-1899)

Geology

It was one thing to declare that we had not yet discovered the trace of a beginning, and another to deny that the earth ever had a beginning. — John Playfair (1748-1819)

It may undoubtedly be said that strata have been always forming somewhere, and therefore at every moment of past time Nature has added a page to her archives. — Sir Charles Lyell (1797-1875)

The birth of a volcanic island is an event marked by prolonged and violent travail; the forces of the earth striving to create, and all the forces of the sea opposing. — Rachel Carson

Physics

Suppose we take a quantity of heat and change it to work. In doing so, we haven't destroyed the heat, we have only transferred it to another place or perhaps changed it into another energy form. — Isaac Asimov

Vacuum I call every place in which a body is able to move without resistance. — Isaac Newton

2. Educators are notorious for their poor writing. Revise this sentence taken from a book:

> Operationally, teaching effectiveness is measured by assessing the levels of agreement between the perceptions of instructors and students on the rated ability of specific instructional behavior attributes which were employed during course instruction.

3.   Simplify these sentences and paragraphs:

Pollutants exist in virtually all sectors of the environment.

It is generally desirable to communicate your thoughts and ideas in a directly forthright manner and style. Toning down your point and tiptoeing around it may, in a large number of circumstances, tempt the reader to tune out and allow her mind to wander.

They are without any nutrients whatsoever.

Some serious problems can be prevented by means of the use of effective writing.

Basically, it would not be reasonable to assume that in the foreseeable future our very unique but nonetheless inefficient experiments will have to be terminated.

The consensus of opinion is that the experiment will not be completely finished until next week.

Your method of mixing together the chemicals is very unique.

It is necessary to have complete documentation of your research.

Pursuant to the regulations promulgated as of this date by this author, endeavor to employ uncomplicated words in writing.

The aquarium was two meters in its length-wise dimension.

I do not have sufficient knowledge of the problem to make a proposal regarding a new analysis.

In order to evaluate the potential significance of certain molecular parameters at the ultracellular and subcellular levels, and so throw some light on the conceivable role of structural configuration in spatial relationships of intracellular and intercellular molecular structures, a comprehensive and integrated approach to the problem of cellular structure and function has been developed. The results, which are in a preliminary stage, are discussed here in quite a bit of detail because of their potential implications in mechanisms of cellular structure and function in a larger context.

Your instructor will not except a report turned in late.

In the event that the chemicals need to be stored prior to their utilization, it is preferred that they be stored in a warehouse before they are unpacked.

Because of the unfortunate fact that it happened to be snowing at that particular point in time, we came to the conclusion that we would refrain from beginning our sampling of the Marlin Prairie.

Experiments are currently in progress to assess the possibility of using the new spectrophotometer.

It consists essentially of two parts.

As a result of the fact that our science students could not be present at our last scheduled lab meeting time, the liberty was taken to cancel that gathering.

Therefore, it is planned to consider the identical research items at the newly scheduled gathering of the students that will gather Tuesday. This gathering will gather in accordance with the same agenda, repeated herein for your convenient reference.

The general feeling of the meeting was that within the basic framework of the experiments a great amount was definitely accomplished and cognitively learned by all participants,and the very unique system has achieved virtually all of our objectives with regard to our experiments.

The answer is in the negative.

4.   Simplify these words and phrases:

assert	a second point is that
transmit	demonstrate
draw to your attention	in view of the fact that
in a number of cases	in the not too distant future
initiate	commence
terminate	seem to suggest
aggregate	converse
have a tendency to	that point in time
as of this date	endeavor
attempt	on two separate occasions
anomalous	utilize
make use of	in most cases
in the majority of cases	is desirous of
had the occasion to be	proved to be
provided that	by the time
at such time as	the question as to whether
during the time that	if it should happen that
until such time as	final outcome
for the purpose of	at the membrane level
general rule	new innovation
positive identification	proposed plan
single unit	component parts
doctorate degree	weather conditions

brief induration

estimated at about

last of all

new all-time record

mutually agreeable

cancel out

connect up

follow after

join together

both together

classify into groups

filled to capacity

rectangular in shape

might possibly

assemble together

attached hereto

enclosed herein

made out of

at a later date

have need for

CHAPTER FOUR

Precision and Clarity

Anything is better than not to write clearly. There is nothing to be said against lucidity, and against simplicity only the possibility of dryness. This is a risk well worth taking when you reflect how much better it is to be bald than to wear a curly wig.
— Somerset Maugham

The habitual use of the active voice . . . makes for forcible writing. . . . [W]hen a sentence is made stronger, it usually becomes shorter. Thus, brevity is a by-product of vigor.
— William Strunk, Jr. and E.B. White, *The Elements of Style*

A clear statement is the strongest argument.
— English proverb

For what good science tries to eliminate, good art seeks to provoke—mystery, which is lethal to the one, and vital to the other.
— John Fowles

It is with words that we do our reasoning, and writing is the expression of our thinking . . . Words and phrases that do not have an exact meaning are to be avoided because once one has given a name to something, one immediately has a feeling that the position has been clarified, whereas often the contrary is true.

— W.I.B. Beveridge, *The Art of Scientific Investigation*

If a reader keeps tripping over strange words, or bumping his head on overhanging clauses or stubbing his toe on concealed antecedents, he tends to give up. The hell with this, he says, and he turns to something else. Our concern here is not with pratfall prose. It is the imperceptible hesitation that matters—the little uncertainty, the small confusion, the modifier that isn't badly lost but only slightly misplaced.

— James Kilpatrick

"Then you should say what you mean," the March Hare went on. "I do," Alice hastily replied; "at least— at least I mean what I say—that's the same thing you know." "Not the same thing a bit!" said the Hatter. "Why you might just as well say that 'I see what I eat' is the same thing as 'I eat what I see'!"

— Lewis Carroll

Some books about scientific writing would have scientists merely count syllables and words to satisfy the writing cliché, "be brief." Although substituting simple words for complex, large words almost always improves communication, it's only a good first step. Concentrating only on simple words will not produce effective writing because counting letters and syllables doesn't explain what makes a sentence awkward.

In scientific writing, precision is the most important goal of language. If your writing does not communicate *exactly* what you think or did, then you have changed your ideas or research. Furthermore, most readers become annoyed with vague writing. Fairly or not, and often without realizing what they're doing, readers conclude that an imprecise writer is lazy and that your ideas aren't worth much of their attention. Vague writing discredits your work. It results from not only a lack of grace, but also a lack of clarity. I argue that these are not separable.

When you choose a precise word, trust it to do its job. If you've picked the right word, adding bankrupt words such as *quite* only impedes communication and defeats your purpose. Similarly, adding redundant modifiers such as *very* provides no compensation when you've chosen the wrong word. However, don't be overly precise; for example, don't spend five pages describing how an electron microscope works if all that you need is a photograph.

Precise and concise writing says exactly what you mean and requires that you choose the right words. This means making every word tell.

Make Every Word Tell

> Scientific articles are most commonly impaired by their authors' misuse of words.
> — John Maddox, former editor of *Nature*

> Vigorous writing is concise. A sentence should contain no unnecessary words, a paragraph no unnecessary sentences, for the same reason that a drawing should have no unnecessary lines and a machine no unnecessary parts. This requires not that the writer make all his sentences short, or that he avoid all detail and treat his subjects only in an outline, but that every word tell.
> — William Strunk

> Words may be either servants or masters. If the former they may safely guide us in the way of truth. If the latter they intoxicate the brain and lead into swamps of thought where there is no solid footing.
> — Bishop George Horne

> Memorable sentences are memorable on account of some single irradiating word.
> — Alexander Smith, *On the Writing of Essays*

> "It's too late to correct it," said the Red Queen; "when you've said a thing, that fixes it, and you must take the consequences."
> — Lewis Carroll

Precise writing is concise; every word tells. Precision is based largely on word choice. As Mark Twain said, "Use the right word, not its second cousin. The difference between the right word and the almost right word is the difference between 'lightning' and 'lightning bug.'" *The United Press International Stylebook* says it better: "A burro is an ass. A burrow is a hole in the ground. As a [writer] you are expected to know the difference."

Archie Bunker and Yogi Berra became famous partly because they often chose the almost right word instead of *the* right word. Choosing the right word is often difficult because many words have several meanings. For example, the 500 most-used words in English have a total of 14,000 meanings—an average of 28 meanings per word.

There are many ways to make every word tell. Having deleted the clutter from your writing is a good start, but you must do more. Specifically, you must know the meaning of every word you use. Don't settle for an *almost* right word in place of *the* right word. If you do, you'll confuse or annoy readers. You might also end up with sentences like these:

Darwin was a child progeny who wrote *Organ of the Species*.

Sir Francis Drake circumcised the world with a 100-foot clipper.

Three kinds of blood vessels are arteries, vanes, and caterpillars.

Although the key words in these sentences are almost correct, they're not *the* correct words. Consequently, the writer appears either careless, confused, or poorly educated. Save yourself embarrassment by buying and using a dictionary and thesaurus (see "Tools of the Trade" on page 47).

Scientific writing must be accurate, precise, and clear, all of which depend on word choice.[1] While precision depends on choosing the right word, clarity often depends on not choosing the wrong word. This requires that you know the meaning and purpose of every word that you use. If you don't know the meaning of words that you use, you risk looking foolish. For example, the authors of an introductory biology book confused the words *mole* and *molecule* in their discussion of cellular respiration, claiming that 36 ATP are the "approximate maximal ATP yield from the complete respiration of 1 mole of glucose." The authors go on to state that each of the 20–30 trillion cells in our bodies cleaves 1–2 billion ATP per minute. Simple arithmetic shows the inconsistency:

If each cell cleaves 1.5 billion ATP per minute, then each cell needs 41,700,000 (1.5 billion ATP/36 ATP per mole of glucose) moles of glucose per minute.

This means that each cell needs 232,000 (41,700 moles of glucose/180 grams per mole) grams (232 kg) of glucose per minute.

[1]Garbled writing has plagued people for ages. According to the Bible, when God wanted to stop people from building the Tower of Babel, God did not zap them with a thunderbolt. Rather, God said, ". . . let us go down, and there confound their language, that they may not understand one another's speech." Apparently God could think of no better way to foil the project than to garble their words.

To supply this much glucose to each of the 25 trillion cells in our body, we would each need to eat about 5,800,000,000,000,000 kg (232 kg of glucose per minute per cell x 25,000,000,000,000 cells) of glucose per minute. That's equal to a staggering 8.35×10^{18} kg of glucose per day, an amount equal to the mass of about 6 trillion blue-whales. Even my huge cousin doesn't eat that much when, to use his colorful words, he roars though the kitchen "hungrier than the gang on Noah's ark."

Know the meaning of all words that you use. Don't use *mole* when you mean *molecule*, *less* when you mean *fewer*, *allude* when you mean *refer*, *imply* when you mean *infer*, *weight* when you mean *mass*, *disinterested* when you mean *uninterested*, and *comprise* when you mean *compose*. Similarly, don't describe a rainy period with a vague phrase such as "unfavorable weather."

VAGUE: Our sampling at Walnut Creek was delayed for a time because of unfavorable weather conditions.

SPECIFIC: Our sampling at Walnut Creek was delayed for a week because of snowstorms.

Appendix 2 provides a list of words frequently misused by scientists. Use these words carefully and, when in doubt, consult a dictionary or usage guide such as the *Harper Dictionary of Contemporary Usage*.

Avoid Doublespeak

Ready-made phrases are the prefabricated strips of words . . . that come crowding in when you do not want to take the trouble to think through what you are saying . . . They will construct your sentences for you—even think your thoughts for you, to a certain extent— and at need they will perform the important service of partially concealing your meaning even from yourself.
— George Orwell

You must learn to talk clearly. The jargon of scientific terminology which rolls off your tongue is mental garbage.
— Martin H. Fischer

The trouble with new or different words to express the old ideas is that the words can give a false sense of progress and

disguise history.
— Thelma Ingles

A good catchword can obscure analysis
for fifty years.
— Wendel L. Wilkie

Whenever ideas fail, men invent words.
— Martin H. Fischer

Doublespeak is a type of writing that pretends to communicate, but really doesn't. Rather, it deliberately retreats from what is tangible, real, and common. For example, on March 24, 1989 the Exxon *Valdez* hit a reef and spilled 11 million gallons of North Slope crude oil into Prince William Sound. David Parish, a spokesperson for Exxon, said that Exxon did not expect major environmental damage from the spill and boldly claimed that Exxon was responsible for cleaning up the spill. However, when Exxon realized the magnitude of the spill—more than 1,000 miles of coastline had been contaminated—they used doublespeak to change their story: Exxon no longer said it would clean up the spill, but instead said it would make the area "environmentally stable." Since no one knows exactly what "environmentally stable" means, Exxon let itself off the hook.

Doublespeak is all around us and is used to deceive, distort, confuse, or mislead us.[2] For example, potholes in streets become "pavement deficiencies"; poor people become "fiscal underachievers"; medical malpractice becomes "diagnostic misadventure"; kickbacks become "rebates"; guards in department stores become "loss prevention specialists"; and gas station attendants become "petroleum transfer engineers." Airlines speak of "water landings" instead of crashes at sea, and the CIA subsidizes "health alteration committees," not assassination squads. Companies don't report losses—they speak of "negative growth," which is analogous to talking about the heat in an ice cube. To paraphrase a famous quote, "Nothing is certain in life except negative patient care outcome and governmental revenue enhancements."

---

[2]Andy Rooney knew the purposes of scientific jargon when he wrote that, "Much of the language of science, medicine, government and law is nonsense designed to exclude outsiders so they won't discover that, basically, the specialists' work is not so complex as it seems." Similarly, George Orwell, in his famous essay "Politics and the English Language," wrote, "the great enemy of clear language is insincerity. When there is a gap between one's real and one's declared aims, one turns as it were instinctively to long words and exhausted idioms, like a cuttlefish squirting out ink." In the nightmarish world of his novel, *1984*, Orwell described how a totalitarian government— "Big Brother"—used a form of doublespeak called Newspeak to create a reality desired by the government. Newspeak epitomized the deceptive power of doublespeak—it diminished thought so much that it allowed people to believe in conflicting ideas. Indeed, the height of doublespeak (and the doublethink that it caused) was the famous phrase, "War is peace." Lest you think that such doublespeak-induced contradictions are restricted only to Orwell's novel, recall that Secretary of State Alexander Haig stated several times that a continued weapons build-up by the United States is "absolutely essential to our hopes for meaningful arms reduction." Or recall that Senator Orrin Hatch said, "Capital punishment is our society's recognition of the sanctity of human life." Such contradictions are doublespeak.

Scientists often chuckle when they hear doublespeak. Words and phrases such as "pseudopseudohypoparathyroidism" and "nonrheumatoid rheumatoid nodules" are whimsical, and few people are fooled by ludicrous phrases such as "oral hygiene appliance" instead of toothbrush, or "personal time control center" instead of wristwatch. However, scientists often use doublespeak. This doublespeak appears as jargon, euphemisms, gobbledygook, and inflated language.[3]

Jargon   The word *jargon* once referred to the unintelligible chatter of birds. Today, jargon is specialized "code" language usually used to impress, not communicate, and is understandable by only a few people. Jargon usually arrives in clusters:

> It is believed that with the parameters that have been imposed by your administrators, a viable research program may be hard to evolve. Net net: If our program is to impact students to the optimum, meaningful interface with your management may be necessitated.

This kind of writing is long-winded and heavy-handed, and is what E.B. White calls "the language of mutilation" because it mutilates your meaning. When you eliminate the jargon, the message becomes more obvious:

> We believe that the limits set by your administrators may prevent an effective research program. If we want to involve our students, we need to talk with your administrators.

Jargon insulates members of a group from the outside world and excludes nonmembers of the group. Jargon is usually pretentious, obscure, and esoteric language used to imply profundity, prestige, or authority. It makes the simple seem complex, the ordinary seem profound, and the obvious seem insightful. For example, you can unclog your drains with a "hydro blastforce cup," (that is, a plunger), and what was once a vacuum cleaner is now Hoover's "Dimension 1000 Electronic Cleaning Machine with Quadraflex Agitator."

Scientists also use unnecessary jargon. For example, some refer to a "fused silicate container" instead of a glass beaker, and a crack in a test tube becomes a "structural discontinuity." Similarly, scientists at the Environmental Protection Agency write about "poorly buffered precipitation" and "atmospheric deposition of anthropogenetically-derived acid substances," when what they really mean is acid rain. Physicians say "utilization of recently introduced therapeutic modalities" instead of "use of new treatments," and "expectorated a hemorrhagic production" instead of "spit out blood." As is its purpose, such writing often isolates you from others. For example, behavioral scientists, instead of writing that "Modern cities are hard to live in," might write, "urban existence in the perpendicular declivities of a stratocosmopolis or megalopolis . . ." Such awful writing prompted writer-historian

---

[3]The National Council of Teachers of English annually awards a Doublespeak Award to officials that are "grossly deceptive, evasive, euphemistic, confusing or self-contradictory." Recent winners of the Doublespeak Award include the Mobil Corporation for claiming that one of its trash bags is "photodegradable" even when the bag is buried in a landfill.

Barbara Tuchman to give an example of the influence of jargon:

> Let us beware the plight of our colleagues, the behavioral scientists, who by use of proliferating jargon have painted themselves into a corner—or isolation ward—of unintelligibility. They know what they mean, but no one else does.

If writing only for people with a background similar to yours, use technical terms. However, when writing to a more general audience, especially decision-makers with limited experience in your field, avoid jargon and use the simplest and most accurate words to express your ideas.

Euphemisms A euphemism is a positive or inoffensive word or phrase used to avoid a harsh or unpleasant reality. People use euphemisms to divert attention from a topic and to avoid communicating effectively.[4] For example, Adolph Hitler referred to murders as "the final solution," and in 1977 the Pentagon tried to obtain funds for a neutron bomb by calling it a "radiation enhancement device." Similarly, the U.S. State Department no longer refers to killings in reports about human rights. Rather, it writes about the "unlawful or arbitrary deprivation of life."

Euphemisms are always used to hide something. For example, dictators have used phrases such as "pacification," "terminating," and "protecting the peace" because they know these phrases do not create an image of bound, blindfolded kneeling prisoners with pistols held to their heads. Similarly, consider all of our euphemistic fig leaves for the word "die": *croak, take the last count, pass away, pass on, pass to one's reward, go beyond, depart this world, expire, come to an untimely end, perish, be taken, give up the ghost, depart this life, kick the bucket,* and *buy the farm*. One scientist even described the death of a laboratory rat by saying that the rat "lost its integrity." Unless religious dictates prevent you from doing so, use the simpler and more dignified word *die*.

Scientists also use euphemisms to deceive readers. For example, scientists seldom write that they kill animals. Rather, they write that animals were "sacrificed," as if they'd been part of some arcane religious ceremony. These writers apparently hope that this will somehow mean that the animals weren't really killed. Similarly, scientists write about "negative patient care outcome" when all they mean is that the patient died, and "national species of special emphasis" when they discuss endangered species that we can kill legally. Avoid euphemisms. Choose simple, direct, more dignified statements.

Gobbledygook During World War II, Congressman Maury Maverick of Texas attended a meeting where the chairman spoke at length about "maladjustments coextensive with problem areas . . . alternative but nevertheless meaningful miminae . . . utilization of factors in which a dynamic democracy can be channelized into both quantitative and qualitative phases." Maverick censured the language by calling it *gobbledygook*, a term that would last far beyond the war.

---

[4]Several years ago the Justice Department filed a suit against the accounting firm Ernst and Whinney for intentionally using "false, misleading, and deceptive" language on a client's tax form. To qualify their client for tax credits, the accountants referred to windows as "decorative features," a fire alarm as a "combustion enunciator," doors as "movable partitions," 92 cubic yards of topsoil as a "planter," and a refrigerated warehouse as a "freezer." In their defense, Ernst and Whinney claimed that they were only trying "to put their client's best foot forward."

Today, gobbledygook is synonymous with "bureaucratese," a writing style that masquerades as serious prose by overwhelming readers with big words, phrases, and sentences. Gobbledygook flourishes in universities and governmental agencies, an observation obvious to Lewis Thomas when he facetiously likened Agricultural Experiment Stations to governmental agencies "devoted to the breeding of new words."[5] For example, consider this statement by Alan Greenspan, then chairperson of President Nixon's Council of Economic Advisors: "It is a tricky problem to find the particular calibration in timing that would be appropriate to stem the acceleration in risk premiums created by falling incomes without prematurely aborting the decline in the inflation-generated risk premiums." *Say what?* A report issued by the U.S. Department of Education claims that "feediness is the shared information between toputness, where toputness is at a time just prior to the inputness." And consider this doozie: "The Arizona Career Ladder Research and Evaluation Project Summated Matrices Depicting Positive Anecdotes Related to Interrelated Organizational Focus." As best I can tell, the translation of this gobbledygook is, "How A Teacher's Salary Affects Student Morale and Performance." Even writing teachers have joined the fray by claiming that students write better if they're taught "concretization of goals, procedural facilitation, and modeling planning." This kind of writing would—to invoke Ernest Hemingway's colorful words—overwhelm even the crudest shit-detector.

Inflated Language   People use inflated language to change everyday observations into things that appear rare and important. Inflated language damages science because its unwieldy style reinforces the mystique of scientific writing being impenetrable. Inflated language is often used by people who fear that the substance of their idea will evaporate if they use simple, clear language. Unable to describe their ideas effectively, these writers retreat behind a cloud of ridiculous words and phrases. Here's how Nobel laureate Richard Feynman described the use of inflated language by NASA officials while investigating the *Challenger* accident:

> At any rate, the engineers all leaped forward. They got all excited and began to describe the problem to me. I'm sure they were delighted, because technical people love to discuss technical problems with technical people . . .
>
> They kept referring to the problem by some complicated name—a "pressure-induced vorticity oscillatory wawa," or something.
>
> I said, "Oh, you mean a whistle."

Feynman, one of the most brilliant physicists of this era, was a passionate advocate of laypeople learning science. Consequently, he opposed and often chided scientist's use of doublespeak (for more about Feynman, see pages 55–67). Here's one of his descriptions of another official at NASA:

[5]Not all governmental administrators endorse the poor writing typical of bureaucrats. For example, soon after taking office as the U.S. Secretary of Commerce, Malcolm Baldridge banned these words and phrases from the department's word processors: *parameter, I would hope, I would like to express my appreciation, As I am sure you know, As you are aware, At the present time, bottom line, enclosed herewith, finalize, hopefully, institutionalize, It is my intention, maximize, Needless to say, new initiatives, prioritize, to impact, optimize, contingent upon, effectuated,* and *utilize.* When someone would type these words and phrases, their screen would flash "Don't use this word."

Mr. Mulloy explains how the seal is supposed to work—in the usual NASA way: he uses funny words and acronyms, and it's hard for anybody else to understand.

Inflated language is often funny. For example, Pentagon writers call a toothpick a "wooden interdental stimulator," and a pencil a "portable hand-held communications inscriber." Others use inflated language to transform car mechanics into "automobile internists," elevator operators into members of the "vertical transportation corps," a black eye into a "circumorbital haematoma," food into "comestibles," used cars into "pre-owned vehicles" or "experienced automobiles," and undertakers into "perpetual rest consultants" who sell "underground condominiums" rather than cemetery lots. Other uses of inflated language aren't so easily deciphered. For example, when Chrysler "initiated a career alternative enhancement program," what it really did was lay off 5,000 workers. And never admit to having lousy handwriting—rather, just say you suffer from "restricted graphomotorial representation."

Although scientists and others scoff at such examples of doublespeak, they're often guilty of using it in their writing—not to reinvent the wheel, but merely to paint it another color. For example, many scientists administer "chemotherapeutic agents" rather than drugs, and report on studies of "hematophagous arthropod vectors," not fleas. Others report that "the biota exhibited a 100% mortality rate" when, in fact, all they meant was that all of the rats died. If you don't work for the Pentagon, follow Mark Twain's advice: "I never use the word *metropolis* because I'm always paid the same amount of money to write *city.*"

Scientists often use doublespeak to make ideas that they think are too simple seem more impressive, much like others use pompous, difficult-to-understand language to protect what they have from others who want it. Doublespeak keeps knowledge from others by hiding information behind language so impenetrable that only a select few can find it. Since science is a search for truth, "scholarly" writing that includes doublespeak hides the truth and does not advance science.

At its worst, doublespeak limits thought and encourages readers to believe in opposing ideas. At its least offensive, doublespeak is inflated language that tries to give importance to common or trivial aspects of science. It is ineffective because it inhibits communication and calls attention to the writing, not to the message. Eliminating doublespeak is especially important for scientists because it is not a matter of having subject and verb agree, but of having words and facts agree. Since doublespeak deceives or distorts the truth, it has no place in scientific writing. Avoid doublespeak by thinking straight and saying exactly what you mean.

Avoid Stacked Modifiers

We use modifiers such as adjectives and adverbs to describe our subjects and actions. Although effective writers use modifiers sparingly, such words are occasionally useful. However, when writers stack two or more modifiers in front of a subject or verb, readers have trouble deciding which word the first modifier is modifying. For example, consider this sentence: "The animals must be given more nourishing food." This is confusing. Does the author mean more food that's nourishing or the same amount of more nutritious food? In the phrase "normal dog hearts," what's normal—the hearts or the dogs? Other examples that have appeared in papers include "proton magnetic resonance spectroscopy literature survey," "rapid gas apparatus deterioration," "isotope dilution assay results," "multiple conductor galvanized angle steel pylon system," "lizard ovary winter lipid level changes," "ankle joint angle measurement," "constant pressure heat capacity temperature maxima," "cyclic ligand planar nitrogen array," "silica gel coated glass fiber paper chromatography," and "average resting and after arginine hydrochloride infusion plasma growth hormone concentration." These phrases are graceless and usually confuse readers. Cramming so much information together, without regard to style or the audience, produces not an idea, but an indigestible lump.

Stacked modifiers can also be entertaining. For example, in 1989 The Mount Sinai Medical Center of New York advertised for a "free radical scientist," apparently hoping to find someone willing to work for nothing. Similarly, a journal article claiming that "flowering plant sperm lack flagella" prompted one of my students to write that "I didn't know that sperm ever flowered."

Stacked modifiers are easily spotted and corrected. Simply count the number of modifiers preceding each noun in a sentence: If there is more than one modifier, rewrite the sentence to avoid the confusion. Although you may lose brevity by making these changes, you'll improve clarity.

Avoid Dangling Modifiers

When a sentence begins with a participial phrase, be sure that the phrase modifies the subject of the sentence. Otherwise, the phrase dangles in front of the wrong word as though it modifies that word instead of the one it should. Dangling modifiers often confuse readers:

> She found nine fish in the water, which was better than usual.
> (What was better than usual, the water or the number of fish?)

> After killing the rat, the diet was tested.
> (Did the diet kill the rat, or did I kill the rat and then test the diet?)

Other times they produce sentences that are more amusing and embarrassing than confusing. For example, consider these gems:

> As a baboon who grew up in the wild, I realized that Tamba had strong sexual needs.

> When dipped in a weak acid, you can see the differential growth of roots.

> Mangy and needing a bath, I let the dog into the lab.

> Being old and dog-eared, I was able to buy the science book for only $1.

After soaking in sulfuric acid, I will rinse the seeds.

Wondering what to do next, the incubator exploded.

When 11 years old, my grandmother died.

Quickly summoning an ambulance, the corpse lay motionless.

Running through the lab, his hat blew off.

After germination, I will grow the plants in the incubator.

King Tut's tomb was unearthed while digging for artifacts.

Looking through his binoculars, the bird flew away.

Another apparently masochistic student wrote, "After 10 minutes in boiling water, I transferred the flask to an ice-bath." Although she provided no details about her injuries, I hope that her grade reflected her dedication to science. Finally, a particularly creative student wrote this sentence in a report: "After closing the incision, the chimp relaxed." I wonder where that chimp went to medical school.

Keep Related Words Together

> Of all the faults found in writing, the
> wrong placing of words is one of the
> most common, and perhaps it leads to
> the greatest number of misconceptions.
> — William Cobbett

The more quickly you get from A to B, the more likely your readers are to see the relationship of A and B. This requires that you keep related words together. If you fail at this, you'll probably confuse the reader. For example, consider this sentence that was included in a laboratory report:

> Here are some suggestions for handling the buffer problems from Sigma Chemical Company.

This sentence confuses readers: Did Sigma provide the suggestions for handling the problems, or did Sigma cause the problems? To eliminate this confusion, rewrite the sentence to say what you mean:

> Sigma Chemical Company sent us some suggestions for solving the buffer problems that we're having.

Poorly organized sentences are often funny. For example, a recruiting brochure for a prominent university claimed that its science department has "a botanical garden, electron microscope, and on-line computer connections in every teaching lab." A botanical garden in every lab? Similarly, a student wrote in her journal that, "In the laboratory is a report written by Bill and an aquarium." Quite a talented aquarium.

Word order is also important. Consider the different meanings that result from reversing two words in the following sentences:

We almost lost our entire sample.

We lost almost our entire sample.

Both of these sentences are grammatically correct, but only one accurately describes the situation. Similarly, the order of words can create confusion or absurdity:

We tested the patients using this procedure.

Randy asked his students while in lab to watch the video.

The marine scientist watched the sea gull wearing a bathing suit.

Watson and Crick completed their paper while traveling to California on the back of a notebook.

The table was used by the scientist with a broken leg.

Gould will discuss how evolution has influenced the use of fossils in the classroom at noon.

The patient left the hospital in good condition.

In barley, a cytoplasmic body containing mitochondria and plastids and recognized as sperm cytoplasm was found outside the egg.

If you occasionally write sentences like these, don't feel too bad—everyone does. For example, this caption appeared beneath a picture on the sports page of the *Washington Star-News* on June 12, 1976:

Pele, the star of Team America, soothes the ankle he sprained in a scrimmage with an ice pack.

Many readers probably wondered why Pele was scrimmaging with an ice pack rather than with other soccer players. A more remarkable caption appeared in the premier issue of *Baylor: The News Journal of Baylor University* in January of 1988:

Former Baylor trustee Mrs. Dorothy Kronzer of Houston greets Jehan Sadat, widow of Egyptian President Anwar Sadat, who was the featured speaker at the fall Baylor University Forum for Distinguished Lecturers.

This was quite a debut for the *News Journal*. Readers must have wondered what strings Baylor University had to pull to get Anwar Sadat as a featured speaker in 1988, seven years after his death. A similar story appeared on the front page of the October 3, 1985 issue of the *Dallas Morning News*:

For the ninth time in less than two months, a young Indian man on the Wind River Reservation in Central Wyoming has committed suicide, officials said Tuesday.

Unfortunately, the article provided no details about how the guy managed to commit suicide nine times. Similarly, a local cafeteria posted a sign warning patrons, "Shoes and a shirt are required to eat in this cafeteria." Apparently my socks can eat anywhere they choose.

Be Careful With Pronouns

You'll confuse readers if you use indefinite pronouns such as "it" and "them" that lack a clear antecedent. For example, the sentence, "Planting marigolds near cucumbers will keep the rabbits away from them" is confusing. Will rabbits stay away from the marigolds, cucumbers, or both? Here are other examples:

Randy asked Darrell if he could dissect the leaf.

When we tried to follow the instructions on the package, we burned it.

Not placing modifiers near words they modify can produce funny—even embarrassing—sentences. Comedians use this trick often. For example, recall Groucho Marx's famous lines from *Animal Crackers*: "One morning I shot an elephant in my pajamas. How he got in my pajamas I don't know." Scientists unknowingly become comedians when they ignore this principle of good writing. For example, few people rushed to respond to this advertisement:

Free information about VD—to get it call 873-2655.

Avoid Clichés

As soon as certain topics are raised, . . . no one seems able to think of turns of speech that are not hackneyed: prose consists less and less of words chosen for the sake of their meaning, and more and more of phrases tacked together like the sections of a prefabricated hen-house.
— George Orwell

A cliché is an expression so overworked that it has become an automatic way of getting around the main business of writing, which is to suit each word to the meaning at hand. The habitual user of cliché brands himself as a lazy writer.
— Frederick Crews

Clichés such as *quantum leap, foregone conclusion, marked contrast, wave of the future, better late than never, few and far between, first and foremost, goes without saying, last but not least, bottom line, tried and true, state-of-the-art, advanced technology, too numerous to mention,* and *by the same token* are trite, overused expressions that cover absence of thought and laziness, and make for dull reading and listening.

Vogue words and clichés mark a copycat. Use your own words, not the clichés of others. Avoid clichés ~~like the plague~~.

Write Positively

The cliché you probably remember about this "rule" is that you should not use two negative words in one sentence—that is, "don't use no double negatives." Although double negatives were used often by Chaucer, Shakespeare, and many other great writers of the past, double negatives are now considered unacceptable. Therefore, avoid expressions such as *haven't scarcely, can't help but,* and *not hardly any.*

Our minds function best when presented with information in a positive form. Therefore, say what is, not what isn't. Describe what happened or what you observed, not what you didn't observe or don't know. For example, instead of writing that the sample was not large enough, write that it was too small. Similarly, write that the rats were always sick, not that they were never healthy. Finally, instead of telling someone, "Pressure must not be lowered until the temperature is not less than 80°C," write, "Keep the pressure constant until the temperature is at least 80°C." Other examples of negative writing common in science papers include the following phrases (negatives are in italics; preferred forms are in roman type):

not many	few
not different	alike; similar
did not remember	forgot
not able	unable; cannot
did not accept	rejected
did not consider	ignored
not known	unknown
not possible	impossible
not aware	unaware
not the same	different
did not	failed to

does not have	lacks
not certain	uncertain
did not allow	prevented
did not stay	left
not on time	late
not sure	unsure

Finding and changing negative phrases is easy. Use the "find" feature of your word-processing program to find every use of "not" in your paper; then ask yourself what word could replace the negative phrase. Such revisions are important because positive writing persuades people, while negative writing nags at and turns people off.

Double negatives usually leave writers dizzy from having to do mental cartwheels. Therefore, ~~don't write in the negative~~ write positively.

Avoid Abbreviations and Foreign Words

Abbreviations often confuse readers if they're not defined. For example, MS can mean mitral stenosis to a cardiologist, multiple sclerosis to a neurologist or virologist, or manuscript to an editor. If you must use abbreviations, be sure to define them.

Most people understand English more readily than they do other languages such as Latin or Greek. For example, we know what *midbrain* means, but are probably not too sure about *mesencephalon*. Therefore, write only in English so that you avoid the pedantic mumbo-jumbo created by foreign words like these:

a fortiori	with stronger reason
a priori	from cause to effect
ab initio	from the beginning
ad hoc	for this particular purpose
ad libitum	at will
infra dignitatem	undignified
loco citato	in the place cited
de facto	in fact, actually
in toto	entirely, completely
non sequitur	it does not follow

per se	of itself
per diem	daily
sine qua non	necessity

Unless instructed to do otherwise, also avoid Latin abbreviations such as:

cf.	confer	compare
et al.	et alii	and other people; use this abbreviation only when instructed to do so by a journal to shorten the list of authors of a reference. For example, ". . . according to methods described by Burton *et al.* (1991)."
etc.	et cetera	and the rest, and so forth; using *etc.* means only that the list is incomplete. For example, consider the sentence, "For lab this week, bring a scalpel, ruler, *etc.*" What else should you bring? A chain-saw? A goat? Use *etc.* only to escape having to repeat all the items of a list already given. Avoid *etc.* at all other times; it suggests you either don't know what you're writing about or that you know but can't be bothered to tell the reader.
e.g.	exempli gratia	for example
et seq.	et sequentes	and the following
ibid.	ibidem	in the same place
i.e.	id est	that is; just say "that is" or "meaning that." Most people don't know what "i.e." means anyway.
infra dig.	infra dignitatem	undignified
loc. cit.	loco citato	in the place cited
N. B.	nota bene	note well
q.v.	quod vide	which see
viz.	videlicet	namely

Define all abbreviations that you use. For example, write "namely" instead of *viz*, and write "about" instead of *circa* or *ca*. If you insist on using foreign words or Latin abbreviations, know their meaning and do not mix them with English in the same phrase.

Express Similar Ideas in Similar Ways

Actors and actresses often bicker about who will get top billing in an upcoming movie—whose name will appear in ten-inch letters and whose will appear in half-inch letters. Something analogous occurs in writing when two or more similar ideas fight for our attention. To avoid confusing readers, writers must express parallel ideas in parallel ways—either with all verbs, all prepositional phrases, or all adjectives. This is called parallelism. For example, consider how this famous quotation loses its punch when its ideas aren't treated equally:

> . . . that all men . . . have certain inalienable rights: . . . life, liberty, and to pursue happiness.

This paragraph included in "More Is Not Merrier" by Andy Rooney shows an effective use of parallelism:

> We've already ruined all the rivers from the Yangtze to the Mississippi. Do you know of a lake you can drink from? Lake water was all drinkable before we started dumping our garbage, our sewage, and our commercial waste in it. Now we've starting ruining our oceans and, big as they are, they'll be seas of slop before long.

Many writers have much trouble with parallelism. Here are examples of those troubles from papers written by science students:

> Einstein enjoyed physics, classical music, and to ride his bicycle.

> I admire dolphins for their intelligence, energy, and because they are beautiful.

> The description of the field site was both accurate and it was easy to read.

> The limitations of this method are its limited sensitivity and that it is dependent on high humidity.

Revising these sentences improves their readability:

> Einstein enjoyed physics, classical music, and riding his bicycle.

> I admire dolphins for their intelligence, energy, and beauty.

> The description of the field site was both accurate and readable.

> The limitations of this method are its limited sensitivity and its dependence on high humidity.

Parallel descriptions of ideas are usually linked with *and* or *or*. Locate and correct them with the search command of your word-processing program. To remember how to write them correctly, recall Lincoln's famous words at Gettysburg: ". . . government of the people, by the people, for the people . . ."[6]

Meet the Expectations of Your Readers

The length-based assignments typical of most English classes have caused most students to adopt a "what do I say next?" approach to writing. This approach usually fails because the length of a paper is unrelated to the *quality* of a paper. To write a paper that helps you learn and communicate effectively you must use a strategy based not on "what do I say next," but rather on meeting readers' expectations. This strategy requires only that the *substance* of your writing (especially its context, action, and emphasis) appear in certain well-defined places within the *structure* of your writing. Although an understanding of readers' expectations is seldom included in writing courses and writing books, it is critical to effective communication. Once you know where your readers look for your message, you'll understand the many options you have to improve the chances of your reader finding your message. Understanding—and therefore being able to predict—how your writing will most likely be interpreted by readers will help move you away from a strategy of "not producing enough words" to a strategy of grasping, developing, and communicating ideas. It will help you write better every time you write, whether it be in a science class or at any of life's many other rhetorical tasks. Moreover, since writing is inextricably linked with thinking, improving either one of these will necessarily improve the other.

If you want readers to grasp what you mean, then you must understand what they need as they read and interpret your paper. This involves understanding what parts of sentences and paragraphs readers look to for emphasis. In speaking, there are many ways to indicate stress. For example, we can alter the volume, speed, or accent of our speech to help readers understand our message. In writing, most emphasis results from the structure of our writing. That structure determines how your paper hangs together, while the substance contains the information that you want to communicate. To show this, consider these two ways of presenting information about the growth of roots:

t (time) = 0.0 days, h (height) = 3.1 mm; t = 6 hours, h = 5.8 mm; t = 12 hours, h = 11 mm; t = 24 hours, h = 20 mm; t = 48 hours, h = 43 mm; t = 96 hours, h = 79 mm

---

[6]The sentence from which these words are taken shows that a long sentence can be powerful. Indeed, the sentence is 82 words long and is one of the most powerful sentences in the English language: "*It is rather for us to be here dedicated to the great task remaining before us—that from these honored dead we take increased devotion to that cause for which they gave the last full measure of devotion—that we here highly resolve that these dead shall not have died in vain—that this nation, under God, shall have a new birth of freedom—and that government of the people, by the people, for the people, shall not perish from the earth.*" Interestingly, the Gettysburg Address, one of the most famous speeches in American history, took only two minutes for Lincoln to recite.

time (hours)	height (mm)
0	3.1
6	5.8
12	11
24	20
48	43
96	79

Although each format presents the same information, the second format is easier to interpret because it gives you an easily perceived context (time) in which you can interpret the important information (height). The context appears on the left in a regular pattern, while the results on the right are in a less obvious pattern, the discovery of which is the point of the chart. You prefer this arrangement because you read from left to right—you prefer to get the familiar information first, followed by the new, important information. Any other arrangement slows your understanding because it puts information where you do not expect it.

Perceiving and absorbing information that we read requires energy. When we read, we have only a limited amount of this energy that we use in a zero-sum game—any energy used to understand the writing's structure is not available to understand the writing's substance. If readers use an excess amount of energy to discern the structure of what you write, such as the meaning of big words or how one sentence relates to a previous sentence, they have little or no energy left to find your message. This defines bad writing: By forcing readers to wade through confusing words and constructions, you unravel the structure of your writing and hide your meaning. This doesn't mean that your writing is impossible to understand; it only means that the reader is less likely to understand what you're trying to say. However, if you put words and connections where readers expect them to appear, the reader can devote more of his or her energy to understanding your ideas. These changes produce subtle and remarkable effects, not only in your thinking, but also in your writing.

Follow the Subject as Soon as Possible With Its Verb

Just as scientists expect certain kinds of information in particular sections of a scientific paper, readers have fixed expectations about where they will find important information in a sentence or paragraph. Consequently, they search for that information in those places. By knowing where these places are, you (rather than the person reading or grading your paper) can identify vagueness in your writing by perceiving problems with the structure of your writing. Consider this example:

The smallest of the URF's (URFA6L), a 207-nucleotide (nt) reading frame overlapping out of phase the NH_2-terminal portion of the adenosinetriphosphatase (ATPase) subunit 6 gene has been identified as the animal equivalent of the recently discovered yeast H^+-ATPase subunit 8 gene. The functional significance of the other URF's has been, on the contrary, elusive. Recently, however, immunoprecipitation experiments with antibodies to purified, rotenone-sensitive NADH-ubiquinone oxido-reductase [hereafter referred to as respiratory chain NADH dehydrogenase or complex I] from bovine heart, as well as enzyme fractionation studies, have indicated that six human URF's (that is, URF1, URF2, URF3, URF4, URF4L, and URF5, hereafter referred to as ND1, ND2, ND3, ND4, ND4L, and ND5) encode subunits of Complex I. This is a large complex that also contains many subunits synthesized in the cytoplasm. . . .

Why is this paragraph hard to read? Most people would say that it requires background knowledge, while others would mention its technical vocabulary. These issues are only a small part of the problem. Here's the passage with its difficult words removed:

The smallest of the URF's, an [A], has been identified as a [B] subunit 8 gene. The functional significance of the other URF's has been, on the contrary, elusive. Recently, however, [C] experiments, as well as [D] studies, have indicated that six human URF's [1-6] encode subunits of Complex I. This is a large complex that also contains many subunits synthesized in the cytoplasm. . . . [7]

This version is more readable, but is still difficult to understand. Although knowing the meanings of "URF" ("uninterrupted reading frame," a segment of DNA that could encode a protein, although no such protein has been identified) and ATPase and NADH oxidoreductase (enzymatic complexes involved in cellular energetics) provides some sense of comfort, it doesn't remedy the confusion. The reader is hindered by more than just the jargon and a lack of background information. What, then, is the problem?

Look again at the first sentence of the original passage. Although it is long (42 words), that's not the real problem; long sentences need not be hard to read. The first sentence of the passage is difficult to read not because it's a long sentence, but because it presents information in places that readers don't expect the information. For example, the sentence's subject ("the smallest") is separated from its verb ("has been identified") by 23 words, more than half the sentence. This violates a fundamental expectation of readers, namely that the subject should be followed immediately by the verb. Anything significant placed between the subject and verb interrupts the reader, causing him or her to lessen its importance. Here's how the writer could have improved the sentence:

The smallest of the URF's is URFA6LK, a 207-nucleotide (nt) reading frame overlapping out of phase the NH_2-terminal portion of the adenosinetrisphosphatase (ATPase) subunit 6 gene; it has also been identified as the animal equivalent of the recently discovered yeast H^+-ATPase subunit 8 gene.

[7]From Gopen, George D. and Judith A. Swan. 1990. "The Science of Scientific Writing." *American Scientist* 78: 550–558.

Similarly, if the intervening material is trivial to the idea, then rewrite the sentence like this:

> The smallest of the URF's (URFA6L) has been identified as the animal equivalent of the recently discovered yeast H^+-ATPase subunit 8 gene.

This sentence gets directly to its point. However, only the author could tell us which of these versions more accurately reflects his or her intentions.

Build Your Writing Around Specific Nouns and Strong Verbs

> [M]ake your words forceful, compact, and energetic . . . . A few strong, carefully selected words will deliver your message with the kick of a mule.
> — Mark S. Bacon, *Write Like the Pros*

> Words and deeds are quite different modes of the divine energy. Words are also actions, and actions are a kind of words.
> — Ralph Waldo Emerson

> I am not built for academic writings. Action is my domain.
> — Gandhi

Strong verbs are the engines that move sentences, no matter if you're talking about a trashy novel, the works of Hemingway, or a scientific masterpiece. All good writers energize their writing by anchoring sentences with strong verbs and by deleting smothering adjectives. Many of these adjectives diminish the meaning of precise words. Others, such as *rather* (*rather* important) and *almost* (*almost* unique) can be contradictory and should therefore be omitted. Finally, others adjectives intensify the meaning of words that are already emphatic. For example, instead of writing that something is *very important*, write that it is *critical* or *crucial*. Or just delete *very* and say that it is *important*. Choose specific nouns that need few adjectives.

Few things improve a sentence more than a well-chosen verb. You'll energize your writing by (1) using verbs that express action, (2) not making nouns out of strong, working verbs, and (3) preferring active voice.

Use verbs to express the action of every clause or sentence. Rambling, unwieldy sentences often hang on wimpish, inert verbs such as *exist*, *occur*, and forms of *to be* and *to have*. These verbs bring the motion of a sentence to a standstill. Replacing inert verbs with action verbs always strengthens your writing. To show this, consider this question:

> What would be the students' reception accorded the introduction of such an experiment?

In earlier chapters you learned how to revise this question by changing the passive verb to an active verb and by removing unnecessary words. This produces the following question:

> How would the students receive such an experiment?

The revision is easier to read than is the original question. Why? The revision is shorter, but its reduced length is a result rather than a cause of the improvement. To understand why the revision is easier to read, study the original question and determine what action is occurring. Action words in the first question are *be, reception, accorded,* and *introduction,* while in the second question *receive* is the only action in the sentence. This is a critical distinction because readers expect to find action in the verb. In the original question, *accorded* sounds like action, but makes no sense as action. This forces readers to divert energy from understanding your message to deciphering its structure. Consequently, when the complexity increases moderately, the chance of misinterpretation or noninterpretation increases dramatically.

Inert verbs that imply rather than state action are common in scientific writing. For example, consider these sentences:

INERT: Ruska performed the development of the electron microscope.

IMPROVED: Ruska <u>developed</u> the electron microscope.

INERT: Watson and Crick made the decision to publish their discovery in *Nature.*

IMPROVED: Watson and Crick <u>decided</u> to publish their discovery in *Nature.*

In these examples, *performed* and *made* imply action, and therefore require other verbs to complete the idea. For example, *performed the development of* means *developed,* and *made the decision* means *decided.*

Improve your sentences by asking yourself these questions: *What's happening here? What did I intend the action to be? Does a verb announce that action?* If not, use a verb to announce the action of the sentence. By doing so, you help ensure that readers understand your message. This is another example of how rewriting is rethinking, and how revision is invention. Indeed, weak verbs allow writers to mechanize their ideas as narration or recitation of facts rather than to explain their thoughts. Force yourself to consider what is happening by stating the action in a strong verb that explains your thoughts.

Don't make nouns out of strong, working verbs. Many scientists make nouns out of strong verbs. This is especially common among writers who are uncomfortable with simple, direct statements. To show this, consider the following sentence:

Authorization for the experiment was given by the professor.

This writer has turned *authorize*, a strong verb, into *authorization*, a weak noun that slows the pace of the sentence. Similar *-ion* nouns such as *determination*, *consideration*, and *conclusion* produce unnecessarily long and dull sentences—long because their weak verbs attract other unnecessary words, and dull because their verbs are abstract. Look at how much more convincing and direct the sentence becomes when you rewrite the sentence around the action verb:

The professor authorized the experiment.

Here are some other examples of sentences that improve when their smothered verb is changed to an action verb:

We conducted a study of the reaction.
We **studied** the reaction.

A need exists for greater proposal selection efficiency.
We must **select** proposals more efficiently.

Martha has expectations of being accepted to medical school.
Martha **expects** to get into medical school.

Our discussion concerned the evolution of life on land.
We **discussed** the evolution of life on land.

Here are some other examples of smothered verbs:

Smothered	Strong or Improved
draw conclusions	conclude
make assumptions	assume
take action	act
make use of	use
bring to a conclusion	conclude
make a determination	determine
have an influence on	influence
give consideration to	consider
arrive at a decision	decide
achieve purification	purify
are found to be in agreement	agree
bring to closure	end/finish
is indicative of	indicates

make mention of	mention
institute an improvement in	improve
present a summary	summarize

Substituting strong verbs for smothered verbs shortens and clarifies sentences by pushing ideas to readers. This explains why strong verbs are a hallmark of effective writing.

Prefer active voice. In writing, voice refers to the relationship of a verb to its subject. If the subject of a verb does the action, the sentence is written in active voice. If the subject receives the action, the sentence is written in passive voice.

Many scientists replace active, personalized storytelling with a passive, abstract style of writing based on *to be* verbs such as *is*, *was*, and *were*. In these sentences, the subject is acted upon. Here are some examples of passive voice:

It was suggested that the experiment be terminated. (Who suggested this? Terminated by whom?)

It was determined that information was insufficient for the scientist to recommend specific action on the question of nutritional needs of the elderly in the designated study.

A cracker is wanted.

Note that the lead verb in the first two sentences is *was* and identifies no one responsible for the action. Passive voice typically produces long sentences that sap readers' strength and often sound pompous. Furthermore, passive voice is usually weak and unconvincing, suggesting that scientists were acted upon rather than that scientists acted. With passive voice, it seems as if the writers are reporting revelation, not information, because passive voice implies that a force—the amorphous "it"—guided their work.

Active voice is a writing style in which the subject does the action. Changing the sentences listed above to active voice produces the following sentences:

I suggested that we end the experiment.

The scientists lacked enough information to recommend nutritional improvements for the elderly included in the study.

Polly wants a cracker.

Some may be shocked by the bluntness of these sentences. Don't worry about people with this concern. Active voice almost always improves writing because it shortens sentences, improves readability, makes analyses more incisive, increases understanding, and more effectively reveals the depth of a scientist's logic. Active voice also produces direct, clear sentences that involve readers. Conversely, passive voice is an impersonal, and usually boring, style of writing.

Most scientists use passive voice either out of habit or to make themselves seem scholarly, objective, or sophisticated. The result, however, is boring writing.

Figure 4–1
The differing effects of active and passive voice.

For example, consider these sentences:

There was considerable erosion of the land by the floods.

Animal movement to less restrictive methods of care may be followed by increased probability of recovery.

A pH meter is capable of detecting the concentration of hydrogen ions in a solution.

The patients were examined by the physicians.

Papers are written by students.

All of these sentences are grammatically correct. However, all are based on weak verbs such as *is, are,* and *were*—derivatives of *to be*. Notice how these sentences improve when you change passive voice to active voice:

The floods eroded the land.

If we treat animals less restrictively, they may recover faster.

A pH meter measures the concentration of hydrogen ions in a solution.

The physicians examined the patients.

Students write papers.

These sentences are shorter, more direct, and communicate better than those written in passive voice.

Contrary to the implication of passive voice, science is a personal activity done

Scientific Writing, Objectivity, and Passive Voice

First-person pronouns such as "I" and "we" disappeared from scientific writing in the United States in the 1920s, when today's inflexible and impersonal style of scientific writing began to dominate science and technology. Since then, scientists have used the anonymity of passive voice to make themselves appear as modest, passive, and objective observers. This greatly diminishes communication. For example, when speaking, most scientists provide information in the way we ordinarily expect to receive it—as a narrative:

> We wanted to understand how penicillin affects growth of bacteria. To do this, we grew bacteria in the presence of varying concentrations of penicillin. We learned that penicillin inhibits growth of bacteria.

Such a narrative isn't something any of us have to think about—we communicate like this all the time. And when we aren't telling these kinds of stories, we're listening to them. Narrative communication is easy for everyone. However, compare this with the abstraction of passive voice:

> The growth of bacteria was studied. Bacteria were grown in the presence of varying concentrations of penicillin. It was learned that bacterial growth is inhibited by penicillin.

Nobody in the real world communicates like this. The insistence of many scientists to write only in passive voice forces readers to shift into a foreign

by people, not machines. Notice how this paragraph from a paper published in *The New England Journal of Medicine* uses active voice to communicate directly with readers (boldface added):

> **Our** findings concerning the overall incidence of acute and chronic GVHD are not appreciably different from those of other groups. Patients with Grades 0 through II acute GVHD fared better than those with more severe involvement, the latter group having a higher incidence of fatal infection. In this series, **we** found no association between acute GVHD and a reduced risk of posttransplantational relapse. However, as in other recent reports, **we** did not find an association between the presence of chronic GHVD and a low relapse rate.

Finally, note how James Watson and Francis Crick used active voice and a simple, personal, get-to-the point writing style to open their monumental paper describing the structure of DNA:

> We wish to suggest a structure for the salt of deoxyribose nucleic acid (D.N.A.).

The notion that "I" and "we" somehow make science undignified is foolish and hobbles science. After all, all research is done by someone, and every paper, book, and essay has someone's name listed as the author. Most editors prefer that writers mode of communication. Consequently, passive voice is hard to write and even harder to read because, especially in reports of experiments, it doesn't reflect what really happened. More importantly, the abstraction of passive voice often provides less information than a narrative using active voice because it removes the feelings, the flavors, the juice, and sometimes even the substance of what we did.

The notion that passive voice ensures objectivity is artificial because objectivity has nothing to do with one's writing style or use of personal pronouns. Objectivity in science results from your choice of subjects, facts that you choose to include or omit, sampling techniques, and how you state your conclusions. Scientific objectivity is a personal trait unrelated to writing. Relying on third-person (for example, "the author") achieves no modesty; moreover, science *is* a personal activity done by people, not machines. Discarding *I* and *we* merely leads to awkward, weak, and indirect writing.

Passive voice can be useful to writers. For example, passive voice is effective when you want to stress "what was done" rather than "who did it," as in the sentence, "Darwin's *The Origin of Species* was published in 1859." Passive voice is also useful when you want to avoid accountability. For example, embarrassed politicians report that "funds were found to be missing," instead of, "I stole the money." Rather than fool anyone, the excessive use of passive voice usually just bores readers. Passive voice is useful for adding variety, softening commands, avoiding responsibility, avoiding embarrassment, and slowing the pace of writing and reading. Unless you have one of these reasons for using passive voice, choose active voice.

use "I" or "we" to describe research. For example, the *American National Standards for the Preparation of Scientific Papers for Written or Oral Presentations*, which includes an impressive list of organizations represented by its views, states, "When a verb concerns action by the author, the first person should be used, especially in matters of experimental design."[8] If the purpose of your paper is to tell what you did and observed, then you should appear at the beginning of sentences and clauses. Insistence on using "the authors" or "the writers of this paper" is pompous, distant, and stuffy.

Reject the argument that scientists should use only passive voice. Indeed, great scientists such as Einstein, Faraday, Watson, Crick, Curie, Darwin, Lyell, Freud, and Feynman communicated their brilliant ideas by preferring active over passive voice. Regardless of why it occurs, abstract writing typical of that based on passive voice is poor writing.

Science is the great adventure of our time. Don't suffocate it with passive, abstract writing.

---

[8]American National Standards Institute (ANSI) 1979. Z39. 16-1979, *American National Standard for the Preparation of Scientific Papers for Written or Oral Presentations.* New York: American National Standards Institute. p. 12.

What Do I Need to Know About Grammar?

Many people are uneasy about anything associated with grammar. Others such as Panini (a 5th century B.C. Indian scholar who wrote one of the world's first grammar books) have claimed that, "Who knows my grammar knows God." Although most writers wouldn't go quite that far, it's clear that grammar means different things to different people. Why all the fuss? What does a scientist need to know about grammar to write well? And what about all of those rules?

Good grammar isn't as easy as rules invented centuries ago, repeated by editors unwilling to determine whether those rules comport with reality, taught by instructors who teach only what textbooks tell them, and ignored by the best writers everywhere. Thoughtlessly following *all* of the rules *all* of the time will offend no one, but will deprive your writing of any flexibility. What good is learning a rule if all you can do is obey it? Indeed, some of the most effective writers are the best violators of rules. These writers usually follow grammatical rules; that's why their occasional violation of a rule is so noticeable. Good writers use such violations for effect in their writing.

Some rules only cause problems if they're followed. For example, forget the adage to "Vary your word choice." Don't try to substitute similar words for the exact word when the exact word is critical to the meaning of a sentence. If you use two words for one concept, you risk having readers think that you mean two concepts. This is why papers written by professional educators often are so confusing. Rather than repeat the word *explanation*, educators often substitute words such as *symbolic modeling, precept, language symbols, words, narrative modeling*, and *instructions*. Rather than provide "elegant variation," this refusal to write precisely merely confuses readers. Several other so-called "rules" lack substance and are ignored by all good writers. For example, ignore the rule about never beginning a sentence with *because, and,* or *but.* Also ignore the notion that you should never split an infinitive. This rule originated with eighteenth-century writers who reasoned that since you can't split the one-word Latin infinitives, you shouldn't split English infinitives. Ignore the "split infinitive rule" if it impedes communication: If you can communicate best by saying *to boldly go*, then say it.

Other rules are imperatives that we violate at the risk of seeming at least careless, at worst illiterate. These rules are observed by even the less-than-best writers. Don't break these rules:

Avoid double negatives.

Example: I have not got no data.

Avoid nonstandard verbs.

Example: I knowed that she was my lab partner.

Avoid double comparatives.

Example: *Paramecium* is more quicker than *Amoeba*.

Do not substitute adjectives for adverbs.

Example: The professor did work real good.

Do not use incorrect pronouns.

Example: Her and me made the media.

Don't let the subject disagree with the verb.

Example: They was in the lab.

What good, then, is grammar? Many scientists view grammar as a series of formal rules that are unnecessary as long as they can "get the message across." This attitude is typical of ineffective writers who don't understand writing, and may suffice as long as you have nothing important to say. However, when you have something important to say, an understanding of grammar is critical because in writing, language is the way words are used. To understand how to use words effectively, you must understand something about grammar because it functions as a series of sign posts that, if followed, organizes your thoughts and improves your writing. Grammar deals with how words function in sentences, and therefore underlies what words mean. Writing is effective only when it communicates, and that occurs most efficiently when you understand a few principles of grammar. Grammatical conventions represent agreement, and agreement stimulates communication.

Words can do many things, and one must understand grammar to use words effectively. Weak grammar puts a greater burden on the reader to grasp your point. And the harder it is for others to get your point, the less likely you are to communicate well, no matter how good your point may be.

Understanding grammar helps you clearly think about and precisely express your ideas—it helps you say what you mean in an orderly and precise way, usually by eliminating common sources of confusion. Although grammar can't substitute for knowing what you want to say, it can save you time by eliminating hazards that block communication and thinking. However, you do not need to understand all of the nuances of grammar to use grammar effectively, just as you do not need to understand how a watch works to tell time.

Place in the Stress Position the Material You Want the Reader to Emphasize

So far in this chapter you've learned the importance of carefully chosen words and phrases for communication. These words and phrases are the ingredients of the next stage of effective writing: strong, concise sentences. Being able to write these sentences depends largely on knowing how to start and end a sentence.

Readers naturally emphasize the material at the end of a sentence. Consequently, they expect the most important idea—the idea that you want to emphasize—in the "stress position" at the end of a sentence. Periods end stress positions, while colons and semicolons indicate secondary stress positions in the sentence. If you fill the stress position with a word that might be emphasized, readers sense that you wanted to emphasize that word. Conversely, if you put emphatic material of a sentence anywhere but in the stress position, readers might find the stress position occupied by material that you didn't want to emphasize. Nevertheless, readers will probably emphasize the imposter material, killing your chances of influencing their interpretation. Alternately, if the stress position is filled with material that can't be logically emphasized, readers must guess from all the other possibilities in the rest of the sentence what you wanted to emphasize. Moreover, the longer the sentence, the greater the chances of your reader choosing the wrong information for emphasis. Such guessing causes you to lose control of the reader's interpretation of your writing.

Put at the end of a sentence the newest or most significant information—the information that you want to stress and information that you will expand on in your next sentence. When you introduce an important term for the first time, design the sentence so that the term appears at the end of the sentence, even if you must invent a sentence just to define or emphasize that term. For example, look at how this paper in *The New England Journal of Medicine* introduces readers to lymphokine-activated killer cells:

> We have previously described a method for generating lymphocytes with antitumor reactivity. The incubation of peripheral-blood lymphocytes with a lymphokine, interleukine-2, generates lymphoid cells that can lyse fresh, noncultured, natural-killer-cell-resistant tumor cells but not normal cells. We have termed these cells lymphokine-activated killer (LAK) cells.

Similarly, look at how Charles Darwin emphasized evolution in the closing words of Chapter 21 of *The Descent of Man*:

> We must, however, acknowledge, as it seems to me, that man with all his noble qualities, still bears in his bodily form the indelible stamp of his lowly origin.

Another simple way to effectively end a sentence is to delete unnecessary words until the stressed information is at the end of the sentence. For example:

> The teaching assistant said that the ocular and objective lens determine magnification of the specimen during use of the microscope.

The teaching assistant said that the ocular and objective lens determine magnification of the specimen.

Knowing to place emphatic information in the stress position of a sentence will help you write more concise, emphatic sentences. It will also help you identify sentences whose length detracts from your message. Longer sentences have more possible stress positions, thus explaining why longer sentences often communicate less effectively than shorter sentences. A sentence is too long when there is more than one candidate for a stress position.

Place Familiar Information at the Beginning of Sentences

> A reader can receive more information
> and understand it better if the informa-
> tion is in an expected and familiar form.
> — John Mitchell

> For me, the big chore is always the
> same—how to begin a sentence, how
> to continue it, how to complete it.
> — Claude Simon, on winning the
> Nobel prize in 1985

Just as readers naturally emphasize the material at the end of a sentence, so too do they search for familiar information at the beginning of a sentence. Readers use the beginning of a sentence to link them with previous information and to provide context for upcoming material. Therefore, use the beginning of a sentence to put the reader in familiar territory; this helps readers see the logic of your argument. If you repeatedly begin sentences with new information and end them with old information, you force readers to carry the new information further into the sentence before it can be put into context. This consumes much of the readers' energy, leaving them with little for understanding your message.

How you begin sentences is critical to how a reader will understand them. The secret to a clear and readable sentence lies in the first five or six words of the sentence. Remember this: If you have to go more than six or seven words into a sentence to get past the subject and verb, or if subject is not one of those words, revise the sentence so that the main ideas appear as subjects and the actions appear as verbs. If you consistently organize the subject around a few concepts and then express that subject's action with a precise verb, your readers will understand what you are writing about. If you do not do this, your writing will seem not just unfocused, but weak and anticlimactic.

Each sentence should teach the reader something new. To do this, design your sentences as good teachers design their lectures—by using simple, straightforward language to expand a previous idea or to connect new ideas with what the reader already knows. Provide context before asking readers to consider anything new. Do this by putting the reader in familiar territory—start sentences with ideas that you have already mentioned, referred to, or implied. For example, notice how this writer uses the phrase *this energy* to show the relationship between the energy in a slice of apple pie and the potential uses of that energy:

> A slice of apple pie contains about 1.5×10^6 J (365 Cal) of energy. **This energy** is enough energy for a woman to run for almost an hour or for a typist to enter about 15,000,000 characters on a manual typewriter.

Note how Rachel Carson linked the sentences in this paragraph from *Silent Spring*, a masterpiece of scientific literature (see also pages 108–114):

> From the green depths of the offshore Atlantic many paths lead back to the coast. They are paths followed by fish; although unseen and intangible, they are linked with the outflow of waters from the coastal rivers. For thousands upon thousands of years the salmon have known and followed these threads of fresh water that lead them back to the rivers, each returning to the tributary in which it spent the first months or years of life. So, in the summer and fall of 1953, the salmon of the river called Miramichi on the coast of New Brunswick moved in from their feeding grounds in the far Atlantic and ascended their native river. In the upper reaches of the Miramichi, in the streams that gather together a network of shadowed brooks, the salmon deposited their eggs that autumn in beds of gravel over which the stream water flowed swift and cold. Such places, the watersheds of the great coniferous forests of spruce and balsam, of hemlock and pine, provide the kind of spawning grounds that salmon must have in order to survive.

Don't confuse restatement with explanation. Avoid starting sentences with phrases such as "*That is*," and delete sentences preceding those beginning with, *In other words*. Similarly, don't start sentences with, *It is interesting to note that*; you'll only tempt the reader to find it dull. Rather, select words to *make* it interesting. Finally, don't say, *I might add that*; just add it.

Use the structure of a sentence to persuade your readers of the relative values of the sentence's contents. Put at the beginning of a sentence the old information that links backward. Similarly, put new information that you want the reader to emphasize at the end of the sentence where the reader naturally exerts the greatest emphasis when reading the sentence. Beginning with the exciting material and ending with what readers already know leaves readers disappointed. Save the best for last—don't start with the ice cream and progress to the broccoli.

A Final Word About Choosing Your Words

> Whatever one wishes to say, there is one
> noun only by which to express it, one
> verb only to give it life, one adjective

only which will describe it. One must
search until one has found them, this
noun, this verb, this adjective, and never
rest content with approximations, never
resort to trickery, however happy, or to
illiteracies, so as to dodge the difficulty.
      — Guy de Maupassant, *Pierre et
Jean*

All words are pegs to hang ideas on.
      — Henry Ward Beecher

As you write, remember that words are the only tool you have. Therefore, use them
originally, directly, and carefully. Value their strength and diversity. And remember—someone is listening.

Francis H.C. Crick

The Uniformity of Biochemistry

Francis Crick (b. 1916) is a scientist who shared the 1962 Nobel prize for Physiology and Medicine with James Watson and Maurice Wilkins for the discovery of the structure of DNA. Crick has written a variety of articles and books, including *What Mad Pursuit: A Personal View of Scientific Discovery* and *The Astonishing Hypothesis.*

You'll learn more about Crick's discovery when you read James Watson's account of their discovery (see 295). Here, Crick writes about the unity of bio-chemistry.

The problem of the origin of life is, at bottom, a problem in organic chemistry—the chemistry of carbon compounds—but organic chemistry within an unusual framework. Living things, as we shall see, are specified in detail at the level of atoms and molecules, with incredible delicacy and precision. At the beginning it must have been molecules that evolved to form the first living system. Because life started on earth such a long time ago—perhaps as much as four billion years ago—it is very difficult for us to discover what the first living things were like. All living things on earth, without exception, are based on organic chemistry, and such chemicals are usually not stable over very long periods of time at the range of temperatures which exist on the earth's surface. The constant buffeting of thermal motion over hundreds of millions of years eventually disrupts the strong chemical bonds which hold the atoms of an organic molecule firmly together over shorter periods; over our own lifetime, for example. For this reason it is almost impossible to find "molecular fossils" from these very early times.

Minerals can be much more stable, at least on a somewhat coarser scale, mainly because their atoms use strong bonds to form regular three-dimensional structures. The failure of a single bond will not disturb the shape of the mineral too much. Fossils are seen in abundance in rocks laid down a little over half a billion years ago, at a time when organisms had evolved sufficiently to develop hard parts. Such fossils are not usually made of the original material of those organisms but consist of mineral deposits which have infiltrated them and taken up their shape. The shape of the soft parts is usually lost, though occasionally traces like wormholes are preserved—footprints on the rocks of time.

Are there any fossils much earlier than this? Careful microscopic examination of very early rocks has shown them to contain small structures which look like the fossilized remnants of very simple organisms, rather similar to some of the unicellular organisms on the earth today. This makes good sense. In the process of evolution we would expect creatures with many cells to develop from earlier ones having only single cells. Although there is still some controversy about the details, the earliest organisms of this type have been dated to about $2\frac{1}{2}$ to $3\frac{1}{2}$ billion years ago. The age of the earth is about $4\frac{1}{2}$ billion years. After the turmoil of its initial formation had subsided there was a period of about a billion years during which life could have evolved from the complex chemistry of the earth's surface, especially in its oceans, lakes, and pools. Of that period we have no fossil record at all, because no preserved parts of the sedimentary rocks from that time have yet been found.

There are only two ways for us to approach this problem. We can try to simulate those early conditions in the laboratory. Since life is probably a happy accident which, even in the extended laboratory of the planet's surface, is likely to have taken many millions of years to occur, it is not too surprising that such research has not yet got very far, though some progress has been made. In addition we can look carefully at all living things which exist today. Because they are all descended from some of the first simple organisms, it might be hoped that they still bear within them some traces of the earliest living things.

At first sight such a hope seems absurd. What could possibly unite the lily and the giraffe? What could a man share with the bacteria in his intestines? A cynic might wonder whether, since all living things eat or are eaten, this at least suggests they have something in common. Remarkably, this turns out to be correct. The unity of biochemistry is far greater and more detailed than was supposed even as little as a hundred years ago. The immense variety of nature—man, animals, plants, microorganisms, even viruses—is built, at the chemical level, on a common ground plan. It is the fantastic elaboration of this ground plan, evolved by natural selection over countless generations, which makes it difficult for us, in our everyday life, to penetrate beneath the outward form and perceive the unity within. In spite of our differences we all use a single chemical language, or, more precisely, as we shall see, *two* such languages, intimately related to each other.

To understand the unity of biochemistry we must first grasp in a very general way what chemical reactions go on within an organism. A living cell can be thought of as a fairly complex, well-organized chemical factory which takes one set of organic molecules—its food—breaks them down, if necessary, into smaller units and then reassorts and recombines these smaller units, often in several discrete steps, to make many other small molecules, some of which it excretes and some of which it uses for further synthesis. In particular, it strings special sets of these small molecules together into long chains, usually unbranched, to make the vital macromolecules of the cell, the three great families of giant molecules: the nucleic acids, the proteins, and the polysaccharides.

The first level of organization we must consider is the lowest of all—that at which atoms are bound together to form small molecules. Now, a single atom is a fairly symmetrical object. Its shape is approximately spherical and if we look at it in a mirror it appears exactly the same, just as a billiard ball would. More intricate structures can have a "handedness"—our own hands are a good example. If we look at a right hand in a mirror we see a left hand, and vice versa. We can oppose our two hands, as in prayer, but this is as if we held a mirror between them. There is no way in which we can exactly superimpose one on the other, even in our imagination.

Some simple organic molecules, such as alcohol, have no "hand"; they are identical to their mirror images, as indeed a cup is. But this is not true of most organic molecules. The sugar on the breakfast table, if looked at in a mirror, becomes a significantly different assembly of atoms. This difference does not matter for *all* types of chemical reaction. If we heated such a molecule and could watch the molecular vibrations increase until one of the bonds broke, we would see that, had we imagined the mirror image of this process, the relative movements of all the atoms would have been identical. The basic reactions of chemistry are symmetrical under reflection to a very high degree of approximation. The difference in the hand only becomes important when two molecules have to fit together. We can see this in the manufacture of a glove. All the components of a glove—the fabric, the sewing thread, even the buttons—are, individually, mirror-symmetric, but they can be put together in two similar but different ways, to make either a right-handed glove or a left-handed one. Obviously we need two sorts because we have two kinds of hands—a good left-handed glove will not fit properly onto a right hand.

The simplest form of asymmetrical molecule of this type arises when a single carbon atom is joined by single bonds to four other *different* atoms, or groups of atoms. This is because the four bonds of the carbon atom do not all lie in the same plane but are spaced out equally in all three dimensions, pointing approximately toward the corners of a regular tetrahedron.

Thus, organic molecules—molecules containing carbon atoms—often have a hand, even though they may be small, but we still have to realize why this matters in a cell. The basic reason is that a biochemical molecule does not exist in isolation. It reacts with other molecules. Almost every biochemical reaction is speeded up by its own special catalyst. A small molecule, to react in this way, has to fit snugly onto the catalyst's surface, and since the small molecule has a hand, the catalyst must also have one. As in the case of a glove, the reaction will not work properly if we try to fit a left-handed molecule into the cavity appropriate for a right-handed one.

Imagine you could watch this minute chemical factory working and could see all the numerous reactions going on, with molecules diffusing rapidly from one place to another, fitting onto the various catalytic molecules, breaking, re-forming, regrouping, and reacting in many different ways. Now imagine you were watching a factory which was the exact mirror image of the first one. Everything would proceed exactly as before, since the laws of chemistry are the same in a mirror. Trouble would arise only if you tried to combine the two, using some components from one system mixed with others from the mirror world.

We can thus see why, in a single organism, the handedness of the many asymmetrical molecules, large and small, must be concordant. Moreover, it is an experimental fact that the asymmetrical molecules on one side of your body have exactly the same hand as those on the other side. But could we not have two distinct types of organisms, one the mirror image of the other, at least as far as its components are concerned? This is what is never found. There are not two separate kingdoms in nature, one having molecules of one hand and the other their mirror images. Glucose has the same hand everywhere. More significantly, the small molecules that are strung together to make proteins—the amino acids—are all L-amino acids (their mirror images are called D-amino acids: L = Levo, D = Dextro) and the sugars in the nucleic acids are also all of one hand. The first great unifying principle of biochemistry is that the key molecules have the same hand in all organisms.

There are many other biochemical features which are astonishingly alike in all cells. The actual metabolic pathways—the precise ways in which one small molecule is converted into another—are often remarkably similar, though not always identical. So are some of the structure features, but the uniformity is even more striking at the deepest levels of organization; striking because there it is both arbitrary and complete.

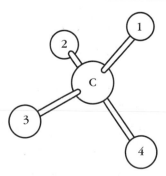

The distribution in space of the four bonds around a single carbon atom.

Much of the structure and the metabolic machinery of the cell are based on one family of molecules, the proteins. A protein molecule is a macro-molecule, running to thousands of atoms. Each protein is precisely made, with every atom in its correct place. Each type of protein forms an intricate three-dimensional structure, peculiar to itself, which allows it to carry out its catalytic or structural function. This three-dimensional structure is formed by folding up an underlying one-dimensional one, based on one or more polypeptide chains, as they are called. The sequence of atoms along this backbone consists of a pattern of six atoms, repeated over and over again. Variety is provided by the very small side-chains which stick out from the backbone, one at every repeat. A typical backbone has some hundreds of them.

Not surprisingly, the synthetic machinery of the cell constructs these polypeptide chains by joining together, end to end, a particular set of small molecules, the amino acids. These are all alike at one end—the part which will form the repeating backbone—but different at the other end, the part which forms the small side-chains. What is surprising is that there are just twenty kinds of them used to make proteins, and this set of twenty is exact-ly the same throughout nature. Yet other kinds of amino acids exist and

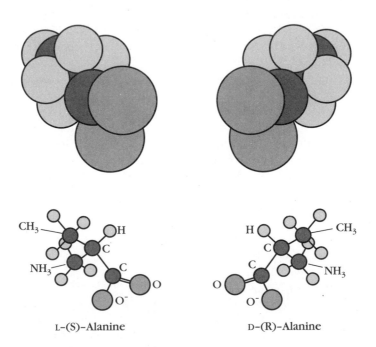

L-(S)-Alanine   D-(R)-Alanine

The two forms of the amino acid alanine. Each is the mirror image of the other. The upper figures use space-filling models; the lower ones use ball-and spoke models. The letters indicate the atoms. The form of alanine found in proteins is L-alanine, the one on the left.

several of them can be found within a cell. Nevertheless, only this particular set of twenty is used for proteins.

A protein is like a paragraph written in a twenty-letter language, the exact nature of the protein being determined by the exact order of the letters. With one trivial exception, this script never varies. Animals, plants, microorganisms, and viruses all use the same set of twenty letters although, as far as we can tell, other similar letters could easily have been employed, just as other symbols could have been used to construct our own alphabet. Some of these chemical letters are obvious choices, since they are small and easily available. Others are less obvious. If every printed text in the world used exactly the same arbitrary set of letters (which, as we know, is far from the case), we would reasonably conclude that the fully developed script had probably originated in one particular place and been passed on by constant copying. It is difficult not to come to the same conclusion for the amino acids. The set of twenty is so universal that its choice would appear to date back to very near the beginning of all living things.

Nature employs a second, very different chemical language which is also fairly uniform. The genetic information for any organism is carried in one of the two closely related families of giant chain molecules, the nucleic acids, DNA and RNA. Each molecule has an immensely long backbone with a regular, repeating structure. Again, a side-group is attached at regular intervals but in this case there are only four types; the genetic language has only four letters. A typical small virus, such as the polio virus, is about five thousand letters long. The genetic message in a bacterial cell usually has a few million letters; man's has several billion, packed in the center of each of our many cells.

One of the major biological discoveries of the sixties was the unraveling of the genetic code, the small dictionary (similar in principle to the Morse code) which relates the four-letter language of the genetic material to the twenty-letter language of protein, the executive language.

To translate the genetic message on a particular stretch of nucleic acid, the sequence of the side-groups is read off by the biochemical machinery in groups of three, starting from some fixed point. Since the nucleic acid language has just four distinct letters, there are sixty-four possible triplets (4 × 4 × 4). Sixty-one of these, codons as they are called, stand for one amino acid or another. The other three triplets stand for "end chain." (The signal for "begin chain" is a little complicated.)

The exact nature of the genetic code is as important for biology as Mendeleev's Periodic Table of the elements is for chemistry, but there is an important difference. The Periodic Table would be the same everywhere in the universe. The genetic code appears rather arbitrary, or at least partly so. Many attempts have been made to deduce the relationship between the two languages from chemical principles, but so far none have been successful. The code has a few regular features, but these might be due to chance.

Even if there existed an entirely separate form of life elsewhere, also

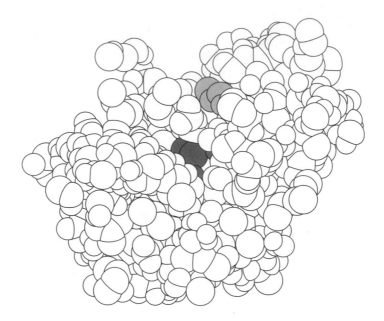

An atomic model of a small protein, the enzyme ribonuclease S. The shaded atoms form part of the active site of the enzyme. The protein would normally be entirely surrounded by water molecules.

based on nucleic acids and protein, I can see no good reason why the genetic code should be exactly the same there as it is here. (The Morse Code, incidentally, is not completely arbitrary. The commonest letters, like *e* and *t*, are allocated the shortest number of dots or dashes.) If this appearance of arbitrariness in the genetic code is sustained, we can only conclude, once again, that all life on earth arose from one very primitive population which first used it to control the flow of chemical information from the nucleic acid language to the protein language.

Thus, all living things use the same four-letter language to carry genetic information. All use the same twenty-letter language to construct their proteins, the machine tools of the living cell. All use the same chemical dictionary to translate from one language to the other. Such an astonishing degree of uniformity was hardly suspected as little as forty years ago, when I was an undergraduate. I find it a curious symptom of our times that those who derive deep satisfaction from brooding on their unity with nature are often quite ignorant of the very unity they are attempting to contemplate. Perhaps in California there already exists a church in which the genetic code is read out every Sunday morning, though I doubt whether anyone would find such a bare recital very inspiring.

We see, then, that one way to approach the origin of life is to try to imagine how this remarkable uniformity first arose. Almost all modern theories and experimental work on life's origin take as their starting point the synthesis of either nucleic acid or protein or both. How could the primitive earth (if indeed life first started on earth) have produced the first relevant macromolecules? We have seen that these chain molecules are made by joining together small subunits end to end. How could the small molecules have been synthesized under early, prebiotic conditions? And how could we decide, even if we could have watched the whole operation in atomic detail, when the system first deserved to be called "living"? To come to grips with this problem we must examine just what attributes we would expect *any* living system to have.

Understanding What You've Read

How is uniformity achieved in biochemistry?

What does Crick tell his readers about himself and his approach to science? Cite examples to support your answer.

Discuss, support, or refute the ideas contained in these quotations by Francis Crick:

> Only time and money stand between us and knowing the composition of every gene in the human genome.

> If you want to understand function, study structure.

> While Occam's razor is a useful tool in the physical sciences, it can be a very dangerous implement in biology. It is thus very rash to use simplicity and elegance as a guide in scientific research.

Exercises

1. You're probably familiar with the opening of Abraham Lincoln's Gettysburg Address:

    > Four score and seven years ago, our fathers brought forth on this continent a new nation conceived in liberty and dedicated to the proposition that all men are created equal.

    But this, according to Oliver Jensen, is what the Gettysburg Address would have sounded like if it had been written by Dwight Eisenhower:

    > I haven't checked these figures but around 87 years ago, I think it was, a number of individuals organized a governmental set-up here in this country, I believe it concerned the Eastern states, with this idea they were following up based on a sort of national independence arrangement and the program that every individual is just as good as every other individual.

    Jensen's parody doesn't say anything more than Lincoln's original, yet it is much less convincing. Had Lincoln used Eisenhower's text at Gettysburg, do you think anyone would have remembered what he'd said? Why? What are the consequences of such writing?

2. Economist John Kenneth Galbraith argues the case against "elitist" writing:

> There are no important propositions that cannot be stated in plain language. The writer who seeks to be intelligible needs to be right; he must be challenged if his argument leads to an erroneous conclusion and especially if it leads to the wrong action. But he can safely dismiss the charge that he has made the subject too easy. The truth is not difficult. Complexity and obscurity have professional value—they are the academic equivalents of apprenticeship rules in the building trades. They exclude the outsiders, keep down the competition, preserve the image of a privileged or priestly class. The man who makes things clear is a scab. He is criticized less for his clarity than for his treachery.

Use evidence to support or attack Galbraith's argument.

3. Edit this sentence from a Nuclear Regulatory Commission report:

> It would be prudent to consider expeditiously the provision of instrumentation that would provide an unambiguous indication of the level of fluid in the reactor vessel.

4. Write a paragraph about the issues raised in this quotation by P. Meredith:

> The great heresy of science has been its adoption of impersonal language. For every observation involves a private world, and science is based on observation . . . we see, then, that to restore the personal pronoun to the language of science is much more than a literary facilitation of readability. It demands the self-understanding of science as an arbitrary enterprise of man. Too long it has masqueraded as the impersonal voice of Nature, another God issuing commandments. And by forcing the restoration we shall force the examination of the values which dominate our choice of axioms. Moreover, this is no mere verbal problem, but a private evaluation of our public human relations. Scientists must restore their own atrophied feelings.

5. A famous writer once claimed, "Passive voice doesn't look you in the eye." What did she mean by this?

6. In 1967 F. Peter Woodford described in delightful detail the poor writing in scientific "journals—even the journals with the highest standards":

> In the linked worlds of experimental science, scientific editing, and science communication many scientists are considering just how serious an effect the bad writing in our journals will have on the future of science.
>
> All are agreed that the articles in our journals—even the journals with the highest standards—are, by and large, poorly written. Some of the worst are produced by the kind of author who consciously pretends to a "scientific scholarly" style. He takes what should be lively, inspiring, and beautiful and, in an attempt to make it seem dignified, chokes it to death with stately abstract nouns; next, in the name of scientific impartiality, he fits it with a complete set of passive constructions to drain away any remaining life's blood or excitement; then he embalms the remains in molasses of polysyllable, wraps the corpse in an impenetrable veil of vogue words, and buries the stiff old mummy with such pomp and circumstance in the most distinguished

journal that will take it. Considered either as a piece of scholarly work or as a vehicle for communication, the product is appalling.[9]

Do you think this criticism is true today? Provide examples to support your answer.

7. Make these sentences and paragraphs more concise. I've included solutions for the first two.

The department deemed it necessary to terminate the professor.

The professor was fired.

It is utterly fruitless to become lachrymost over precipitately decanted lacteal fluid.

Don't cry over spilled milk.

We could, of course, proceed on the assumption that an animal's behavior is innate.

The desirability of abbreviating the incubation period was worth mentioning by the professor.

Assessment of the experiment should precede implementation of the program.

Randy's data failed to impact positively on me.

The purpose of this experiment is to offer the student the ways and resources to assist those in charge of developing protocols and procedures to increase the effective utilization of the amino acid analysis information system.

Health educators serve as change agents: they facilitate lifestyle alterations in recipients of their services.

Feathered bipeds of similar plumage will live gregariously. (Members of avian species of identical plumage congregate.)

A revolving volcanic conglomeration accumulates no congeries of small bryophytic herbage.

The temperature of the aqueous content of an unremittingly ogled vessel will never attain 212 degrees Fahrenheit.

Abstentation from uncertain undertakings precludes a potential escalation of remuneration.

Ingestion of an apple (the pome fruit of any tree of the genus, *Malus*, said fruit being usually round in shape and red in color) on a diurnal basis will with absolute certainty keep a primary member of the disease prevention establishment absent from one's local environment.

Scintillate, scintillate, astra minific.

Neophytic serendipity.

Sorting by mendicants must be extirpated.

[9]Woodford, F. Peter. 1967. "Sounder Thinking Through Clearer Writing." *Science* 156: 743.

As a case in point, other authorities have proposed that slumbering canines are best left in a recumbent position.

There is no utility in belaboring a deceased equine.

I do not want to read this obsolete operator's manual.

Hoping for a better crop, the field was irrigated weekly.

Being confined to a small cage, the zoologist tried to enrich the baby baboon's food.

While looking through the microscope, a new idea popped into my head.

Since hiring the technician, our data have improved.

Reaching the heart, a bypass was performed on the blocked arteries.

At the age of 12 my mother entered me in a science fair.

To become a science professor, two degrees must be earned.

To learn the technique the first three exercises must be completed.

Jimmy was unsuccessful in his attempt to pass the test.

The absorption coefficient of the chlorophyll must be determined over the sample range by the students in the classroom.

The students reached a decision to terminate the experiment after extensive deliberations.

The new computer is quieter, faster, and will break down less often than the current computer.

These data provide support for your conclusions.

The experiment will be done by the graduate students.

The decision was made to repeat the experiment.

A series of standards, varying in concentration over the range corresponding to approximately 0.5 to 4 times the standard, is prepared and analyzed under the same conditions and during the same period of time as the unknown samples.

Included in these plans is the construction of a high quality, multiuser, easy access, biotechnical center for optimization of research output.

Before analysis of the samples, the analytical system is allowed to equilibrate until a steady baseline data reading is attained and reached on the recorder.

The replacement of a lens is a job that can be handled by the least experienced student.

The end of a sentence is the place where the most important information should be placed.

It can be concluded that my data are worthless.

It was suggested by the students that lab reports be accepted late.

In the next chapter the functions of the nucleus are explained.

8.   Write a short essay describing how this *The Far Side* Cartoon relates to science.

Great moments in evolution.

Figure 4–2
The Far Side. Gary Larson (@ Universal Press Syndicate).

9.   Discuss, support, or refute the ideas contained in these quotations:

Biology

 Plants, instead of affecting the air in the same manner as animal respiration, reverse the effect of breathing and tend to keep the atmosphere sweet and wholesome.
— Joseph Priestley (1733–1804)

The structure of tissues and their functions, are two aspects of the same thing. — Alexis Carroll

Heredity is just environment stored. — Luther Burbank (1849–1926)

Chemistry

Chemistry without catalysis, would be a sword without a handle, a light without brilliance, a bell without sound. — Alwyn Mittasch (1869–1953)

I have procured air [oxygen] . . . between five and six times as good as the best common air that I have ever met with. — Joseph Priestley (1733–1804)

Geology

We are as much gainers by finding a new property in the old earth as by acquiring a new planet. — Ralph Waldo Emerson (1803–1882)

One naturally asks, what was the use of this great engine set at work ages ago to grind, furrow, and knead over, as it were, the surface of the earth? We have our answer in the fertile soil which spreads over the temperate regions of the globe. The glacier was God's great plough. — Jean Louis Agassiz (1807–1873)

Physics

Every body continues in its state of rest or of uniform motion in a straight line, except in so far as it is compelled to change that state by forces impressed upon it. — Isaac Newton (1642–1727)

Physics regards matter solely as a vehicle of energy—physics may be regarded as the science of energy, precisely as chemistry may be regarded as the science of matter. — G.F. Barker

In some sort of crude sense which no vulgarity, no humor, no overstatement can quite extinguish, the physicists have known sin; and this is a knowledge which they cannot lose. — Julius Robert Oppenheimer (1904–1967)

Science has nothing to be ashamed of, even in the ruins of Nagasaki. — Jacob Bronowski (1908–1974)

The conception of the atom stems from the concepts of subject and substance: there has to be "something" to account for any action. The atom is the last descendant of the concept of the soul. — Friedrich Nietzsche (1844–1900)

CHAPTER FIVE

Cohesion

Science is built up with facts, as a house is with stones. But a collection of facts is no more a science than a heap of stones is a house.
> — Henri Poincaré

If your research is strong, then everyone should know it. So state what you did up front—the first paragraph, even the first sentence—and leave the dilly-dallying for people who don't have anything to report.
> — Erich Kuhnardt

The best argument is that which seems merely an explanation.
> — Dale Carnegie

So far I've discussed clarity as if scientists write only individual sentences. Simple, concise, and precise sentences increase communication, but can also produce dull prose—noble and virtuous, but dull. Matching characters and actions to subjects and verbs creates clarity, but this clarity is a local clarity restricted to only one sentence. This is another instance where mere rules such as "Be clear" will fail you; indeed, readers may understand all of your words and sentences, but may not know what you're trying to say. Since clear sentences can confuse readers if the sentences are not in the proper context, good writing requires more than accuracy and local clarity. Writing an effective paragraph requires a continual compromise between local clarity and cohesion, with priority given to the cohesiveness that helps readers see the paragraph's message.

The most important concerns of a writer are not individual topics of individual sentences, but the cumulative effect—the communication—of a cohesive sequence of sentences. These sequences of sentences are paragraphs, the major units of scientific writing. Paragraphs are logically constructed passages having one theme. Concise, effective sentences are a must, because if one sentence is weak, language falters and the reader stumbles.

The cliché stressed in most writing books to cover cohesion is, "Each paragraph should communicate one idea." Yes, a paragraph *should* communicate an idea. But how do you ensure that? Since it takes more than one sentence to describe most ideas, implementing this principle requires that a writer know how to arrange sentences into an effective paragraph. This involves understanding how readers read sentences—not individually, but in sequences.

Readers divide paragraphs into two parts: (1) a summary of the paragraph's take-home message, and (2) points that stress the take-home message. The take-home message is best placed in a short opening sentence, with the stress toward the end of the sentence. The rest of the paragraph is where the writer develops, and where readers look for, new ideas. Notice how the first sentence of this paragraph introduces evolution, while later sentences develop the idea:

> Clark's practice of carefully mapping every fossil made it possible to follow the evolutionary development of various types through time. Beautiful sequences of antelopes, giraffes and elephants were obtained; new species evolving out of old and appearing in younger strata. In short, evolution was taking place before the eyes of the Omo surveyors, and they could time it. The finest examples of this process were in several lines of pigs which had been common at Omo and had developed rapidly. Unsnarling the pig story was turned over to paleontologist Basil Cook. He produced family trees for pigs whose various types were so accurately dated that pigs themselves became measuring sticks that could be applied to fossils of questionable age in other places that had similar pigs.[1]

[1]Johanson, Donald C., and Maitland A. Edey. 1981. *Lucy: The Beginnings of Humankind*. New York: Simon and Schuster. pp. 116–117.

Although effective paragraphs are made of simple, straightforward sentences, a paragraph is much more than a group of related sentences. Rather, a well written paragraph shows relationships and makes a point. To show relationships you must link the ideas of the paragraph. You'll do this most effectively by (1) organizing sentences in a logical order, and (2) linking these sentences with transitions that lead the reader to your conclusion.

Organizing Sentences in a Logical Order

The cohesion of a paragraph depends on the arrangement of concise sentences. This arrangement must reflect the logic behind your argument and develop your idea. Here are some suggestions for writing effective paragraphs:

The most prominent and important part of a paragraph is its first sentence. Use the first sentence to tell readers what the paragraph is about. Don't make readers be detectives—tell them the significance of your work. Use your first sentence or paragraph to tell readers that the butler did it, and the rest of the paragraph or paper to describe how it happened. Again note the opening sentence of Watson's and Crick's monumental paper from *Nature*:

> We wish to suggest a structure for the salt of deoxyribose nucleic acid (D.N.A.).

Develop the paragraph's idea with data, examples, comparisons, and analogies. Watson and Crick emphasized the importance of their discovery with a great understatement:

> This structure has novel features which are of considerable scientific interest.

Delete sentences that don't relate to the paragraph's topic sentence. By doing so, you'll stick to the point of the paragraph and make it easier to understand.

Link each sentence to the ones preceding and following so that they produce a coherent idea. In the following paragraph from *On the Origin of Species*, Charles Darwin's logic dictates the order of sentences. He expresses some thoughts with simple sentences or independent clauses, while expressing others as dependent clauses. Finally, he uses transitions to show relations between the sentences and clauses, concessions and exceptions, correlations and disjunctions:

> The truth of the principle, that the greatest amount of life can be supported by great diversification of structure, is seen under many natural circumstances. In an extremely small area, especially if freely open to immigration, and where the contest between individual and individual must be severe, we always find great diversity in its inhabitants. For instance, I found that a piece of turf, three feet by four in size, which had been

exposed for many years to exactly the same conditions, supported twenty species of plants, and these belonged to eighteen genera and to eight orders, which shows how much these plants differed from each other. So it is with the plants and insects on small and uniform islets; and so in small ponds of fresh water. Farmers find that they can raise most food by a rotation of plants belonging to the most different orders: nature follows what may be called a simultaneous rotation. Most of the animals and plants which live close round any small piece of ground, could live on it (supposing it not to be in any way peculiar in its nature), and may be said to be striving to the utmost to live there; but, it is seen, that where they come into the closest competition with each other, the advantages of diversification of structure, with the accompanying differences of habitat and constitution, determine that the inhabitants, which thus jostle each other most closely, shall, as a general rule, belong to what we call different genera and orders.

Writing is clear when readers can quickly get your message and can follow your supporting arguments. Anything else is bad writing. Therefore, write so that readers get your message quickly and on first reading. Don't hesitate to use comparisons to help readers visualize your ideas. For example:

A virus is to a person as a person is to the earth. Likewise, the size of a human cell is to that of a person as a person's size is to that of Rhode Island; an atom is to a person as a person is to the earth's orbit around the sun; and a proton is to a person as a person is to the distance to Alpha Centauri.

End the paragraph with a short, direct, concluding sentence, and start a new paragraph to announce a change of subject. For example, in *In the Shadow of Man*, Jane Goodall describes a baby chimpanzee named Pom who, unlike other baby chimps, always stayed close to her mother. She closes the chapter's last paragraph with this emphatic sentence:

Obviously she was terrified of being left behind.

Start a new paragraph when you start describing a new idea. This usually corresponds to places you would say, "Oh, by the way . . ." or "Here's something else . . ."

Remember that readers grasp material most readily when it is presented in units of 75 to 200 words (five to twelve sentences).

If your text is more than about six paragraphs long, consider using headings and subheadings to divide the text. Use these subheadings to tell readers what's ahead.

Writers must communicate with their readers. To communicate, a writer must show, and to show, a writer must provide details. General statements provide an overview, but they alone are unconvincing. For example, a general statement like, "Microbes are interesting" only prompts readers to ask for details—*how* are microbes interesting? Examples and details give meaning to generalizations. Readers remember details and examples: they want to hear about miles per gallon, not that a car "runs good." Therefore, instead of saying that a piece of apple pie contains "a lot of energy," write that it "contains about 1.5×10^6 J (365 Cal) of energy, enough energy for a woman to run for almost an hour or for a typist to enter about 15,000,000 characters on a manual typewriter."

Be as specific as possible. Just as you are annoyed by vague sentences such as "Dr. DeBakey inserted a valve in the vicinity of the heart," so too will readers be annoyed if you do not give them enough details about your subject. Notice how these generalizations improve when you add details:

Energetics

GENERAL: A piece of apple pie contains a lot of energy.

SPECIFIC: The energy in a piece of apple pie (1.5×10^6 J) is equivalent to the energy content of 11.2 ounces of TNT and is enough energy to fuel a one-hour run by a woman. The 3,000 calories of food energy needed daily by most college students are contained in about nine pieces of apple pie and are equivalent to the energy in six pounds of TNT.

Humans

GENERAL: Fetuses grow fast just before birth.

SPECIFIC: Fetuses grow fastest in the last three months before birth. If infants continued to grow at that rate, they would be 18' 4" tall when they reached age ten.

Dinosaurs

GENERAL: Dinosaurs were large but had small brains.

SPECIFIC: The *Stegosaurus*, a playful 25' long dinosaur, had a brain the size of a walnut.

Pollution

GENERAL: Plants can help solve the pollution problem.

SPECIFIC: It takes more than 100,000 trees to cancel the pollution of one jet flying round-trip from New York to Los Angeles.

Anchor general statements with details and examples. For example, write "eight" instead of "several," and instead of writing that "some salt was added to the medium," write that "I added 10.2 g of NaCl to the medium." Use numbers and exact words wherever possible.

We remember best what we hear first and last. This is why effective paragraphs need a topic sentence summarizing their point and why effective essays need a summarizing sentence or paragraph to convey their message. It's also why most famous quotations are usually from the beginning or end of speeches or books. For example, *government of the people, by the people, for the people . . .* comes from the end of Lincoln's Gettysburg Address, and *To be or not to be* opens one of Shakespeare's works. Similarly, consider the opening line of Dickens's *A Tale of Two Cities: It was the best of times; it was the worst of times. . .* Most people have not read *A Tale of Two Cities*, yet are familiar with its opening line. Indeed, since being published in 1859, this line is now overused to the point of being a cliché. You'd be hard-pressed to find anyone who can quote any other lines from the 600-page book.[2]

Vary the Length of Your Sentences

Even when you follow all of these suggestions, you may end up with a monotonous paragraph. This monotony results not from the writing being incorrect, but rather from all of the sentences having a similar length. Indeed, a string of short sentences makes writing choppy, and a string of long sentences makes a reader drowsy. To avoid this monotony and to communicate most effectively, write sentences short enough to be held in your reader's mind. Do this by building paragraphs with sentences having an average length less than 17 words—that's about the average sentence-length in *Newsweek* (the average length of sentences in most "scholarly journals" exceeds 25 words).[3] However, don't make *all* of your sentences less than 20 words long. If you do, you may communicate adequately, but you'll be like a singer who uses only one octave: You'll be able to carry the tune, but you'll be unable to produce much variety.

[2]The only other lines that someone might be able to quote from *A Tale of Two Cities* are these: *It is a far, far better thing that I do now than I have ever done. It is a far better rest that I go to than I have ever known.* Not surprisingly, these lines end the novel.

[3]Sentences of many popular (Ann Landers), famous (Winston Churchill), and inspiring (John Steinbeck) writers average about 15 words, as do sentences in magazines such as *Reader's Digest, Atlantic, Harper's, and Saturday Review.* Consider these analyses of *Wise Blood* (by Flannery O'Connor) and *Of Mice and Men* (by John Steinbeck):

	Wise Blood	*Of Mice and Men*
Average sentence length	15.3 words	15.2 words
Longest sentence	40 words	35 words
Shortest sentence	5 words	6 words

The average sentence-length usually increases with one's education. For example, sentences written by fourth graders average 11.1 words, those written by ninth graders average 17.3 words, and those written by college juniors average 21.5 words. Most college graduates (but ones lacking good training in effective writing) write sentences averaging about 25 words.

Mark Twain was a master of varying the length of sentences to produce rhythmic writing. Here's how he varied the number of words in the opening paragraph of *The Adventures of Huckleberry Finn*: 25, 14, 12, 3, 21, 33, 28, 8, 12, 35, 51, 16, 33, 4.

Vary the length of your sentences. Do this by introducing a topic with a short, direct sentence. Develop the topic with longer sentences, and conclude with short, direct sentences. Concentrate on the first and last words of sentences—these are the words that stick most with readers. Also remember that short sentences can carry tremendous punch; that's why they dominate good writing. Appreciate and use simple declarative sentences.

Make Smooth Transitions

Arranging sentences in what you feel is a logical order may still leave readers confused. This confusion results from your having written separate, albeit well-written, sentences rather than a cohesive paragraph. Consequently, each sentence appears independent of others rather than as a part of an argument. In short, you've not linked the sentences' ideas, and therefore your logic is hidden. Each sentence yanks and surprises readers because they have no hint as to what's coming. Your transitions are either absent or too abrupt.

Walter Campbell has likened a good transition to a bridge, which "must be level, and alike at both ends." There are many ways to effectively link ideas. One way is to use a paragraph's last sentence to close one idea and introduce the next. A common way of doing this is to repeat an important word in the last sentence of one paragraph in the first sentence of the next. For example, notice how these writers use the words testing and tests to bridge these paragraphs:

> A company considers enrolling one of its brightest young female employees in an expensive training program. Before making the investment, it asks the woman to undergo scientific testing with the latest diagnostic techniques, including genetic screening.
>
> The tests reveal an unexpected problem. . . .

Another way to link paragraphs involves ending one paragraph with a question and starting the following paragraph with the question's answer. For example, notice how Peter Raven uses a question, answer, and repeated idea to link these paragraphs in his article entitled "The Less-Noticed Worldwide Revolution:"

> Many parents across the country reportedly have been stunned to discover that their teenagers are baffled about the significance of events in Eastern Europe and the Soviet Union over the past several months. How, these parents wonder, can young people be so ignorant about such historic occurrences?
>
> I share their concern, but I think many parents are as culpable as their children when it comes to awareness of another, equally important upheaval taking place in the world today. I refer to the revolution in science, which is likely to change the course of human history as profoundly as anything in today's political arena.
>
> The recent transformation of the scientific sciences . . .

Rules for Writing

Never, ever use repetitive redundancies.

Never use a long word when a diminutive one will do.

Use parallel structure when you write and in speaking.

Place pronouns as close as possible, especially in long sentences—such as those of 10 or more words—to their antecedents.

Proofread your paper carefully to find any words that you out.

Do not let a colon separate: the main parts of a sentence.

Whenever possible, avoid unnecessary, or superfluous, commas.

Delete unnecessary, excess words that are not needed.

Avoid run on sentences they are hard to read.

Put an apostrophe where its needed.

Do not put statements in negative form.

Do not use a hyphen when it is un-necessary.

Do not use a semicolon where; it is not needed.

Put an apostrophe where its needed. Use apostrophe's correctly.

Avoid overuse, of commas.

Don't abbrev.

Have a good reason for Capitalizing a word.

Consult a dictionary for prosper spelling.

After studying these rules, dangling modifiers will be easy to correct.

Avoid using trendy words whose parameters are not viable.

Remember that verbs has to agree with their subjects.

Each pronoun should agree with their antecedent.

Don't use no double negatives.

Remember to never split an infinitive.

DO NOT OVERCAPITALIZE.

No sentence fragments. Enough of this.

Here's another example of linking paragraphs with a question and an answer:

> National surveys assure us that public understanding of science has never been poorer while our national need for a scientifically literate public has never been greater. We are faced with an endless array of issues—AIDS, pesticides, ecological Armageddon, space stations and Stealth bombers—that are inextricably entwined with science and technology. How are we to manage these issues if we do not understand them?
>
> Scientists and science communicators must do more to enhance "science literacy" among their fellow Americans through books, television programs, museums and, perhaps most productively, our nation's schools and colleges.

You'll also link paragraphs effectively if you state an idea or observation in one paragraph and use the following paragraph to explain the significance of that idea or observation. For example:

> Presidential elections and other dramatic news may capture the headlines, but one of the most profound events of our time has been the dramatic increase in life expectancy. Americans now live about 25 years longer than they did a century ago.
>
> That is an essential fact to keep in mind when one considers whether it is ethical to carry out experiments on animals, a controversy that has existed for more than a century. The debate has become more charged over the past decade, with some radical animal-rights activists breaking into laboratories to "liberate" animals.

Notice also how this writer used *controversy* and *the debate* to link the sentences of the second paragraph. Here's another example:

> Imagine you are an ambulance driver and someone's life depends on your finding an address—but you have no map. That's the dilemma faced by medical researchers, who must search for the cures to serious genetic diseases without knowing where the relevant genes are located on the human chromosomes.
>
> The result is that research today proceeds much more slowly than it might on heart disease, cancer, certain kinds of Alzheimer's disease, cystic fibrosis and some 3,000 other disorders with a genetic component. . . .
>
> Like the ambulance driver searching blindly along city streets, these researchers could do their work far more effectively—and save many more lives—if they had a decent road map.

An even simpler way is to bluntly tell readers what's coming. Here's how Charles Darwin did this:

> We will now discuss in a little more detail the struggle for existence.

You can also lead readers by linking the ideas of successive sentences with transitions that tell readers what to expect. For example, *therefore* spotlights a conclusion, while *however* and *but* warn readers that you're changing direction. Without these transitions, sentences remain independent and readers have no time to grasp the full meaning of one idea before the next is thrust upon them.

Ironically, the inability to connect ideas is one of the most common symptoms of poor writing, yet is one of the easiest problems to solve. Remember that *transitions connect related ideas* by showing relationships, by leading the reader to your conclusions, and by telling the reader about the next idea—what's going to happen, when it has started, and when it is finished. Use these words and phrases to make smooth transitions in your writing:

Add information with words such as *furthermore, also, similarly,* and *moreover.*

Compare with words such as *similarly, like,* and *likewise.*

Change directions, contrast ideas, and **point out exceptions** with words and phrases such as *but, however, on the contrary, even though, nonetheless, conversely, though, nevertheless, on the other hand, although, alas,* and *yet.*

Caution readers with words such as *however* and *but.*

Illustrate with phrases such as *for example, namely,* and *for instance.*

Qualify a statement with *yet* and *still.*

Concede with words and phrases such as *although, even though, since,* and *though.*

Subordinate ideas with words such as *because, although, while, where,* and *if.*

Indicate time sequences with words such as *first, next, later, finally, when, as, then,* and *while.*

Indicate location with phrases such as *in the next room, down the hall,* and *in the zoology lab.*

Alert the reader of a change with words such as *but, yet, however, nevertheless, still,* and *thus.* These words alert readers that their expectations are not likely to be met, thereby preparing them for new information.

Link equal ideas with words such as *and, either/or,* and *neither/nor.*

Indicate cause and effect relationships with words such as *therefore, so, because, thus, hence,* and *consequently.*

Conclude with words such as *therefore, thus, accordingly, moreover,* and *consequently.*

Transitions create dialogue and discussion while keeping the reader oriented to what's going on. Thus, they usually occur at the beginning of sentences and near the middle of paragraphs where examples, comparisons, and cautions are most likely to occur. Without connectors, writing usually crumbles into staccato-like "Me

Tarzan. You Jane" prose that annoys, bores, and eventually loses the reader. However, be sure to use connectors correctly: Don't write "but" unless you actually show a contrast or change direction. Signaling a relationship that doesn't exist confuses readers.

Just as few things can improve a sentence more than a well-chosen verb, so too can few things improve a paragraph's cohesiveness more than logic and effective transitions. However, other things also affect cohesion when you write about science. These include trying not to hedge, writing fairly, punctuation, and knowing when to stop writing.

Try Not to Hedge

> It is a strange model and embodies several unusual features. However, since DNA is an unusual substance, we are not hesitant in being bold.
> — James Watson, writing to a friend one month before the public announcement of the structure of DNA—a discovery many scientists now regard as the most significant since Mendel's

Careful scientists always have some degree of doubt about their conclusions. To express this doubt, science has its own idiom of caution and confidence. None of us wants to sound like either a smug dogmatist or an uncertain milquetoast. How successfully we walk the line between seeming timidity and arrogance depends in part on how well we hedge with words such as *almost, possibly*, and *perhaps*. Hedge words give us room to backpedal and to make exceptions. However, they can also make us appear wishy-washy. So what's a scientist to do?

Don't mumble. Once you've decided what you want to say, come right out and say it. State points as emphatically as you can support them—no more, no less. Never hesitate to suspend judgment (and do more experiments) when your evidence is inconclusive; such a measured agnosticism is a valuable reflex for a scientist. However, remember that you can't expect others to make up their mind about your work if *you* haven't made up your mind about your work. Therefore, try to come to a firm conclusion. Use hedge words and phrases such as *often, sometimes, virtually, apparently, seemingly, in some ways, to a certain extent, may, might, perhaps, suggest, is likely to, could, may be possible*, and *probably* only if they're essential. Hedge words such as these are often referred to as "weasel words" because they're used by people trying to weasel their way out of taking a stand. If they're overused, readers feel that the writer is unsure of everything in the paper and refuses to be held accountable. At worst, overusing weasel words makes you appear deceitful, and at best they make you appear wishy-washy. For exam-

The Hidden Meanings of Some Hedge Words

It's always wise to state your conclusions as emphatically as you can support them. When you're unsure, don't hedge; just state your level of uncertainty and the reasons behind it as straightforwardly as possible. However, hedging has its place, especially when you need to cover problems in effort or logic. With many people, many hedge words and phrases have taken on a new meaning. For example, consider this list compiled by an unknown author as a guide for interpreting reports:

It has long been known that . . .

I haven't bothered to look up the original reference.

Of great theoretical and practical importance . . .

It's interesting to me.

Nine of the samples were chosen for detailed study . . .

The results from other samples didn't make sense so I ignored them.

Typical results are shown . . .

The best results are shown.

These results will be presented at a later date . . .

I might get around to this sometime.

An exhaustive review of the literature shows that . . .

I found a 1996 paper that says . . .

It is believed that . . .

I think that . . .

It is generally believed that . . .

A couple of people think that . . .

It is clear that much additional work is required before a complete understanding of this work will be achieved.

I don't understand it.

Thanks are due to Joyce Corban for assistance with the experiment and to Jerry Hubschman for valuable discussions.

Corban did the work and Hubschman told me what it meant.

ple, consider this sentence:

> It seems that it could possibly be wise to follow this procedure if a better one is not proposed very soon.

This student uses a quadruple hedge (*seems, could, possibly, if*) before ironically trying to intensify his idea by adding *very.*

Every qualifier whittles away at the reader's trust in you. Don't diminish this trust with a flood of hedge words. Be bold, not *kind of* bold. Do not hesitate to emphatically state conclusions supported by your results. Also remember that many phrases used for emphasis can project arrogance. For example, phrases such as *As everyone knows, It is true that, It is clear that, As any fool can see, Of course, This clearly demonstrates that, As you should know,* and *It is obvious that,* rather than ward off challengers, often insult readers. At best, these phrases usually mean little more than "believe me," and they become background static that robs your writing of precision and clarity. At worst, they make you seem arrogant and defensive.

Accepting evasiveness, like accepting pretentiousness, pollutes writing. Don't use a squid's approach to writing and hide behind a sea of ink. Similarly, don't shirk your scientific responsibility of making a judgment. Give readers a take-home message that's logical, convincing, and based on evidence. If your evidence is sufficient to delete qualifiers such as *may,* then delete them. If not, ~~it is quite possible that~~ you *may* need to do more research before publishing your work.

Write Fairly

> For I am well aware that scarcely a single point is discussed in this volume on which facts cannot be addressed, often apparently leading to conclusions opposite to those at which I have arrived. A fair result can be obtained only by fully stating and balancing the facts and arguments on both sides of each question . . .
> — Charles Darwin, *On The Origin of Species*

> He who knows only his own side of the case knows little of that.
> — John Stuart Mill

Write fairly by acknowledging sources of error, assumptions inherent in your argument, and other possible interpretations of your results. Demonstrate fairness and objectivity not by using passive voice, but by basing your arguments on evidence and logic. Consider all options and approaches and be sure that nothing is implied or left to the readers' imagination. Cite relevant papers even if they don't support your argument. If your argument is sound and you've included evidence to *show* the reader that it's sound, there's no danger that the reader will turn on you. The only danger is that readers will discover the conflict themselves and wonder why it never occurred to you.

Do not assign human traits to other organisms or things. For example, do not write that results suggest or that data indicate. These are human activities that results and data cannot do. Similarly, do not write teleological sentences such as "Giraffes evolved long necks because they wanted to reach leaves high in trees" or "Birds have wings for flying and a beak for feeding." These sentences state a result as if it were an explanation, goal, cause, or purpose.

Remove biases toward race, disabilities, and sex from your writing.[4] For example, whether used as a prefix (*manpower*) or as a suffix (*chairman*), *man* evokes a strong reaction in many readers. Since *man* can refer both to a male and to both sexes, it produces ambiguity that many readers find unsettling or even offensive. Revise sentences that contain potentially offensive words; you lose nothing and gain much when you do this.[5] For example:

BIASED: Select a spokesman to present the results of your experiment.

NEUTRAL: Select someone to present the results of your group.

Know When to Stop Writing

Not knowing how or when to stop writing is as much a liability in writing as it is in speaking. Consider the simple advice of The King to White Rabbit:

. . . go on till you come to the end; then stop.   — Lewis Carroll

---

[4]Many women become brilliant scientists despite the societal and economic barriers to their success. For example, Gerty Cori won a Nobel Prize in 1947 for her studies of sugar metabolism in humans, Dorothy Hodgkin won a Nobel Prize in 1964 for her work with vitamin B_{12} and penicillin, Barbara McClintock in 1983 for discovering mobile elements in genetics, and Rita Levi-Montaleini in 1986 for discovering nerve-growth factor. Many of these discoveries have been based on work of other women. For example, Marie Sklodowska Curie discovered two elements (radium and polonium) and opened the Atomic Age by showing that radiation is an atomic property. (See pages 326–328.) In 1903 she became the first woman to win a Nobel Prize and later, in 1911, the first person to win two Nobel Prizes. Her daughter, Irene Joliot-Curie, won a Nobel Prize in 1935 for discovering artificial isotopes. Twenty-two years later, Rosalyn Yallow won a Nobel Prize for developing the radioimmunoassay, a technique sensitive enough to measure a teaspoon of insulin in a lake 62 miles long, 62 miles wide, and 30 feet deep. Yallow's work was based on artificial isotopes made possible by Irene Curie. Unfortunately, the great work of many women scientists has also been overlooked. For example, Watson and Crick's Nobel Prize in 1962 was based on analyses and interpretations of data by Rosalind Franklin. Similarly, Candace Pert discovered endorphins, but the Lasker Prize for their discovery went to Solomon Snyder, the director of her lab.

[5]The English "he" meaning he or she resulted from a conspiracy by a group of eighteenth century grammarians. Until that time, "they" was acceptable as a singular pronoun (for example, Shakespeare wrote, "God send everyone their heart's desire."), and the grammarians' recommendation that "he" should be used when gender was unknown was even endorsed by an Act of Parliament in 1850. However, it was not always followed: in 1879 the Massachusetts Medical Society refused to admit a woman physician because their rule book referred only to "he."

Heed the King's advice: Stop when you've made all of your points. Repetition won't strengthen your argument. When you've said everything highlighted in your outline, and said it well, stop writing. However, remember the words of E.B. White: "When you say something, make sure you have said it. The chances of your having said it are only fair."

There are many ways to end a paper. Look at how Stephen Jay Gould closed an essay on the probability of finding extraterrestrial life:

> Ultimately, however, I must justify the attempt at such a long shot simply by stating that a positive result would be the most cataclysmic event in our entire intellectual history. Curiosity impels, and makes us human. Might it impel others as well?

Compare this with what a confused student wrote when he didn't know quite when to stop writing about heredity:

> Heredity means that if your grandfather didn't have any children, then your father probably wouldn't have any, and neither would you, probably.

If you knew where you were going with your paper, you'll know when you arrive and when you should stop writing. You'll be able to answer yes to all of these questions:

Have you said what you meant to say?

Have you told your readers everything they need to know?

Have you answered an important scientific question?

Are the conclusions that you want readers to draw from your evidence clear?

Have you written your paper precisely, clearly, and concisely?

Have you arranged your arguments logically and supported your arguments with evidence?

Have you presented information in digestible chunks?

When you've answered these questions, stop writing.

Punctuate Your Writing Correctly

> No matter what any of the grammar books or English teachers say, punctuation is an arbitrary matter. It should be used to make sentences clear.
> — Andy Rooney

> SALLY: Show me where you sprinkle in
> the little curvy marks.
> CHARLIE BROWN: Commas.
> SALLY: Whatever.
> — Charles Schulz

Punctuation is for the reader, not the writer: It should clarify the meaning of your writing. Thus, punctuation is to reading as traffic signs and signals are to driving: It tells readers what's coming, when to stop, and when to go. It also tell readers how sentences relate to each other.

Common sense will help you punctuate your writing, but it is no substitute for understanding a few principles of punctuation. Choosing correct punctuation requires as much discretion as choosing words. Although there are about 30 different punctuation marks, you need to know about only a few to write effectively.

Period (.)

The primary misuse of periods is that most writers don't reach them soon enough. Remember that short sentences dominate effective writing. Use a period after all declarative sentences and abbreviations. For example,

I am not in the least afraid to die.   — last words of Charles Darwin

I don't want it.   — last words of Marie Curie, when offered a pain-killing injection

I cannot.   — last words of Louis Pasteur, when offered a glass of milk

Don't overdo it.   — message given by H.L. Mencken to an obituary writer before Mencken's death

Colon (:)

A colon is a mark of anticipation. It has a special function: that of delivering what was invoiced in the preceding words. A colon introduces a clause that explains an idea in the first part of the sentence and tips off the reader about what's next, such as a list, a quotation, or a move from the general to the specific. For example,

We studied three kinds of marsupials: wombats, bandicoots, and kangaroos.

My brain: it's my second favorite organ.   — Woody Allen

That is right: I have now done.   — last words of English chemist and scientist Joseph Priestley, after finishing a few corrections in a manuscript

The wisdom of the wise and the experience of the ages are perpetuated by quotations. — Benjamin Disraeli

The difference between intelligence and education is this: intelligence will make you a good living.   — Charles F. Kettering

Semicolon (;)

A semicolon is sort of a half-period: It indicates a pause that is stronger than the pause indicated by a comma but weaker than the pause indicated by a period. It indicates that you're weighing two sides of the same problem. A semicolon behaves like a supercomma, separating two or more related clauses and those with internal punctuation where another comma would be confusing. The semicolon is a convenient device for grouping related elements for which the period is too strong and the comma too weak. For example,

> Cnidarians have no brain, and cnidarian behavior seems to be completely rigid; no one has yet trained a jellyfish.

> There is no known cure for AIDS; it kills thousands of Americans every year.

> Anatomy is to physiology as geography to history; it describes the theatre of events.
> — Jean Fernel

> No amount of experimentation can ever prove me right; a single experiment can prove me wrong.   — Albert Einstein

The semicolon is one of the least-appreciated punctuation marks. It suggests a close relationship between two independent clauses; it signals contrast without the stopping power of a period; it helps to amplify a point; it nudges the reader onward with a pleasant feeling of anticipation; it tells readers that there is more to come.

Hyphens (-)

Use hyphens in compound adjectives. For example,

> 100-foot-tall tree

> 20-day-old mouse

Also use hyphens to indicate single-bonds, linked amino-acids, and linked nucleotides in polynucleotides. For example,

> CH_3-CH_2OH

> Leucine-Proline-Serine-Alanine

> pG-A-C-C-T-T-G-C-Gp

Dashes (–)

Dashes are used to interrupt thought, to amplify an idea, or to insert ideas for clarity, emphasis, or explanation. For example,

> In the division of coelomate animals into protostomes and deuterostomes, three phyla—Phoronida, Bryozoa, and Barchipopda—are difficult to assign.

Facts are not science—as the dictionary is not literature.   — Martin H. Fischer

Familiarity breeds contempt – and children.   — Mark Twain

Quotation Marks (")

Quotation marks are used to indicate exact words, to enclose titles, or to indicate a word that is special or being defined. Closing quotation marks go outside of periods and commas:

"God does not cast the die."   — Albert Einstein

"Famous remarks are very seldom quoted correctly."   — Simeon Strunsky

Parentheses ( )

Parentheses are used to insert explanatory or supplemental information or a reference. For example,

We purified the antigens according to methods described by Stinner (1996).

The turquoise streaks in blue cheese and Roquefort are mycelia of certain species of *Penicillium* (Deuteromycota).

Comma (,)

Commas are the most frequently used and abused punctuation. Many writers avoid commas, while others paint their writing with them. Commas mark a short pause among items that might be confusing if run together. If you use a comma to signal a short pause, you'll usually be correct.

When you begin a sentence with a modifying clause (as I just did), put a comma after the clause. These clauses usually start with words such as *when, as, because, if, since,* and *although.* For example,

Because they shoot a jet of water through the excurrent siphon when molested, tunicates are also called sea squirts.

As cruel a weapon as the cave man's club, the chemical barrage has been hurled against the fabric of life.   — Rachel Carson, *Silent Spring*

Commas are also used to insert a word or phrase that interrupts the sentence:

Earth, the only truly closed ecosystem any of us knows, is an organism. — Lewis Thomas

Man is developed from an ovule, about 1/25th of an inch in diameter, which differs in no respect from the ovules of other animals.   — Charles Darwin

to separate consecutive adjectives:

". . . small, green, and flat leaves. . ."

to separate introductory phrases:

> In the main, opera in English is just about as sensible as baseball in Italian.
> — H.L. Mencken

to separate independent clauses linked by conjunctions such as *and, but, or,* and *for* :

> I am about to, or, I am going to die. Either expression is used. — last words of Dominique Bouhors, a French grammarian

> A doctor can bury his mistakes, but an architect can only advise his client to plant vines.
> Frank Lloyd Wright

> I have read all of Darwin's papers, but this one interests me most.

> Medical scientists are nice people, but you should not let them treat you. — August Bier

> Opportunities are usually disguised as hard work, so most people don't recognize them.
> — Ann Landers

> Science is the father of knowledge, but opinion breeds ignorance. — Hippocrates

> We can lick gravity, but sometimes the paperwork is overwhelming. — Wernher Von Braun

and to address someone directly.

> Nurse, it was I who discovered that leeches have red blood. — Baron George Cuvier, on his deathbed when the nurse came to apply leeches

> Winston, if you were my husband, I should flavor your coffee with poison. — Lady Astor

> Madam, if I were your husband, I should drink it. — Winston Churchill

The comma is the most common and versatile punctuation mark, and can greatly affect meaning.[6] You'll best grasp its many uses by noting how good writers use it.

Exclamation Marks (!)

Exclamation marks indicate surprise, disbelief, or other strong emotions.

> But I have to! So little done! So much to do! — last words of Alexander Graham Bell, when asked not to hurry his dictation

---

[6]In the 1984 presidential election, Republicans argued for two days about whether this statement in their platform should contain a comma: "The Republicans oppose any attempt to raise taxes which would harm the recovery." Without a comma after *taxes,* they left the door open to increase taxes that they felt would not harm the recovery. However, adding a comma after *taxes* meant that they opposed any taxes because any taxes would harm the recovery.

Chores on the Word Farm

After reading all of this chapter's suggestions for improving punctuation, grammar, and style, you may want a few simple guidelines to help with your writing. Almost 50 years ago, George Orwell provided some advice that I can't improve:

> Never use a metaphor, simile, or other figure of speech which you are used to seeing in print.
>
> Never use a long word where a short one will do.
>
> If it is possible to cut a word out, always cut it out.
>
> Never use the passive voice where you can use the active [voice].
>
> Never use a foreign phrase, a scientific word or a jargon word if you can think of an everyday English equivalent.
>
> Break any of these rules sooner than say anything outright barbarous.

> Cut out all those exclamation marks. An exclamation mark is like laughing at your own joke.   — F. Scott Fitzgerald

Like a loud voice, exclamation marks annoy readers when the marks are overused. Most good writers consider exclamation marks a poor substitute for a well-chosen word. Use exclamation marks sparingly, if at all, in your writing. Use truth, logic, and words for emphasis and understatement.

Question Marks (?)

A question mark follows a direct question and takes the place of one's voice being raised at the end of a question:

> What can be more important than the science of life to any intelligent being who has the good fortune to be alive?   — Isaac Asimov

> Of all these questions the one he asks most insistently is about man. How does he walk? How does the heart pump blood? What happens when he yawns and sneezes? How does a child live in the womb? How does he die of old age? Leonardo discovered a centenarian in a hospital in Florence and waited gleefully for his demise so that he could examine his veins.   — Sir Kenneth Clark, referring to Leonardo da Vinci

Simple, well-written sentences are easy to punctuate. However, others are not. For example, try to figure this one out:

> That which is is that which is not is not is not that it it is

That sentence is from *Flowers of Algernon* and is punctuated like this:

That which is, is. That which is not, is not. Is not that it? It is!

If you have a question about punctuation, consult one of the books listed in Appendix 1. Also remember that although punctuation is important, it alone can't save a poorly written paper. Don't use punctuation as a Band-Aid when basic surgery is needed. If you're struggling with punctuation, you're probably fighting a losing battle. Since, as Aristophanes said, "You cannot teach a crab to walk straight," scrap the sentence and start over.

Have Someone Else Read What You've Written

It is impossible to write one's best if
nobody else ever has a look at it.
— C.S. Lewis

Find the grain of truth in criticism—
chew it and swallow it.
— D. Sutten

To avoid criticism, do nothing, say nothing, be nothing.
— Elbert Hubbard

Few things will improve your writing as much as constructive criticism. Therefore, always have someone read what you've written. Ask for their criticisms and be willing to consider what they say. Although the criticisms may sometimes sting, they'll help you produce an effective paper.

Michael Faraday

The Chemical History of a Candle

One day, Sir, you may tax it.
— Michael Faraday to Mr.
Gladstone, the Chancellor of the
Exchequer, about the practical
value of electricity

Michael Faraday (1791–1867) was born into the
English working class. Faraday had little formal education, and worked as an errand boy with a local
bookseller and bookbinder (he learned his trade
well; some of his volumes survive). Surrounded by
books, Faraday became a voracious reader, especially about chemistry and physics. One day, Faraday
read the article "Electricity" in the *Encyclopedia
Brittanica*; from that moment he devoted his life to
science.

Faraday made several major discoveries. In
1821, at the age of 29, Faraday pioneered the electric motor with a demonstration of electromagnetic
rotation. Two years later, he liquefied chlorine, a
water-soluble gas that would later be used in water
purification, bleaches, and warfare. That same year,
Faraday pioneered mechanical refrigeration when
he discovered that certain gases under constant
pressure will condense until they cool. However,
Faraday's most important discovery came in 1831
when he wound an iron ring with two coils of
wire—one connected to a voltaic battery, the other
leading away to convert the vibrations to a current.
Faraday discovered that if the current was interrupt-
ed, a current flowed in a second wire wrapped
around the rod; that is, there was a transfer of electrical energy between two circuits.

Although Faraday's discovery was important, he
did not see the potential for stepping down power

from a high-voltage line for use with an alternating current which regularly reverses its direction. Moreover, he was not a mathematician; he needed the work of another British scientist, James Clerk Maxwell, to convert his ideas into quantitative results. Maxwell's subsequent prediction that electromagnetic waves exist led to discovery of radio waves by German physicist Heinrich Hertz.

Although Faraday is most remembered for his work on electricity and magnetism, he would have been famous from his work on chemistry alone. Indeed, Faraday's work provided the laws of electrochemistry as well as the first clue to the existence of a unit of electricity.

Faraday, like many of his contemporaries, believed intensely that scientists should share their work with the public. Thus, in December of 1827, he inaugurated a series of lectures for young people at the Royal Institution. Those lectures are still presented and often appear on television. Faraday's famous article "The Chemical History of a Candle," reprinted here, started as one of those lectures.

A Candle: The Flame—Its Sources—Structure—Mobility—Brightness

I propose, in return for the honor you do us by coming to see what are our proceedings here, to bring before you, in the course of these lectures, the Chemical History of a Candle. I have taken this subject on a former occasion, and, were it left to my own will, I should prefer to repeat it almost every year, so abundant is the interest that attaches itself to the subject, so wonderful are the varieties of outlet which it offers into the various departments of philosophy. There is not a law under which any part of this universe is governed which does not come into play and is touched upon in these phenomena. There is no better, there is no more open door by which you can enter into the study of natural philosophy than by considering the physical phenomena of a candle. I trust, therefore, I shall not disappoint you in choosing this for my subject rather than any newer topic, which could not be better, were it even so good.

And, before proceeding, let me say this also: that, though our subject be so great, and our intention that of treating it honestly, seriously, and philosophically, yet I mean to pass away from all those who are seniors among us. I claim the privilege of speaking to juveniles as a juvenile myself. I have done so on former occasions, and, if you please, I shall do so again. And, though I stand here with the knowledge of having the words I utter given to the world, yet that shall not deter me from speaking in the same familiar way to those whom I esteem nearest to me on this occasion.

And now, my boys and girls, I must first tell you of what candles are made. Some are great curiosities. I have here some bits of timber, branches of trees particularly famous for their burning. And here you see a piece of that very curious substance, taken out of some of the bogs in Ireland, called *candle-wood*, a hard, strong, excellent wood, evidently fitted for good work as a register of force, and yet, withal, burning so well that where it is found they make splinters of it, and torches since it burns like a candle, and gives a very good light indeed. And in this wood we have one of the most beautiful illustrations of the general nature of a candle that I can possibly give. The fuel provided, the means of bringing that fuel to the place of chemical action, the regular and gradual supply of air to that place of action—heat and light—all produced by a little piece of wood of this kind, forming, in fact, a natural candle.

But we must speak of candles as they are in commerce. Here are a couple of candles commonly called dips. They are made of lengths of cotton cut off, hung up by a loop, dipped into melted tallow, taken out again and cooled, then redipped, until there is an accumulation of tallow around the cotton. In order that you may have an idea of the various characters of these candles, you see these which I hold in my hand—they are very small and very curious. They are, or were, the candles used by the miners in coal mines. In olden times the miner had to find his own candles, and it was supposed that a small candle would not so soon set fire to the fire-damp in the coal mines as a large one; and for that reason, as well as for economy's sake, he had candles made of this sort—20, 30, 40, or 60 to the pound. They have been replaced since then by the steel-mill, and then by the Davy lamp, and other safety-lamps of various kinds. I have here a candle that was taken out of the *Royal George*, it is said, by Colonel Pasley. It has been sunk in the sea for many years, subject to the action of salt water. It shows you how well candles may be preserved; for, though it is cracked about and broken a good deal, yet when lighted it goes on burning regularly, and the tallow resumes its natural condition as soon as it is fused.

Mr. Field, of Lambeth, has supplied me abundantly with beautiful illustrations of the candle and its materials; I shall therefore now refer to them. And, first, there is the suet—the fat of the ox—Russian tallow, I believe, employed in the manufacture of these dips, which Gay Lussac, or some one who intrusted him with his knowledge, converted into that beautiful substance, stearin, which you see lying beside it. A candle, you know, is not now a greasy thing like an ordinary tallow candle, but a clean thing, and you may almost scrape off and pulverize the drops which fall from it without soiling any thing. This is the process he adopted: The fat or tallow is first boiled with quick-lime, and made into a soap, and then the soap is decomposed by sulphuric acid, which takes away the lime, and leaves the fat rearranged as stearic acid, while a quantity of glycerin is produced at the same time. Glycerin—absolutely a sugar, or a substance similar to sugar—comes out of the tallow in this chemical change. The oil is then pressed out of it; and you see here this series of pressed cakes, showing how beautifully the impurities are carried out by the oily part as the pressure goes on increasing, and at last you have left that substance, which is melted, and cast into candles as here represented. The candle I have in my hand is a stearin candle, made of stearin from tallow in the way I have told you. Then here is a sperm candle, which comes from the purified oil of the spermaceti whale. Here, also, are yellow bees-wax and refined bees-wax, from which candles are made. Here, too, is that curious substance called paraffine, and some paraffine candles, made of paraffine obtained from the bogs of Ireland. I have here also a substance brought from Japan since we have forced an entrance into that out-of-the-way place—a sort of wax which a kind friend has sent me, and which forms a new material for the manufacture of candles.

And how are these candles made? I have told you about dips, and I will show you how moulds are made. Let us imagine any of these candles to be made of materials which can be cast. "Cast!" you say. "Why, a candle is a thing that melts, and surely if you can melt it you can cast it." Not so. It is wonderful, in the progress of manufacture, and in the consideration of the means best fitted to produce the required result, how things turn up which one would not expect beforehand. Candles can not always be cast. A wax candle can never be cast. It is made by a particular process which I can illustrate in a minute or two, but I must not spend much time on it. Wax is a thing which, burning so well, and melting so easily in a candle, can not be cast. However, let us take a material that can be cast. Here is a frame, with a number of moulds fastened in it. The first thing to be done is to put a wick through them. Here is one—a plaited wick, which does not require snuffing—supported by a little wire. It goes to the bottom, where it is pegged in; the little peg holding the cotton tight, and stopping the aperture so that nothing fluid shall run out. At the upper part there is a little bar placed across, which stretches the cotton and holds it in the mould. The tallow is then melted, and the moulds are filled. After a certain time, when the moulds are cool, the excess of tallow is poured off at one corner, and then cleaned off altogether, and the ends of the wick cut away. The candles alone then remain in the mould, and you have only to upset them, as I am doing, when out they tumble, for the candles are made in the form of cones, being narrower at the top than at the bottom; so that, what with their form and their own shrinking, they only need a little shaking and out they fall. In the same way are made these candles of stearin and of paraffine. It is a curious thing to see how wax candles are made. A lot of cottons are hung upon frames, as you see here, and covered with metal tags at the ends to keep the wax from covering the cotton in those places. These are carried to a heater, where the wax is melted. As you see, the frames can turn round; and, as they turn, a man takes a vessel of wax and pours it first down one, and then the next, and the next, and so on. When he has gone once around, if it is sufficiently cool, he gives the first a second coat, and so on until they are all of the required thickness. When they have been thus clothed, or fed, or made up to that thickness, they are taken off and placed elsewhere. I have here, by the kindness of Mr. Field, several specimens of these candles. Here is one only half finished. They are then taken down and well rolled upon a fine stone slab, and the conical top is moulded by properly shaped tubes, and the bottoms cut off and trimmed. This is done so beautifully that they can make candles in this way weighing exactly four or six to the pound, or any number they please.

We must not, however, take up more time about the mere manufacture, but go a little farther into the matter. I have not yet referred you to luxuries in candles (for there is such a thing as luxury in candles). See how beautifully these are colored; you see here mauve, Magenta, and all the chemical colors recently introduced, applied to candles. You observe, also, different forms employed. Here is a fluted pillar most beautifully shaped; and I have also here some candles sent me by Mr. Pearsall, which are ornamented with designs upon them, so that, as they burn, you have, as it were, a glowing sun above, and a bouquet of flowers beneath. All, however, that is fine and beautiful is not useful. These fluted candles, pretty as they are, are bad candles; they are bad because of their external shape. Nevertheless, I show you these specimens, sent to me from kind friends on all sides, that you may see what is done and what may be done in this or that direction; although, as I have said, when we come to these refinements, we are obliged to sacrifice a little in utility.

Now as to the light of the candle. We will light one or two, and set them at work in the performance of their proper functions. You observe a candle is a very different thing from a lamp. With a lamp you take a little oil, fill your vessel, put in a little moss or some cotton prepared by artificial means, and then light the top of the wick. When the flame runs down the cotton to the oil, it gets extinguished, but it goes on burning in the part above. Now I have no doubt you will ask how it is that the oil which will not burn of itself gets up to the top of the cotton, where it will burn. We shall presently examine that; but there is a much more wonderful thing about the burning of a candle than this. You have here a solid substance with no vessel to contain it; and how is it that this solid substance can get up to the place where the flame is? How is it that this solid gets there, it not being a fluid? or, when it is made a fluid, then how is it that it keeps together? This is a wonderful thing about a candle.

We have here a good deal of wind, which will help us in some of our illustrations, but tease us in others; for the sake, therefore, of a little regularity, and to simplify the matter, I shall make a quiet flame, for who can study a subject when there are difficulties in the way not belonging to it? Here is a clever invention of some costermonger or street-stander in the market-place for the shading of their candles on Saturday nights, when they are selling their greens, or potatoes, or fish. I have very often admired it. They put a lamp-glass round the candle, supported on a kind of gallery, which clasps it, and it can be slipped up and down as required. By the use of this lamp-glass, employed in the same way, you have a steady flame, which you can look at, and carefully examine, as I hope you will do, at home.

You see then, in the first instance, that a beautiful cup is formed. As the air comes to the candle, it moves upward by the force of the current which the heat of the candle produces, and it so cools all the sides of the wax, tallow, or fuel as to keep the edge much cooler than the part within; the part within melts by the flame that runs down the wick as far as it can go before it is extinguished, but the part on the outside does not melt. If I made a current in one direction, my cup would be lop-sided, and the fluid would consequently run over; for the same force of gravity which holds worlds together holds this fluid in a horizontal position, and if the cup be not horizontal, of course the fluid will run away in guttering. You see, therefore, that the cup is formed by this beautifully regular ascending current of air playing upon all sides, which keeps the exterior of the candle cool. No fuel would serve for a candle which has not the property of giving this cup, except such fuel as the Irish bogwood, where the material itself is like a sponge and holds its own fuel. You see now why you would have had such a bad result if you were to burn these beautiful candles that I have shown you, which are irregular, intermittent in their shape, and can not, therefore, have that nicely-formed edge to the cup which is the great beauty in a candle. I hope you will now see that the perfection of a process—that is, its utility—is the better point of beauty about it. It is not the best looking thing, but the best acting thing, which is the most advantageous to us. This good-looking candle is a bad-burning one. There will be a guttering round about it because of the irregularity of the stream of air and the badness of the cup which is formed thereby. You may see some pretty examples (and I trust you will notice these instances) of the action of the ascending current when you have a little gutter run down the side of a candle, making it thicker there than it is elsewhere. As the candle goes on burning, that keeps its place and forms a little pillar sticking up by the side, because, as it rises higher above the rest of the wax or fuel, the air gets better round it, and it is more cooled and better able to resist the action of the heat at a little distance. Now the greatest mistakes and faults with regard to candles, as in many other things, often bring with them instruction which we should not receive if they had not occurred. We come here to be philosophers, and I hope you will always remember that whenever a result happens, especially if it be new, you should say, "What is the cause? Why does it occur?" and you will, in the course of time, find out the reason.

Then there is another point about these candles which will answer a question—that is, as to the way in which this fluid gets out of the cup, up the wick, and into the place of combustion. You know that the flames on these burning wicks in candles made of bees-wax, stearin, or spermaceti, do not run down to the wax or other matter, and melt it all away, but keep to their own right place. They are fenced off from the fluid below, and do not encroach on the cup at the sides. I can not imagine a more beautiful example than the condition of adjustment under which a candle makes one part subserve to the other to the very end of its action. A combustible thing like that, burning away gradually, never being intruded upon by the flame, is a very beautiful sight, especially when you come to learn what a vigorous thing flame is—what power it has of destroying the wax itself when it gets hold of it, and of disturbing its proper form if it come only too near.

But how does the flame get hold of the fuel? There is a beautiful point about that—*capillary attraction*. "Capillary attraction!" you say—"the attraction of hairs." Well, never mind the name; it was given in old times, before we had a good understanding of what the real power was. It is by what is called capillary attraction that the fuel is conveyed to the part where combustion goes on, and is deposited there, not in a careless way, but very beautifully in the very midst of the centre of action, which takes place around it. Now I am going to give you one or two instances of capillary attraction. It is that kind of action or attraction which makes two things that do not dissolve in each other still hold together. When you wash your hands, you wet them thoroughly; you take a little soap to make the adhesion better, and you find your hand remains wet. This is by that kind of attraction of which I am about to speak. And, what is more, if your hands are not soiled (as they almost always are by the usages of life), if you put your finger into a little warm water, the water will creep a little way up the finger, though you may not stop to examine it. I have here a substance which is rather porous—a column of salt—and I will pour into the plate at the bottom, not water, as it appears, but a saturated solution of salt which can not absorb more, so that the action which you see will not be due to its dissolving any thing. We may consider the plate to be the candle, and the salt the wick, and this solution the melted tallow. (I have colored the fluid, that you may see the action better.) You observe that, now I pour in the fluid, it rises and gradually creeps up the salt higher and higher; and provided the column does not tumble over, it will go to the top. If this blue solution were combustible, and we were to place a wick at the top of the salt, it would burn as it entered into the wick. It is a most curious thing to see this kind of action taking place, and to observe how singular some of the circumstances are about it. When you wash your hands, you take a towel to wipe off the water; and it is by that kind of wetting, or that kind of attraction which makes the towel become wet with water, that the wick is made wet with the tallow. I have known some careless boys and girls (indeed, I have known it happen to careful people as well) who, having washed their hands and wiped them with a towel, have thrown the towel over the side of the basin, and before long it has drawn all the water out of the basin and conveyed it to the floor, because it happened to be thrown over the side in such a way as to serve the purpose of a siphon. That you may the better see the way in which the substances act one upon another, I have here a vessel made of wire gauze filled with water, and you may compare it in its action to the cotton in one respect, or to a piece of calico in the other. In fact, wicks are sometimes made of a kind of wire gauze. You will observe that this vessel is a porous thing; for if I pour a little water on to the top, it will run out at the bottom. You would be puzzled for a good while if I asked you what the state of this vessel is, what is inside it, and why it is there? The vessel is full of water, and yet you see the water goes in and runs out as if it were empty. In order to prove this to you I have only to empty it. The reason is this: the wire, being once wetted, remains wet; the meshes are so small that the fluid is attracted so strongly from the one side to the other as to remain in the vessel, although it is porous. In like manner, the particles of melted tallow ascend the cotton and get to the top; other particles then follow by their mutual attraction for each other, and as they reach the flame they are gradually burned.

Here is another application of the same principle. You see this bit of cane. I have seen boys about the streets, who are very anxious to appear like men, take a piece of cane, and light it, and smoke it, as an imitation of a cigar. They are enabled to do so by the permeability of the cane in one direction, and by its capillarity. If I place this piece of cane on a plate containing some camphene (which is very much like paraffine in its general character), exactly in the same manner as the blue fluid rose through the salt will this fluid rise through the piece of cane. There being no pores at the side, the fluid can not go in that direction, but must pass through its length. Already the fluid is at the top of the cane; now I can light it and make it serve as a candle. The fluid has risen by the capillary attraction of the piece of cane, just as it does through the cotton in the candle.

Now the only reason why the candle does not burn all down the side of the wick is that the melted tallow extinguishes the flame. You know that a candle, if turned upside down, so as to allow the fuel to run upon the wick, will be put out. The reason is, that the flame has not had time to make the fuel hot enough to burn, as it does above, where it is carried in small quantities into the wick, and has all the effect of the heat exercised upon it.

There is another condition which you must learn as regards the candle, without which you would not be able fully to understand the philosophy of it, and that is the vaporous condition of the fuel. In order that you may understand that, let me show you a very pretty but very commonplace experiment. If you blow a candle out cleverly, you will see the vapor rise from it. You have, I know, often smelt the vapor of a blown-out candle, and a very bad smell it is; but if you blow it out cleverly, you will be able to see pretty well the vapor into which this solid matter is transformed. I will blow out one of these candles in such a way as not to disturb the air

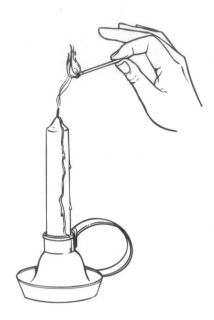

around it by the continuing action of my breath; and now, if I hold a lighted taper two or three inches from the wick, you will observe a train of fire going through the air till it reaches the candle. I am obliged to be quick and ready, because if I allow the vapor time to cool, it becomes condensed into a liquid or solid, or the stream of combustible matter gets disturbed.

Now as to the shape or form of the flame. It concerns us much to know about the condition which the matter of the candle finally assumes at the top of the wick, where you have such beauty and brightness as nothing but combustion or flame can produce. You have the glittering beauty of gold and silver, and the still higher lustre of jewels like the ruby and diamond; but none of these rival the brilliancy and beauty of flame. What diamond can shine like flame? It owes its lustre at night-time to the very flame shining upon it. The flame shines in darkness, but the light which the diamond has is as nothing until the flame shines upon it, when it is brilliant again. The candle alone shines by itself and for itself, or for those who have arranged the materials. Now let us look a little at the form of the flame as you see it under the glass shade. It is steady and equal, and its general form is that which is represented in the diagram, varying with atmospheric disturbances, and also varying according to the size of the candle. It is a bright oblong, brighter at the top than toward the bottom, with the wick in the middle, and, besides the wick in the middle, certain darker parts toward the bottom, where the ignition is not so perfect as in the part above. I have a drawing here, sketched many years ago by Hooker, when he made his investigations. It is the drawing of the flame of a lamp, but it will apply to the flame of a candle. The cup of the candle is the vessel or lamp; the melted spermaceti is the oil; and the wick is common to both. Upon that he sets this little flame, and then he represents what is true, a certain quantity of matter rising about it which you do not see, and which, if you have not been here before, or are not familiar with the subject, you will not know of. He has here represented the parts of the surrounding atmosphere that are very essential to the flame, and that are always present with it. There is a current formed, which draws the flame out; for the flame which you see is really drawn out by the current, and drawn upward to a great height, just as Hooker has here shown you by that prolongation of the current in the diagram. You may see this by taking a lighted candle, and putting it in the sun so as to get its shadow thrown on a piece of paper. How remarkable it is that the thing which is light enough to produce shadows of other objects can be made to throw its own shadow on a piece of white paper or card, so that you can actually see streaming round the flame something which is not part of the flame, but is ascending and drawing the flame upward. Now I am going to imitate the sunlight by applying the voltaic battery to the electric lamp. You now see our sun and its great luminosity; and by placing a candle between it and the screen, we get the shadow of the flame. You observe the shadow of the candle and of the wick; then there is a darkish part, as represented in the diagram, and then a part which is more distinct. Curiously enough, however, what we see in the shadow as the darkest part of the flame is, in reality, the brightest part; and here you see streaming upward the ascending current of hot air, as shown by Hooker, which draws out the flame, supplies it with air, and cools the sides of the cup of melted fuel.

I can give you here a little farther illustration, for the purpose of show-ing you how flame goes up or down according to the current. I have here a flame—it is not a candle flame—but you can, no doubt, by this time gener-alize enough to be able to compare one thing with another: what I am about to do is to change the ascending current that takes the flame upward into a descending current. This I can easily do by the little apparatus you see before me. The flame, as I have said, is not a candle flame, but it is produced by alcohol, so that it shall not smoke too much. I will also color the flame with another substance, so that you may trace its course; for, with the spirit alone, you could hardly see well enough to have the opportunity of tracing its direction. By lighting this spirit of wine we have then a flame produced, and you observe that when held in the air it naturally goes upward. You understand now, easily enough, why flames go up under ordi-nary circumstances: it is because of the draught of air by which the com-bustion is formed. But now, by blowing the flame down, you see I am enabled to make it go downward into this little chimney, the direction of the current being changed. . . . You see, then, that we have the power in this way of varying the flame in different directions.

There are now some other points that I must bring before you. Many of the flames you see here vary very much in their shape by the currents of air blowing around them in different directions; but we can, if we like, make flames so that they will look like fixtures, and we can photograph them—indeed, we have to photograph them—so that they become fixed to us, if we wish to find out every thing concerning them. That, however, is not the only thing I wish to mention. If I take a flame sufficiently large, it does not keep that homogeneous, that uniform condition of shape, but it breaks out with a power of life which is quite wonderful. I am about to use another kind of fuel, but one which is truly and fairly a representative of the wax or tallow of a candle. I have here a large ball of cotton, which will serve as a wick. And, now that I have immersed it in spirit and applied a light to it, in what way does it differ from an ordinary candle? Why, it differs very much is one respect, that we have a vivacity and power about it, a beauty and a life entirely different from the light presented by a candle. You see those fine tongues of flame rising up. You have the same general disposition of the mass of the flame from below upward, but, in addition to that, you have this remarkable breaking out into tongues which you do not perceive in the case of a candle. Now, why is this? I must explain it to you, because, when you understand that perfectly, you will be able to follow me better in what I have to say hereafter. I suppose some here will have made for themselves the experiment I am going to show you. Am I right in supposing that any body here has played at snapdragon? I do not know a more beautiful illustration of the philosophy of flame, as to a certain part of its history, than the game of snapdragon. First, here is the dish; and let me say, that when you play snapdragon properly you ought to have the dish well warmed; you ought also to have warm plums, and warm brandy, which, however, I have not got. When you have put the spirit into the dish, you have the cup and the fuel; and are not the raisins acting like the wicks? I now throw the plums into the dish, and light the spirit, and you see those beautiful tongues of flame that I refer to. You have the air creeping in over the edge of the dish forming these tongues. Why? Because, through the force of the current and the irregularity of the action of the flame, it can not flow in one uniform stream. The air flows in so irregularly that you have what would otherwise be a single image broken up into a variety of forms, and each of these little tongues has an independent existence of its own. Indeed, I might say, you have here a multitude of independent candles. You must not imagine, because you see these tongues all at once, that the flame is of this particular shape. A flame of that shape is never so at any one time. Never is a body of flame, like that which you just saw rising from the ball, of the shape it appears to you. It consists of a multitude of different shapes, succeeding each other so fast that the eye is only able to take cognizance of them all at once. In former times I purposely analyzed a flame of that general character, and the diagram shows you the different parts of which it is composed. They do not occur all at once; it is only because we see these shapes in such rapid succession that they seem to us to exist all at one time.

It is too bad that we have not got farther than my game of snapdragon;

but we must not, under any circumstances, keep you beyond your time. It will be a lesson to me in future to hold you more strictly to the philosophy of the thing than to take up your time so much with these illustrations.

Understanding What You've Read

Faraday goes to great pains to debunk the idea that science is mysterious. How does he do this? Had you attended this lecture, what would have been your impressions of the nature of science?

What are some of the examples that Faraday uses to make his points? Why are they important?

This article shows Faraday's direct approach to science. Indeed, he even described experiments that failed, always with the goal of leading the reader to his conclusions. Do scientists still write like this? Why or why not?

Faraday was unusual in his day in that he was paid for his work; that is, he was a professional scientist, despite having no formal qualification for such a position. Could someone with comparable qualifications succeed in science today? Why or why not?

What Faraday did for electricity and magnetism, Einstein later did for gravitation. Explain.

Faraday's work provided the basic components to establish an electrical industry. What other developments led to that industry?

Charles Darwin

Keeling Islands: Coral Formations

Earlier in this book you were introduced to Charles Darwin (1809–1882), a British scientist who became famous for his ideas about evolution. Although society is still coming to terms with those monumental ideas, most people do not know that Darwin was also an excellent geologist.

This article, first published in 1837, eventually appeared as a book entitled *The Structure and Distribution of Coral Reefs*.

I will now give a very brief account of the three great classes of coral-reefs; namely, Atolls, Barrier, and Fringing-reefs, and will explain my views on their formation. Almost every voyager who has crossed the Pacific has expressed his unbounded astonishment at the lagoon-islands, or as I shall for the future call them by their Indian name of atolls, and has attempted some explanation. Even as long ago as the year 1605, Pyrard de Laval well exclaimed, "C'est une merveille de voir chacun de ces atollons, environné d'un grand banc de pierre tout autour, n'y ayant point d'artifice humain." The sketch of Whitsunday Island in the Pacific, copied from Capt. Beechey's admirable *Voyage*, gives but a faint idea of a faint idea of the singular aspect of an atoll: it is one of the smallest size, and has its narrow islets united together in a ring. The immensity of the ocean, the fury of the breakers, contrasted with the lowness of the land and the smoothness of the bright green water within the lagoon, can hardly be imagined without having been seen.

The earlier voyagers fancied that the coral-building animals instinctively built up their great circles to afford themselves protection in the inner parts; but so far is this from the truth, that those massive kinds, to whose growth on the exposed outer shores the very existence of the reef depends, cannot live within the lagoon, where other delicately-branching kinds flourish. Moreover, on this view, many species of distinct genera and families are supposed to combine for one end; and of such a combination, not a single instance can be found in the whole of nature. The theory that has been most generally received is, that atolls are based on sub-marine craters; but when we consider the form and size of some, the number, proximity, and relative positions of others, this idea loses its plausible character: thus, Suadiva Atoll is 44 geographical miles in diameter in one line, by 34 miles in another line; Rimsky is 54 by 20 miles across, and it has a strangely sinuous margin; Bow Atoll is 30 miles long, and on an average only 6 in width; Menchicoff Atoll consists of three atolls united or tied together. This theory, moreover, is totally inapplicable to the northern Maldiva Atolls in the Indian Ocean (one of which is 88 miles in length, and between 10 and 20 in breadth), for they are not bounded like ordinary atolls by narrow reefs, but by a vast number of separate little atolls; other little atolls rising out of the great central lagoon-like spaces. A third and better theory was advanced by Chamisso, who thought that from the corals growing more vigorously where exposed to the open sea, as undoubtedly is the case, the outer edges would grow up from the general foundation before any other part, and that this would account for the ring or cup-shaped structure. But we shall immediately see, that in this, as well as in the crater-theory, a most important consideration has been overlooked, namely, on what have the reef-building corals, which cannot live at a great depth, based their massive structures?

Numerous soundings were carefully taken by Captain Fitz-Roy on the steep outside of Keeling Atoll, and it was found that within ten fathoms, the prepared tallow at the bottom of the lead, invariably came up marked with the impressions of living corals, but as perfectly clean as if it had been dropped on a carpet of turf; as the depth increased, the impressions became less numerous, until at last it was evident that the bottom consisted of a smooth sandy layer: to carry on the analogy of the turf, the blades of grass grew thinner and thinner, till at last the soil was so sterile, that nothing sprang from it. From these observations, confirmed by many others, it may be safely inferred that the utmost depth at which corals can construct reefs is between 20 and 30 fathoms. Now there are enormous areas in the Pacific and Indian Oceans, in which every single island is of coral formation, and is raised only to that height to which the waves can throw up fragments, and the winds pile up sand. Thus the Radack group of atolls is an irregular square, 520 miles long and 240 broad; the Low Archipelago is elliptic-formed, 840 miles in its longer, and 420 in its shorter axis: there are other small groups and single low islands between these two archipelagoes, making a linear space of ocean actually more than 4000 miles in length, in which not one single island rises above the specified height. Again, in the Indian Ocean there is a space of ocean 1500 miles in length, including three archipelagoes, in which every island is low and of coral formation. From the fact of the reef-building corals not living at great depths, it is absolutely certain that throughout these vast areas, wherever there is now an atoll, a foundation must have originally existed within a depth of from 20 to 30 fathoms from the surface. It is improbable in the highest degree that broad, lofty, isolated, steep-sided banks of sediment, arranged in groups and lines hundreds of leagues in length, could have been deposited in the central and profoundest parts of the Pacific and Indian Oceans, at an immense distance from any continent, and where the water is perfectly limpid. It is equally improbable that the elevatory forces should have uplifted throughout the above vast areas, innumerable great rocky banks within 20 to 30 fathoms, or 120 to 180 feet, of the surface of the sea, and not one single point above that level; for where on the whole face of the globe can we find a single chain of mountains, even a few hundred miles in length, with their many summits rising within a few feet of a given level, and not one pinnacle above it? If then the foundations, whence the atoll-building corals sprang, were not formed of sediment, and if they were not lifted up to the required level, they must of necessity have subsided into it; and this at once solves the difficulty. For as mountain after mountain, and island after island, slowly sank beneath the water, fresh bases would be successively afforded for the growth of the corals. It is impossible here to enter into all the necessary details, but I venture to defy any one to explain in any other manner, how it is possible that numerous islands should be distributed throughout vast areas—all the islands being low—all being built of corals, absolutely requiring a foundation within a limited depth from the surface.

Before explaining how atoll-formed reefs acquire their peculiar structure, we must turn to the second great class, namely, barrier-reefs. These either extend in straight lines in front of the shores of a continent or of a large island, or they encircle smaller islands; in both cases, being separated from the land by a broad and rather deep channel of water, analogous to the lagoon within an atoll. It is remarkable how little attention has been paid to encircling barrier-reefs; yet they are truly wonderful structures. The following sketch represents part of the barrier encircling the island of Bolabola in the Pacific, as seen from one of the central peaks. In this instance the whole line of reef has been converted into land; but usually a snow-white line of great breakers, with only here and there a single low islet crowned with cocoa-nut trees, divides the dark heaving waters of the ocean from the light-green expanse of the lagoon-channel. And the quiet waters of this channel generally bathe a fringe of low alluvial soil, loaded with the most beautiful productions of the tropics, and lying at the foot of the wild, abrupt, central mountains.

Encircling barrier-reefs are of all sizes, from three miles to no less than forty-four miles in diameter; and that which fronts one side, and encircles both ends, of New Caledonia, is 400 miles long. Each reef includes one, two, or several rocky islands of various heights; and in one instance, even as many as twelve separate islands. The reef runs at a greater or less distance from the included land; in the Society Archipelago generally from one to three or four miles; but at Hogoleu the reef is 20 miles on the southern side, and 14 miles on the opposite or northern side, from the included islands. The depth within the lagoon-channel also varies much; from 10 to 30 fathoms may be taken as an average; but at Vanikoro there are spaces

Figure 1
Bolabola

no less than 56 fathoms or 336 feet deep. Internally the reef either slopes
gently into the lagoon-channel, or ends in a perpendicular wall sometimes
between two and three hundred feet under water in height: externally the
reef rises, like an atoll, with extreme abruptness out of the profound depths
of the ocean. What can be more singular than these structures? We see an
island, which may be compared to a castle situated on the summit of a lofty
submarine mountain, protected by a great wall of coral rock, always steep
externally and sometimes internally, with a broad level summit, here and
there breached by narrow gateways, through which the largest ships can
enter the wide and deep encircling moat.

As far as the actual reef of coral is concerned, there is not the smallest
difference, in general size, outline, grouping, and even in quite trifling
details of structure, between a barrier and an atoll. The geographer Balbi
has well remarked, that an encircled island is an atoll with high land rising
out of its lagoon; remove the land from within, and a perfect atoll is left.

But what has caused these reefs to spring up at such great distances
from the shores of the included islands? It cannot be that the corals will not
grow close to the land; for the shores within the lagoon-channel, when not
surrounded by alluvial soil, are often fringed by living reefs; and we shall
presently see that there is a whole class, which I have called Fringing-reefs
from their close attachment to the shores both of continents and of islands.
Again, on what have the reef-building corals, which cannot live at great
depths, based their encircling structures? This is a great apparent difficulty,
analogous to that in the case of atolls, which has generally been overlooked. It will be perceived more clearly by inspecting the following sections, which are real ones, taken in north and south lines, through the
islands with their barrier-reefs, of Vanikoro, Gambier, and Maurua; and they
are laid down, both vertically and horizontally, on the same scale of a quarter of an inch to a mile.

It should be observed that the sections might have been taken in any
direction through these islands, or through many other encircled islands,

and the general features would have been the same. Now bearing in mind that reef-building coral cannot live at a greater depth than from 20 to 30 fathoms, and that the scale is so small that the plummets on the right hand show a depth of 200 fathoms, on what are these barrier-reefs based? Are we to suppose that each island is surrounded by a collar-like submarine ledge of rock, or by a great bank of sediment, ending abruptly where the reef ends? If the sea had formerly eaten deeply into the islands, before they were protected by the reefs, thus having left a shallow ledge round them under water, the present shores would have been invariably bounded by great precipices; but this is most rarely the case. Moreover, on this notion, it is not possible to explain why the corals should have sprung up, like a wall, from the extreme outer margin of the ledge, often leaving a broad space of water within, too deep for the growth of corals. The accumulation of a wide bank of sediment all round these islands, and generally widest where the included islands are smallest, is highly improbable, considering their exposed positions in the central and deepest parts of the ocean. In the case of the barrier-reef of New Caledonia, which extends for 150 miles beyond the northern point of the island, in the same straight line with which it fronts the west coast, it is hardly possible to believe, that a bank of sediment could thus have been straightly deposited in front of a lofty island, and so far beyond its termination in the open sea. Finally, if we look to other oceanic islands of about the same height and of similar geological constitution, but not encircled by coral-reefs, we may in vain search for so trifling a circumambient depth as 30 fathoms, except quite near to their shores; for usually land that rises abruptly out of water, as do most of the encircled and non-encircled oceanic islands, plunges abruptly under it. On what then, I repeat, are these barrier-reefs based? Why, with their wide and

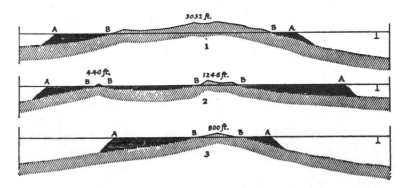

Figure 2
1. Vanikoro. 2. Gambier Islands. 3. Maurua. The horizontal shading shows the barrier-reefs and lagoon-channels. The inclined shading above the level of the sea (AA), shows the actual form of the land; the inclined shading below this line, shows its probable prolongation under water.

deep moat-like channels, do they stand so far from the included land? We shall soon see how easily these difficulties disappear.

We come now to our third class of Fringing-reefs, which will require a very short notice. Where the land slopes abruptly under water, these reefs are only a few yards in width, forming a mere ribbon or fringe round the shores: where the land slopes gently under the water the reef extends further, sometimes even as much as a mile from the land; but in such cases the soundings outside the reef always show that the submarine prolongation of the land is gently inclined. In fact the reefs extend only to that distance from the shore, at which a foundation within the requisite depth from 20 to 30 fathoms is found. As far as the actual reef is concerned, there is no essential difference between it and that forming a barrier or an atoll: it is, however, generally of less width, and consequently few islets have been formed on it. From the corals growing more vigorously on the outside, and from the noxious effect of the sediment washed inwards, the outer edge of the reef is the highest part, and between it and the land there is generally a shallow sandy channel a few feet in depth. Where banks of sediment have accumulated near to the surface, as in parts of the West Indies, they sometimes become fringed with corals, and hence in some degree resemble lagoon-islands or atolls; in the same manner as fringing-reefs, surrounding gently-sloping islands, in some degree resemble barrier-reefs.

No theory on the formation of coral reefs can be considered satisfactory which does not include the three great classes. We have seen that we are driven to believe in the subsidence of these vast areas, interspersed with low islands, of which not one rises above the height to which the wind and waves can throw up matter, and yet are constructed by animals requiring a foundation, and that foundation to lie at no great depth. Let us then take an island surrounded by fringing-reefs, which offer no difficulty in their structure; and let this island with its reef, represented by the unbroken lines in the woodcut, slowly subside. Now as the island sinks down, either a few feet at a time or quite insensibly, we may safely infer, from what is known of the conditions favourable to the growth of coral, that the living masses, bathed by the surf on the margin of the reef, will soon regain the surface. The water, however, will encroach little by little on the shore, the island becoming lower and smaller, and the space between the inner edge of the reef and the beach proportionally broader. A section of the reef and island in this state, after a subsidence of several hundred feet, is given by the dotted lines. Coral islets are supposed to have been formed on the reef; and a ship is anchored in the lagoon-channel. This channel will be more or less deep, according to the rate of subsidence, to the amount of sediment accumulated in it, and to the growth of the delicately branched corals which can live there. The section in this state resembles in every respect one drawn through an encircled island: in fact, it is a real section (on the scale of .517 of an inch to a mile) through Bolabola in the Pacific. We can now at once

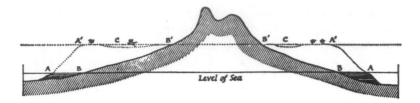

Figure 3

AA. Outer edges of the fringing-reef, at the level of the sea. BB. The shores of the fringed island. A'A'. Outer edges of the reef after its upward growth during a period of subsidence, now converted into a barrier, with islets on it. B'B'. The shores of the now encircled island. CC. Lagoon-channel.

see why encircling barrier-reefs stand so far from the shores which they front. We can also perceive, that a line drawn perpendicularly down from the outer edge of the new reef, to the foundation of solid rock beneath the old fringing-reef, will exceed by as many feet as there have been feet of subsidence, that small limit of depth at which the effective corals can live: the little architects having built up their great wall-like mass, as the whole sank down, upon a basis formed of other corals and their consolidated fragments. Thus the difficulty on this head, which appeared so great, disappears.

If, instead of an island, we had taken the shore of a continent fringed with reefs, and had imagined it to have subsided, a great straight barrier, like that of Australia or New Caledonia, separated from the land by a wide and deep channel, would evidently have been the result.

Let us take our new encircling barrier-reef, of which the section is now represented by unbroken lines, and which, as I have said, is a real section through Bolabola, and let it go on subsiding. As the barrier-reef slowly sinks down, the corals will go on vigorously growing upwards; but as the island sinks, the water will gain inch by inch on the shore—the separate mountains first forming separate islands within one great reef—and finally, the last and highest pinnacle disappearing. The instance this takes place, a perfect atoll is formed: I have said, remove the high land from within an encircling barrier-reef, and an atoll is left, and the land has been removed. We can now perceive how it comes that atolls, having sprung from encircling barrier-reefs, resemble them in general size, form, in the manner in which they are grouped together, and in their arrangement in single or double lines; for they may be called rude outline charts of the sunken islands over which they stand. We can further see how it arises that the atolls in the Pacific and Indian oceans extend in lines parallel to the generally prevailing strike of the high islands and the great coast-lines of those oceans. I venture, therefore, to affirm, that on the theory of the upward growth of the corals during the sinking of the land, all the leading features in those wonderful structures, the lagoon-islands or atolls, which have so long excited the attention of voyagers, as well as in the no less wonderful barrier-reefs, whether encircling small islands or stretching for hundreds of miles along the shores of a continent, are simply explained.

It may be asked, whether I can offer any direct evidence of the subsidence of barrier-reefs or atolls; but it must be borne in mind how difficult it must ever be to detect a movement, the tendency of which is to hide under water the part affected. Nevertheless, at Keeling Atoll I observed on all sides of the lagoon old cocoa-nut trees undermined and falling; and in one place the foundation-posts of a shed, which the inhabitants asserted had stood seven years before just above high-water mark, but now was daily washed by every tide: on inquiry I found that three earthquakes, one of them severe, had been felt here during the last ten years. At Vanikoro, the lagoon-channel is remarkably deep, scarcely any alluvial soil has accumulated at the foot of the lofty included mountains, and remarkably few islets have been formed by the heaping of fragments and sand on the wall-like barrier-reef; these facts, and some analogous ones, led me to believe that this island must lately have subsided and the reef grown upwards: here again earthquakes are frequent and very severe. In the Society Archipelago, on the other hand, where the lagoon-channels are almost choked up, where much low alluvial land has accumulated, and where in some cases long islets have been formed on the barrier-reefs—facts all showing that the islands have not very lately subsided—only feeble shocks are most rarely felt. In these coral formations, where the land and water seem struggling for mastery, it must be ever difficult to decide between the effects of a change in the set of the tides and of a slight subsidence: that many of these reefs and atolls are subject to changes of some kind is certain; on some atolls the islets appear to have increased greatly within a late period; on others they have been partially or wholly washed away. The inhabitants of parts of the Maldiva Archipelago know the date of the first formation of some islets; in other parts, the corals are now flourishing on water-washed reefs, where holes made for graves attest the former existence of inhabited land. It is difficult to believe in frequent changes in the tidal currents of an open ocean; whereas, we have in the earthquakes recorded by the natives on some atolls, and in the great fissures observed on other atolls, plain evidence of changes and disturbances in progress in the subterranean regions.

It is evident, on our theory, that coasts merely fringed by reefs cannot have subsided to any perceptible amount; and therefore they must, since the growth of their corals, either have remained stationary or have been upheaved. Now it is remarkable how generally it can be shown, by the presence of upraised organic remains, that the fringed islands have been elevated: and so far, this is indirect evidence in favour of our theory. I was particularly struck with this fact, when I found to my surprise, that the descriptions given by MM. Quoy and Gaimard were applicable, not to reefs in general as implied by them, but only to those of the fringing-class; my surprise, however, ceased when I afterwards found that, by a strange chance, all the several islands visited by these eminent naturalists could be shown by their own statements to have been elevated within a recent geological era.

Not only the grand features in the structure of barrier-reefs and of atolls, and of their likeness to each other in form, size, and other characters, are explained on the theory of subsidence—which theory we are independently forced to admit in the very areas in question, from the necessity of finding bases for the corals within the requisite depth—but many details in structure and exceptional cases can thus also be simply explained. I will give only a few instances. In barrier-reefs it has long been remarked with surprise, that the passages through the reef exactly face valleys in the included land, even in cases where the reef is separated from the land by a lagoon-channel so wide and so much deeper than the actual passage itself, that it seems hardly possible that the very small quantity of water or sediment brought down could injure the corals on the reef. Now, every reef of the fringing-class is breached by a narrow gateway in front of the smallest rivulet, even if dry during the greater part of the year, for the mud, sand, or gravel, occasionally washed down, kills the corals on which it is deposited. Consequently, when an island thus fringed subsides, though most of the narrow gateways will probably become closed by the outward and upward growth of the corals, yet any that are not closed (and some must always be kept open by the sediment and impure water flowing out of the lagoon-channel) will still continue to front exactly the upper parts of those valleys, at the mouths of which the original basal fringing-reef was breached.

We can easily see how an island fronted only on one side, or on one side with one end or both ends encircled by barrier-reefs, might after long-continued subsidence be converted either into a single wall-like reef, or into an atoll with a great straight spur projecting from it, or into two or three atolls tied together by straight reefs—all of which exceptional cases actually occur. As the reef-building corals require food, are preyed upon by other animals, are killed by sediment, cannot adhere to a loose bottom, and may be easily carried down to a depth whence they cannot spring up again, we need feel no surprise at the reefs both of atolls and barriers becoming in parts imperfect. The great barrier of New Caledonia is thus imperfect and broken in many parts; hence, after long subsidence, this great reef would not produce one great atoll 400 miles in length, but a chain or archipelago of atolls, of very nearly the same dimensions with those in the Maldiva Archipelago. Moreover, in an atoll once breached on opposite sides, from the likelihood of the oceanic and tidal currents passing straight through the breaches, it is extremely improbable that the corals, especially during continued subsidence, would ever be able again to unite the rim; if they did not, as the whole sank downwards, one atoll would be divided into two or more. In the Maldiva Archipelago there are distinct atolls so related to each other in position, and separated by channels either unfathomable or very deep (the channel between Ross and Ari Atolls is 150 fath-

oms, and that between the north and south Nillandoo Atolls is 20 fathoms
in depth), that it is impossible to look at a map of them without believing
that they were once more intimately related. And in this same archipelago,
Mahlos-Mahdoo Atoll is divided by a bifurcating channel from 100 to 132
fathoms in depth, in such a manner, that it is scarcely possible to say
whether it ought strictly to be called three separate atolls, or one great atoll
not yet finally divided.

I will not enter on many more details; but I must remark that the curious structure of the northern Maldiva Atolls receives (taking into consideration the free entrance of the sea through their broken margins) a simple
explanation in the upward and outward growth of the corals, originally
based both on small detached reefs in their lagoons, such as occur in common atolls, and on broken portions of the linear marginal reef, such as
bounds every atoll of the ordinary form. I cannot refrain from once again
remarking on the singularity of these complex structures—a great sandy and
generally concave disk rises abruptly from the unfathomable ocean, with its
central expanse studded, and its edge symmetrically bordered with oval
basins of coral-rock just lipping the surface of the sea, sometimes clothed
with vegetation, and each containing a lake of clear water!

One more point in detail: as in two neighbouring archipelagoes corals
flourish in one and not in the other, and as so many conditions before enu-
merated must affect their existence, it would be an inexplicable fact if, during the changes to which earth, air, and water are subjected, the reef-building corals were to keep alive for perpetuity on any one spot or area. And
as by our theory the areas including atolls and barrier-reefs are subsiding,
we ought occasionally to find reefs both dead and submerged. In all reefs,
owing to the sediment being washed out of the lagoon or lagoon-channel
to leeward, that side is least favourable to the long-continued vigorous
growth of the corals; hence dead portions of reef not unfrequently occur on
the leeward side; and these, though still retaining their proper wall-like
form, are now in several instances sunk several fathoms beneath the surface. The Chagos group appears from some cause, possibly from the subsidence having been too rapid, at present to be much less favourably circum-
stanced for the growth of reefs than formerly: one atoll has a portion of its
marginal reef, nine miles in length, dead and submerged; a second has only
a few quite small living points which rise to the surface; a third and fourth
are entirely dead and submerged; a fifth is a mere wreck, with its structure
almost obliterated. It is remarkable that in all these cases, the dead reefs
and portions of reef lie at nearly the same depth, namely, from six to eight
fathoms beneath the surface, as if they had been carried down by one uniform movement. One of these "half-drowned atolls," so called by Capt.
Moresby (to whom I am indebted for much invaluable information), is of
vast size, namely, ninety nautical miles across in one direction, and seventy
miles in another line; and is in many respects eminently curious. As by our
theory it follows that new atolls will generally be formed in each new area

of subsidence, two weighty objections might have been raised, namely, that atolls must be increasing indefinitely in number; and secondly, that in old areas of subsidence each separate atoll must be increasing indefinitely in thickness, if proofs of their occasional destruction could not have been adduced. Thus have we traced the history of these great rings of coral-rock, from their first origin through their normal changes, and through the occasional accidents of their existence, to their death and final obliteration.

In my volume on "*Coral Formations*" I have published a map, in which I have coloured all the atolls dark-blue, the barrier-reefs pale-blue, and the fringing-reefs red. These latter reefs have been formed whilst the land has been stationary, or, as appears from the frequent presence of upraised organic remains, whilst it has been slowly rising: atolls and barrier-reefs, on the other hand, have grown up during the directly opposite movement of subsidence, which movement must have been very gradual, and in the case of atolls so vast in amount as to have buried every mountain-summit over wide ocean-spaces. Now in this map we see that the reefs tinted pale and dark blue, which have been produced by the same order of movement, as a general rule manifestly stand near each other. Again we see that the areas with the two blue tints are of wide extent; and that they lie separate from extensive lines of coast coloured red, both of which circumstances might naturally have been inferred, on the theory of the nature of the reefs having been governed by the nature of the earth's movement. It deserves notice, that in more than one instance where single red and blue circles approach each other, I can show that there have been oscillations of level; for in such cases the red or fringed circles consist of atolls, originally by our theory formed during subsidence, but subsequently upheaved; and on the other hand, some of the pale-blue or encircled islands are composed of coral-rock, which must have been uplifted to its present height before that subsidence took place, during which the existing barrier-reefs grew upwards.

Authors have noticed with surprise, that although atolls are the commonest coral-structures throughout some enormous oceanic tracts, they are entirely absent in other seas, as in the West Indies: we can now at once perceive the cause, for where there has not been subsidence, atolls cannot have been formed; and in the case of the West Indies and parts of the East Indies, these tracts are known to have been rising within the recent period. The larger areas, coloured red and blue, are all elongated; and between the two colours there is a degree of rude alternation, as if the rising of one has balanced the sinking of the other. Taking into consideration the proofs of recent elevation both on the fringed coasts and on some others (for instance, in South America) where there are no reefs, we are led to conclude that the great continents are for the most part rising areas; and from the nature of the coral-reefs, that the central parts of the great oceans are sinking areas. The East Indian Archipelago, the most broken land in the world, is in most parts an area of elevation, but surrounded and penetrated, probably in more lines than one, by narrow areas of subsidence.

I have marked with vermilion spots all the many known active volcanoes within the limits of this same map. Their entire absence from every one of the great subsiding areas, coloured either pale or dark blue, is most striking; and not less so is the coincidence of the chief volcanic chains with the parts coloured red, which we are led to conclude have either long remained stationary, or more generally have been recently upraised. Although a few of the vermilion spots occur within no great distance of single circles tinted blue, yet not one single active volcano is situated within several hundred miles of an archipelago, or even small group of atolls. It is, therefore, a striking fact that in the Friendly Archipelago, which consists of a group of atolls upheaved and since partially worn down, two volcanoes, and perhaps more, are historically known to have been in action. On the other hand, although most of the islands in the Pacific which are encircled by barrier-reefs, are of volcanic origin, often with the remnants of craters still distinguishable, not one of them is known to have ever been in eruption. Hence in these cases it would appear, that volcanoes burst forth into action and become extinguished on the same spots, according as elevatory or subsiding movements prevail there. Numberless facts could be adduced to prove that upraised organic remains are common wherever there are active volcanoes; but until it could be shown that in areas of subsidence, volcanoes were either absent or inactive, the inference, however probable in itself, that their distribution depended on the rising or falling of the earth's surface, would have been hazardous. But now, I think, we may freely admit this important deduction.

Taking a final view of the map, and bearing in mind the statement made with respect to the upraised organic remains, we must feel astonished at the vastness of the areas, which have suffered changes in level either downwards or upwards, within a period not geologically remote. It would appear, also, that the elevatory and subsiding movements follow nearly the same laws. Throughout the spaces interspersed with atolls, where not a single peak of high land has been left above the level of the sea, the sinking must have been immense in amount. The sinking, moreover, whether continuous, or recurrent with intervals sufficiently long for the corals again to bring up their living edifices to the surface, must necessarily have been extremely slow. This conclusion is probably the most important one which can be deduced from the study of coral formations;—and it is one which it is difficult to imagine, how otherwise could it ever have been arrived at. Nor can I quite pass over the probability of the former existence of large archipelagoes of lofty islands, where now only rings of coral-rock scarcely break the open expanse of the sea, throwing some light on the distribution of the inhabitants of the other high islands, now left standing so immensely remote from each other in the midst of the great oceans. The reef-constructing corals have indeed reared and preserved wonderful memorials of the subterranean oscillations of level; we see in each barrier-reef a proof that the land has there subsided, and in each atoll a monument over an island now lost. We may thus, like unto a geologist who had lived his ten thousand years and kept a record of the passing changes, gain some insight into the great system by which the surface of this globe has been broken up, and land and water interchanged.

Understanding What You've Read

Darwin's theory of subsidence has been proven correct, except that we now know that instead of the whole sea floor sinking, the islands press down upon the sea floor, creating a saucer-like depression. How does Darwin prepare readers for his theory of subsidence for reef formation? Would you have been convinced by Darwin's theory had you read this article in 1837?

Darwin often ends his essays with sincere and restrained expressions of awe. Does he do that in this essay? How does this compare with Rachel Carson's (see pages 108–114) or Lewis Thomas's (see pages 253–256) style of expressing emotion?

Darwin probably would not have come to the same conclusions had he not been familiar with the work of geologist Charles Lyell. Write an essay about a scientific topic of your choice showing how historical influences affected its empirical features.

Exercises

1. Rachel Carson's *The Sea Around Us* won a National Book Award and went into eleven printings the year it was published (1951). Here is one of its paragraphs:

    Between the sunlit surface waters of the open sea and the hidden hills and valleys of the ocean floor lies the least known region of the sea. These deep, dark waters, with all their mysteries and their unsolved problems, cover a very considerable part of the earth. The whole world ocean extends over about three-fourths of the surface of the globe. If we subtract the shallow areas of the continental shelves and the scattered banks and shoals, where at least the pale ghost of sunlight moves over the underlying bottom, there still remains about half the earth that is covered by miles-deep, lightless water, that has been dark since the world began.[7]

    What gives this paragraph its cohesion and rhythm?

2. Eliminating unnecessary words and phrases often produces a series of short, choppy sentences. To make your paper readable, you must merge these sentences into cohesive sentences and paragraphs. Combine each of the following sets of sentences into a well-written sentence or paragraph:

[7]Carson, Rachel L. 1951. *The Sea Around Us.* New York: Oxford University Press, p. 37.

All of the glassware should be cleaned and rinsed.
The glassware is not disposable.
The cleaning and rinsing should be thorough.
Wash the glassware with nitric acid.
The acid should be a solution of 50%.
All nondisposable glassware should be thoroughly cleaned and rinsed with a 50%
solution of nitric acid.

The scientist is studying the effects of alcohol on the circulatory systems of rats.
The scientist is Dr. Dennis Clark.

The project will study DNA in chimpanzees.
The chimpanzees are native to Africa.
DNA is the genetic building-block.

The 1991 guidelines of the EPA are available.
The guidelines are for air quality.
The levels for air quality are tolerable levels.
The levels are for each type of pollutant in the air.
The levels vary with toxicity of each pollutant.

3. Improve the readability and clarity of these sentences and paragraphs:

Enclosed please find material which is descriptive of the graduate programs in the Department of Geology at Auburn University. I would deeply appreciate your posting of this material and calling it to the attention of any students who might have an interest in our program.

We suggest it is quite possible that the solutions may have been contaminated.

It is possible that death may occur.

Often, bacteria cannot be identified with an optical microscope.

Our proposal follows the sequential itemization of points occurring elsewhere in your RFP, wherever possible, to facilitate your review.

More than mere numbers are required for this experiment.

There has been an affirmative decision for program termination.

An evaluation of the experiment by us will assure greater efficiency in utilization of experimental animals.

Pursuant to the recent memorandum issued 21 June 1991, because of financial exigencies, it is incumbent upon us all to endeavor to make maximal utilization of telephonic communication in lieu of personal visitation.

Imagine a picture of someone engaged in the activity of trying to learn how to operate a pH meter.

The instructions must be followed in an accurate manner.

As per our aforementioned discussion, I am herewith enclosing a report of our results.

In my personal opinion, we should basically listen to and think over in a punctilious manner each and every suggestion that is offered to us.

4. Read a science-related article from a recent issue of Scientific American. Reduce each paragraph to one sentence.

5. Discuss, support, or refute the ideas expressed in these quotations:

Biology

Freedom to breed will bring ruin to all. — Garrett Hardin

Ontogeny recapitulates phylogeny. — Ernst Heinrich Haeckel

The main conclusion arrived at in this work, namely, that man is descended from some lowly organized form, will, I regret to think, be highly distasteful to many. But there can hardly be a doubt that we are descended from barbarians. — Charles Darwin

Chemistry

The significant chemicals of living tissue are rickety and unstable, which is exactly what is needed for life. — Isaac Asimov

Geology

Daily it is forced home on the mind of the geologist that nothing, not even the wind that blows, is so unstable as the level of the crust of this Earth. — Charles Darwin (1809–1882)

Few scientists are willing to jeopardize their research funds by publicly criticizing EPA's interpretation of the scientific record. — Roy Gould

The facts proved by geology are briefly these: that during an immense, but unknown period, the surface of the earth has undergone successive changes; land has sunk beneath the ocean, while fresh land has risen up from it; mountain chains have been elevated; islands have been formed into continents, and continents submerged till they have become islands; and these changes have taken place, not once merely, but perhaps hundreds, perhaps thousands of times. — Alfred Russel Wallace (1823–1913)

Physics

It [the earth] alone remains immovable, whilst all things revolve round it. — Pliny the Elder (23–79)

Since nothing stands in the way of the movability of the earth, I believe we must now investigate whether it also has several motions, so that it can be considered one of the planets. That it is not the center of all the revolutions is proven by the irregular motions of the planets, and their varying distances from the earth, which cannot be explained as concentric circles with the earth at the center. — Nicholas Copernicus (1473–1543)

The total energy of the universe is constant and the total entropy is continually increasing. — Isaac Newton (1642–1727)

An elementary particle is not an independently existing, unanalyzable entity. It is, in essence, a set of relationships that reach outward to other things. — William Stapp

UNIT TWO

WRITING FOR
YOUR
AUDIENCE

Anyone who writes about science must know about
science, which cuts down competition considerably.
  — Isaac Asimov

CHAPTER SIX

Writing for a

General Audience

I believe—as Galileo did when he wrote his two greatest works as dialogues in Italian rather than didactic treatises in Latin, as Thomas Henry Huxley did when he composed his masterful prose free of jargon, as Darwin did when he published all his books for general audiences—that we can still have a genre of science books suitable for and accessible alike to professionals and interested laypeople.
    — Stephen Jay Gould, *Wonderful Life: The Burgess Shale and the Nature of History*

Anyone who can't explain their work to a fourteen-year-old is a charlatan.
    — Irving Langmuir

The average scientist is basically toilet-trained to the point where if what he does is comprehensible to the general public, it means he's not a good scientist.
    — Paul Ehrlich

In 1924, Albert Einstein and Indian physicist Satyendra Nath Bose predicted that low temperatures would cause atoms in a dilute, noninteracting gas to condense to the point where they would fall into the quantum state, essentially behaving like a single atom. That prediction was proven accurate when, on June 5, 1995, a group of scientists created the so-called Bose-Einstein condensate. The researchers chilled rubidium-87 to a few billionths of a degree above absolute zero, at which point the gas became so dense that the atoms came to a near standstill. As the atoms' movement slowed, the quantum-mechanical waves that describe each atom merged until the entire gas was locked in the same quantum state. In effect, the atoms lost their separate identities and became one.

The production of the first Bose-Einstein condensate was announced as the cover story of the July 14, 1995 issue of *Science*. In the opening pages of the journal, the discovery was announced with the headline, "Physicists Create New State of Matter"; the formal article, which appeared later in the journal, was titled, "Observation of Bose-Einstein Condensation in a Dilute Atomic Vapor." Other publications announced the discovery in different ways. Here are some of those headlines, along with the articles' opening sentences:

Akron Beacon Journal: **Scientists In Colorado Create Matter That Has Never Existed Before**

In a lab in Colorado, inside a jar cooled to the lowest temperature ever reached on Earth or anywhere else, scientists have created a form of matter that has never existed before anywhere in the universe—something they have dubbed a "superatom."

USA TODAY: **Coldest Temperature Ever, Hot Scientific Breakthrough**

Beating the heat in a big way, scientists have chilled atoms to the lowest temperature ever achieved—creating an entirely new state of matter.

Science News: **Physics "Holy Grail" Finally Captured**

A team of scientists in Colorado has done something really cool.

Clearly, these headlines and opening sentences differ markedly from those in *Science*. The differences result from each article being written for different types of audiences.

Many scientists have long understood the importance of communicating with the public. Benjamin Franklin wrote for the *Pennsylvania Gazette;* Michael Faraday (see pages 200–211) delivered a series of public lectures and wrote for the general public; J.B.S. Haldane edited a newspaper; and Richard Feynman (see pages 55–67) made television programs. These and other famous scientists did so because they knew that science is part of our culture and heritage; it is common property and needs the public's support. They also had a strong faith in the educability of the public.

Many scientists continue to write for nonscientists. Scientists often write articles for newspapers and popular magazines, as well as essays for their students, colleagues, and peers. The secret to writing effectively for a general audience is to demystify science. Do this by connecting the science to the everyday world and by showing readers that science is a human activity—an enterprise managed and mismanaged by men and women. As such, it involves luck and hunches. It also changes people's lives.

Magazines such as *Natural History, Discover*, and *Smithsonian* publish articles written for nonscientists. These articles resemble and are often excellent models for term papers and essays. In these articles, scientists define terms, describe simple concepts, include images that readers understand, and discuss examples outside of science to communicate with readers. The writing is simple and logical, thereby helping convince readers that the conclusions are inevitable.

Analogies and examples, especially those with human interest, help a general audience understand what you're saying. Whatever the topic, you need to explain the unknown in terms of the known. Einstein was a master of this: He made his work accessible to nonscientists by writing about passing ships, flying birds, and other everyday experiences. For example, consider his definition of relativity: "When a man sits with a pretty girl for an hour, it seems like a minute. But let him sit on a hot stove for a minute, and it's longer than an hour. That's relativity."

Nonscientists have no use for the dull, jargon-laced writing of most scientists—they'll quickly pitch your writing aside if it doesn't meet their needs. You'll best attract the eye of nonscientists if you explain science in everyday language. For example, saying that "this substance has the consistency of pancake syrup" communicates ideas that would require many paragraphs to describe. By making science relevant, you make it meaningful to a general audience.

There are many ways to write about science for a general audience. Whatever your approach, make sure that your paper has an introduction, a body, and a conclusion (Figure 6-1).

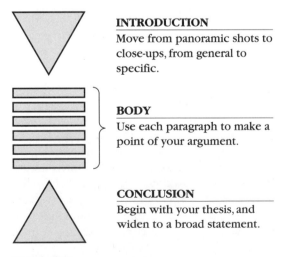

INTRODUCTION
Move from panoramic shots to close-ups, from general to specific.

BODY
Use each paragraph to make a point of your argument.

CONCLUSION
Begin with your thesis, and widen to a broad statement.

Figure 6–1
Model for the design of an essay.

Introduction   Introduce your paper in the first paragraph or two. Open with general statements that catch the reader's attention. Then progress to a focusing statement of the purpose of your paper, much like photographers begin their videos with panoramic shots before zeroing in on close-ups.

Body   Develop your arguments in the body of your paper. The number of paragraphs in the body of your paper depends on the number of points that you want to make.

Conclusion   Use the conclusion of your paper to tie everything together. Begin the conclusion of your paper by stating the paper's thesis. Widen the thesis to make a broad, concluding statement.

The paradox of writing about scientific discoveries is that while you have the advantage of the subject's novelty, you have the disadvantage of presenting your readers with unfamiliar information. The most effective way of solving this paradox is to use familiar information to lead your readers to unfamiliar information. This strategy always makes new ideas less intimidating.

Here are a few of the many ways to write effectively for a general audience:

Analogies

Scientists have always used analogies to communicate their ideas. For example, J.B.S. Haldane wrote a famous essay in which he likened the production of hot gas in a bomb to that of steam in a kettle, and the changes which occur in a bird each year to those which take place in people once in a lifetime at puberty. Similarly, Charles Darwin's monumental *On the Origin of Species* opened with an extended analogy involving domestic animals. Darwin and Haldane, like all effective writers, understood the power of analogies for presenting new ideas.

Use analogies to link a scientific idea with a common experience. For example, analogies such as, "The eye is like a camera, with the retina being the film," communicate quickly and effectively with readers. Such analogies are easily developed into more detailed explanations:

> The human eye functions like a camera. The iris and the pupil of the eye function as the lens and shutter of the camera by controlling the depth of focus. The depth of focus depends on the size of the hole through which the photographer looks: The smaller the hole, the greater the depth of focus and the sharper the picture. And just like a camera's diaphragm, the pupil of the eye also shrinks or enlarges according to available light. When we squint, turn toward or away from bright light, or put on sunglasses, we mimic a photographer's adjustment of the camera's shutter-speed to let more or less light into the camera.

Consider how Richard Feynman uses numerical analogies to explain the significance of the magnitude of electrical forces:

If you were standing at arm's length from someone and you had one percent more electrons than protons, the repelling force would be incredible. How great? Enough to lift the Empire State Building? No! To lift Mount Everest? No! The repulsion would be enough to lift a weight equal to that of the entire earth.[1]

Here's how Paul and Anne Ehrlich use analogies (and sarcasm) to show the absurdity of the claim that "the world's food resources theoretically could feed 40 billion people:"

In one sense they were right. It's "theoretically possible" to feed 40 billion people—in the same sense that it's theoretically possible for your favorite major-league baseball team to win every single game for fifty straight seasons, or for you to play Russian roulette ten thousand times in a row with five out of six chambers loaded without blowing your brains out.[2]

Although such analogies communicate effectively when used properly, remember that they clarify, but do not prove, a point. If two things are alike in the two ways shown by the analogy, they're not necessarily alike in all ways.

Human Interest

This writer uses a "Look at me" opening to capture readers' attention long enough to lead them to read further:

"My mother almost never drank," says Nancy Wexler, president of the Hereditary Diseases Foundation, "yet one day, as she was crossing the street, a policeman said, 'Aren't you ashamed to be drunk so early in the morning!'"

Those words terrified her mother, for her father and her three older brothers had died of a frightening ailment called Huntington's disease, Wexler says, which made them lose their balance, twitch uncontrollably, and eventually lose their minds. Mrs. Wexler thought she had been spared: She was 53, past the age when symptoms of the disease usually begin. None of her family or friends had noticed the slight weave in her walk, for it had come about very gradually. But those startling and brutal words from the policeman sent her rushing to consult with her doctor.

When the doctor confirmed that she was doomed to become totally incapacitated and die within 10 to 20 years, her husband, a Los Angeles psychoanalyst, called their two daughters home from graduate school. He had to explain not only that their mother had Huntington's but also that they, too, stood a 50–50 chance of getting the disease.[3]

[1]Feynman, Richard P. 1964. *The Feynman Lectures on Physics*. Reading, MA: Addison-Wesley Publishing Co. II, 1.

[2]Ehrlich, Paul. R. and Anne H. Ehrlich. 1990. *The Population Explosion*. New York: Simon and Schuster,. p. 19.

[3]Pines, Maya. 1984. "In the Shadow of Huntington's." *Science 84* 5: 32.

News

The article, "Not Just a Big, Boring Rock—a Space Rock," published in the February 1994 issue of *Discover*, opens with news:

> The space probe *Galileo* is now hurtling toward a rendezvous with Jupiter in December 1995.

In *National Geographic*, Esmond Martin starts his article with an interesting piece of news:

> Sometime in the late 1970s half of the white rhinoceros population of Uganda suddenly disappeared—a single rhino, probably shot by a gang of poachers. At the time, nobody realized that it was one of only two left in the country.[4]

In this article from *Science News*, the writer saves the news for the last sentence of the paragraph:

> In a mid-19th-century monastery garden, Gregor Mendel's experiments with smooth and wrinkled peas revealed the rules by which parents pass on traits to their offspring. But a theory called genomic imprinting is putting a new wrinkle into Mendelian genetics: A gene's expression may depend on which parent contributed it.
>
> In the latest finding, . . .

Background Information

Other authors begin paragraphs with background information that leads to a question:

> The sight of a flock of birds migrating south in the fall or north in the spring hardly ever fails to evoke a sense of wonder. The flight may be the orderly aerobatics of a V of Canada geese or the ragtag progress of a group of starlings. Whatever its details, the overwhelming impression conveyed to the observer is that of a powerful inner impulse. The birds do not hesitate in their flight; they travel smoothly and unerringly toward a goal far out of the viewer's sight. Where does the impulse come from that guides the birds toward warmer climates in winter and brings them back to their northern breeding grounds in the spring?[5]

[4]Martin, Esmond B. 1984. "They're Killing off the Rhino." *National Geographic* (March 1984): 404.

[5]Gwinner, Eberhard. 1986. "Internal Rhythms in Bird Migration." *Scientific American* 254: 84.

Questions

Still others begin their papers with questions followed by answers. Here's how John Allen Paulos (author of *Innumeracy*) opened his "My Turn" column entitled "Orders of Magnitude" in *Newsweek*:

> Quick: how fast does human hair grow in miles per hour? What is the volume of all the human blood in the world? If you don't know, it's no surprise; even math students sometimes don't, either. The answers are perhaps intriguing: hair grows a little faster than 10 to the minus-8th miles an hour (that is, a decimal point followed by 7 zeros and a 1); the totality of human blood will fill a cube 800 feet on a side, or cover New York's Central Park to a depth of about 20 feet.

Having gotten readers' attention with these questions and answers, Paulos then makes his point:

> What does surprise me, however, is how often I find adults who have no ideas of easily imagined numbers: the population of the United States, say, or the approximate distance from the East Coast to the West. Many otherwise sophisticated people have no feel for magnitude, no grasp of large numbers like the federal deficit or small probabilities like the chances of ingesting cyanide-laced painkillers. This disability, which the computer scientist Douglas Hofstadter calls innumeracy, is so widespread that it can lead to bad public policies, poor personal decisions—even a susceptibility to pseudoscience.

Contrast and Origin

William Brown combines contrast with origin to describe selection before 1900:

> For thousands of years Indians of the Western Hemisphere grew corn, varieties pollinated by the wind and bred largely by chance. Despite their lack of scientific insight, they transformed a wild grass from Mexico into one of the world's most productive plants. Sixteenth- and 17th-century farmers continued the practice of corn improvement. Distinct varieties were developed by selecting the best ears at the time of harvest and using seed from those ears to produce the next year's crop. This kind of selection continued until about 1900 and resulted in scores of high-yielding, randomly pollinated varieties. Then, in the course of just a few years, scientists applied genetics to corn breeding—and brought about a transformation of agriculture in this century.
>
> The development of hybrid corn resulted from the exploitation of a phenomenon known as heterosis or hybrid vigor. This increased yield, vigor, and rate of growth of plants comes from the mating of unrelated parents. Many early botanists and horticulturists, including Charles Darwin, had previously observed this phenomenon. But it was geneticist George Harrison Shull who developed the heterosis concept as it is applied today. He and E. M. East, a

contemporary whose experiments at the Connecticut Agricultural Experiment Station in New Haven closely paralleled Shull's, were the first to isolate pure strains of corn. These were then crossed to produce the reliable vigor of hybrid corn.[6]

Comparisons

Here are examples of how to use comparisons to communicate effectively:

> The Moon is 240,000 miles away from the Earth. How far is that? It's two and a half times the distance from Greenland to South Africa. It's the distance you'd cover if you walked for five years.

> Atoms are small—so small that when chemists count out quantities they work in "moles," each of which contains six hundred thousand billion billion atoms. If chemists need six hundred thousand billion billion atoms before they can handle them they must be pretty small . . . but how many is six hundred thousand billion billion? It's actually about the number of apples you'd need to fill a bag as big as the Earth.

Everyday Life

In this essay entitled "Following Aspirin's Trail," Steve Olson uses specific information about our medicine cabinets to grab our attention:

> Of all the drugs in a medicine cabinet, none is as familiar as aspirin. Most of us pop an aspirin as blithely as we take a shower or get a haircut. Americans swallow some 20 tons of the stuff everyday—enough to fill four good-sized dump trucks.
>
> But even though aspirin has been around since 1899, only in the last few years have scientists begun to uncover how it works. . . .

In "The Space Between the Stars," published in the December 1994 issue of *Discover*, Bob Berman describes a common scene: the night sky.

> This blackest month, when the stars emerging from early darkness recall the masterful illumination of a Rembrandt, we can inspect the canvas itself—the sky between the stars. Here is an emptiness rich in both mystery and misconception.

David Baltimore links our everyday lives to a less common bit of information—the discovery of the structure of DNA:

---

[6]Brown, William. 1984. "Twenty Discoveries that Changed our Lives." *Science 84* 5: 65, 69, 73, 76, 79, 83, 99, 111, 115, 121, 127, 131–132, 141, 149, 153.

Science, for all its mystery, is woven into the fabric of our daily lives. We cultivate plants and raise animals to eat; we burn wood to keep warm; we wear leather, cotton, or wool. Over the centuries, man has become increasingly adept at harnessing the biologic world, but before the 1950s our knowledge of how living things work was at best superficial. That has changed. One discovery stands out as the primary generator of our new understanding of biologic systems and our power to manipulate them: the 1953 elucidation of the structure of DNA by James D. Watson and Francis Crick.

Several Points of View

In "A Charming Resistance to Parasites," Marlene Zuk states several points of view in the opening paragraphs:

> By 1889, British naturalist Alfred Russell Wallace had concurred with many of Darwin's ideas about evolution. He balked, however, at the notion that males of many animal species have evolved bright colors and complicated mating displays simply as a result of female's preference for them, and wrote:
>
>> . . . it may also be admitted. . . that the female is pleased or excited by the display. But it by no means follows that slight differences in the shape, pattern or colours of the ornamental plume are what lead a female to give the preference to one male over another. . . A young man, when courting, brushes or curls his hair, and has his moustache, beard of whiskers in perfect order, and no doubt his sweetheart admires them; but this does not prove that she marries him on account of these ornaments, still less that hair, beard, whiskers and moustache were developed by the continued preferences of the female sex.
>
> Darwin disagreed and maintained that "the power to charm the female has sometimes been more important than the power to conquer other males in battle." Most evolutionary scientists would now agree with Darwin, but they still disagree over exactly what females find charming and how their tastes could cause the development of such bizarre ornaments as a peacock's tail or the elongated wattles of a turkey. Recently, even the possible role of tiny blood parasites has been brought into the controversy.[7]

Misconceptions

In "The Space Between the Stars," Bob Berman addresses a common misconception about the night sky:

[7]Zuk, Marlene. 1984. "A Charming Resistance to Parasites." *Natural History* 93: 28

Can you see millions of stars with the naked eye? No—the count is surprisingly low.

Similarly, Carl Zimmer refutes another misconception in his opening lines of "Inconstant Field," an article that appeared in the February 1994 issue of *Discover:*

Although it has long served as a fixed reference for navigators, Earth's magnetic field is anything but static. Over the course of decades and centuries, in what is called secular variation, the pattern of the field drifts randomly, such that at a given geographic location the direction a compass needle points may change by tens of degrees.

Yves Dunant and Maurice Israel address a misconception as they begin their paper in *Scientific American:*

For many years there was a wide consensus among neuroscientists that acetylcholine is released from small, spherical organelles called synaptic vesicles, which are found inside the nerve terminal. It was thought that when the nerve terminal is stimulated, the vesicles fuse with the terminal membrane and release their contents into the space between the neuron and the tissue with which it communicates.

Our recent investigations contradict this simple picture. They suggest that although the vesicles do indeed store acetylcholine and play a role in its regulation within the cell, the acetylcholine released by the nerve terminal does not originate in the vesicles. Instead the released acetylcholine is derived directly from the cytoplasm, which makes up the ground material inside the neuron. The releasing mechanism appears to be operated by a compound, most likely a protein, that is embedded in the membrane of the nerve cell. The protein may act as a valve, enabling acetylcholine to pass through membrane.[8]

This article entitled "Butterfly Fallacy" (from the April 11, 1991 issue of *USA Today*) presents a scientist's data to overturn a popular misconception about butterflies:

For years, science students have been taught that the delectable viceroy butterfly resembles its foul-tasting cousin, the monarch, to fool predators. Now David Ritland and Lincoln Brower of the University of Florida say the viceroy tastes lousy, too. Their study in the British journal *Nature* details how viceroys, monarchs and other butterflies known to be a good meal were stripped of their telltale wings and offered to 16 wild blackbirds. Only 41% of the viceroys were eaten vs. 98% of the tasty species. "It kind of shows how some of the obvious things we've assumed have never been tested," says Jim Miller, curator of moths and butterflies at the American Museum of Natural History in New York.

---

[8]Dunant, Yves and Maurice Israel. 1985. "The Release of Acetylcholine." *Scientific American* 252: 58.

Readability Formulas

Several people have tried to devise ways to measure the readability of writing. The most popular method is the "fog index" devised by Robert Gunning in 1944. The fog index of a paragraph or essay is the grade-level of education needed by the reader to understand the article. For example, an essay having a fog index of 12 could be read by a twelfth grader, but not a ninth grader. Similarly, a paragraph having a fog index of 16 could be read by a college senior, but probably not by a high school senior.

The fog index of a paragraph or essay is based on two important traits of writing: average sentence length and the percentage of difficult words. These features of writing must be matched to your audience if you expect your audience to read and understand what you've written. This is an important function of an effective editor. For example, most magazines have a consistent writing style:

Magazine	Average Sentence Length	% Polysyllabic Words
True Confessions	13	4
Reader's Digest	15	8
Time	17	9
Newsweek	17	9
Science	28	14

Other popular magazines such as *Harpers* and *Ladies' Home Journal* all have average sentence lengths less than 22 words and fewer than 12% polysyllabic words. This does not mean that all of the articles in these magazines sound alike; it merely means that the editor of each magazine matches the reading level of each article with the magazine's audience. Most scientific journals have a fog index that varies between 18 and 25, thus explaining why they are often so hard to read.

Here's how to determine the fog index of what you write:

- Choose a paragraph that you'd like to evaluate. Count the number of words and sentences in the essay.

- Divide the number of words by the number of sentences. This is the average sentence length of the paragraph.

- Count the number of words having three or more syllables. Do not include proper names, combined words (for example, bookkeeper), or verbs created by adding endings such as -ed (for example, created).

- Divide this number of words by the total number of words in the paragraph. This is the percentage of polysyllabic words.

- Add the average sentence length plus the percent polysyllabic words and multiply by 0.4. This is the fog index of the writing.

As an example, consider this tongue-in-cheek essay description of some scientists' insistence on poor writing:

> It is the opinion of the writer that it is the appropriate moment to re-examine the style of writing which might most effectively be used by members of the science profession. It is also the writer's belief that the long-lasting tradition about the inappropriateness of the active voice and the personal pronoun for technical writing has made for a great deal of inefficiency. This kind of writing has been exemplified in the past by numerous national publications. It would appear that an application of the principles of science to the problem wound be beneficial and it would seem the result might be that such a style would be eliminated.[*]

The numbers of words in each sentence of this paragraph are 31, 33, 14, 17, and 14. (Note that the last sentence has two independent clauses and is therefore counted as two sentences.) The total number of words in the paragraph (109) divided by the number of sentences (5) gives an average sentence length of 22 words. Furthermore, there are 20 words having three or more syllables. This means that 18% (20/109) of the words in the paragraph are polysyllabic.

The average sentence length plus the percentage of polysyllabic words is 22 + 18 = 40. This number multiplied by 0.4 produces a fog index of 16, which corresponds to the reading level of a college senior. It is absurd to present the paragraph's simple idea at such a high level of reading difficulty. However, that effect was intentional. A translation of the paragraph reads like this:

> I think it's about time we stop insisting on impersonal style for scientists. I think that our national publications could set a good example in breaking this strait jacket. After all, the scientist wants efficiency and a "common-sense" approach to his professional work. Why not encourage him to apply this practical method to his writing? If he does, he'll save time, money, material, and his reader's temper.

The fog index of this paragraph is 12.5. Not surprisingly, it's also easier to read and understand.

Readability formulas such as the one for determining the fog index can help you match your writing to your audience. However, blindly following any formula or rule will not make you an effective writer. Moreover, formulas such as that for calculating a fog index cannot evaluate the precision or style of your writing. Use such formulas to check your work *after* you've written something, not to pattern or control your ideas before you write. Effective writing is lively— don't kill it with this or any other "system."

[*]Edited from Shurter, Robert L. 1952. "Let's Take the Strait Jacket off Technical Style." *Mechanical Engineering.* 74: 664.

Stating the Problem

These two authors concisely state the problem they studied:

> A seed falls, and finding the weather hospitable, it swells, splits its protective coat, and sends a shot and root out into the air. Suddenly and without fanfare, the shoot curves upward and heads for the sky, while the root turns downward and angles its way into the soil.
>
> This uncanny ability to sense gravity and act accordingly, called gravitropism, has long fascinated botanists. But so far no one has been able to explain how plants get their directions straight.[9]

> The neuronal mechanisms of information storage remain one of the principal challenges in contemporary neuroscience. Over the past decade, however, the development of effective model systems has significantly advanced our understanding of the cellular basis of nonassociative learned behaviors. This progress has resulted largely from the exploitation of "simple" invertebrate models, but few effective systems are available for cellular analysis of associative learning or of learning in invertebrates.[10]

Descriptions

Scientists often use descriptive writing to present their ideas. Here's how Walter Sullivan uses this style of writing to describe recent discoveries on the nature of the Milky Way, the center of our galaxy:

> A voyager to the heart of the galaxy would find the sky almost blindingly bright with stars. According to recent observations by the Very Large Array of radio telescopes in New Mexico of the area within 200 light years of the Milky Way's center, the traveler would enter a region of gigantic, parallel gaseous filaments, arcing around the core, each about 100 light years long. Additional clouds reaching from there toward the core are cut by strange, narrow "threads."

Here's how Neil de Grasse Tyson opens "The Milky Way Bar," an article that appeared in *Natural History* in August, 1995:

> As you probably know, our solar system is part of a large, flattened, spiral-shaped disk of several hundred billion stars known as the Milky Way. In its center—protruding slightly above and below the plane—is the galactic bulge, a dense collection of 10 billion stars. Taken as a whole, the shape of the Milky Way greatly resembles that of a fried egg."

[9]Fellman, Bruce. 1984. "How Do Plants Know Which End Is Up?" *Science 84* 5: 26.

[10]Gold, M.R. and G.H. Cohen. 1981. "Modifications of the Discharge of the Vagal Cardiac Neurons During Learned Heart Rate Change." *Science* 214: 345.

The following excerpt from a National Cancer Institute publication uses a simple, straightforward style to describe the pancreas:

> The pancreas is a thin, lumpy gland about six inches long that lies behind the stomach. Its broad right end, called the head, fills the loop formed by the duodenum (the first part of the small intestine). The midsection of the pancreas is called the body, and the left end is called the tail. The pancreas produces two kinds of essential substances. Into the bloodstream it releases insulin, which regulates the amount of sugar in the blood. Into the duodenum it releases pancreatic juice containing enzymes that aid in the digestion of food.
>
> As pancreatic juice is formed, it flows through small ducts (tubes) into the main pancreatic duct that runs the full length of the gland. At the head of the pancreas, the main duct joins with the common bile duct, and together they pass through the wall of the duodenum. The common bile duct carries bile (a yellowish fluid that aids in the digestion of fat) from the liver and gallbladder to the duodenum.[11]

Interesting Phenomenon

The article entitled "The Photon's Progress" includes this incredible claim:

> A typical photon born at the center of the sun takes a good 170,000 years to reach the surface.

Similarly, this author uses a simple writing style to describe an interesting phenomenon—how a snake smells with its tongue:

> For years it was thought that the snake's forked tongue was used to feel along in the dark, like the whiskers of a cat. Another popular belief was that it was poisonous. The forked tongue, which the snake can flick in and out without opening its mouth, through a gap in its lips, is actually an organ of smell.
>
> The tongue flicks out and waves around to attract a few odor molecules floating in the air. The two forks of the tongue are inserted briefly into two cavities in the top of the snake's mouth called Jacobson's organ. Sensors inside this organ immediately analyze the molecules and send the appropriate message to the snake's brain. The tongue can detect smells that signal danger, approaching prey, or the presence of a mate long before the snake can see or hear the source of the odors.[12]

---

[11] *What You Need to Know About Cancer of the Pancreas.* Washington, D.C.: National Cancer Institute, 1979.

[12] Thomas, Warren D. and Daniel Kaufman. 1990. *Dolphin Conferences, Elephant Midwives, and Other Astonishing Facts About Animals.* Los Angeles: Jeremy P. Tarcher, Inc.

Definitions

Science contains many unusual terms, and writers must define these terms if they are to use them to communicate effectively with their readers. General audiences do not need a complete technical definition of a term to understand its meaning. Therefore, an adequate definition of penicillin for a general audience might be "a drug used to cure bacterial infections." The audience would be confused by this technically more accurate information presented in *Van Nostrand's Scientific Encyclopedia*:

> Penicillin is an antibacterial substance produced by microorganisms of the *Penicillium chrysogenum* group, principally *Penicillium notatum* NRRL832 for deep or submerged fermentation and NRRL 1249, B21 for surface culture. Penicillin is antibacterial toward a large number of gram positive and some gram negative bacteria and is used in the treatment of a variety of infections.

Notice how Acker and Hartsel, in their article entitled "Fleming's Lysozyme," effectively define difficult terms by first offering a common meaning:

> Bacterial anatomists are indebted to the late Sir Alexander Fleming for a sensitive chemical tool with which they have been studying bacteria, dissolving away the cell wall, and exposing the cell body or cytoplasm within. In 1922 at St. Mary's Hospital in London, six years before his epochal discovery of penicillin, Fleming found "a substance present in the tissues and secretions of the body, which is capable of rapidly dissolving certain bacteria." Because of its resemblance to enzymes and its capacity to dissolve, or lyse, the cells, he called it "lysozyme."[13]

As mentioned above, you may want to define some terms by discussing the origin of the term. Here is an example:

> The word *science* is derived from two Greek words, *bios*, "*life*" and *logos*, "word" or "discourse," and so "science." In its narrower sense, science may be defined as the science of life, that is, the science which treats the theories concerning the nature and origin of life; in a broader sense, science is the sum of zoology (Greek *zoon*, "animal," + *logos*, "science"), or the science of animals, and botany (Greek *botone*, "plant"), or the science of plants. It is in this broad sense that the word is used in this volume.[14]

The first sentence of this paragraph gives the derivation of the word, and the second sentence provides details of the wider definition and its Greek origins. The final sentence is a transition between that definition and what will be discussed in the rest of the article.

There are many ways to write effectively for nonscientists. Choose the method best suited for your audience, purpose, and subject.

---

[13]Acker, Robert and S.E. Hartsel. 1960. "Fleming's Lysozyme." *Scientific American*, June.

[14]Rice, Edward L. 1935. *An Introduction to Biology*. Boston: Ginn and Co.

James Watson

The Double Helix

James Watson (b. 1928) was a boy-wonder who
entered college when he was 15 and received a
Ph.D. in genetics when he was 20. Soon thereafter,
he teamed with Francis Crick to discover the structure of DNA—work for which in 1962 they received
the Nobel Prize in Physiology and Medicine. Watson
described his view of that collaboration and research
in *The Double Helix*, published in 1968.
Throughout the book, Watson mirrors himself as an
arrogant, western cowboy.

Bragg was in Max's office when I rushed in the next day to blurt out
what I had learned. Francis was not yet in, for it was a Saturday morning and he was still home in bed glancing at the *Nature* that had come in
the morning mail. Quickly I started to run through the details of the B form,
making a rough sketch to show the evidence that DNA was a helix which
repeated its pattern every 34 Å along the helical axis. Bragg soon interrupt-
ed me with a question, and I knew my argument had got across. I thus
wasted no time in bringing up the problem of Linus, giving the opinion that
he was far too dangerous to be allowed a second crack at DNA while the
people on this side of the Atlantic sat on their hands. After saying that I
was going to ask a Cavendish machinist to make models of the purines and
pyrimidines, I remained silent, waiting for Bragg's thoughts to congeal.

To my relief, Sir Lawrence not only made no objection but encouraged
me to get on with the job of building models. He clearly was not in sympa-
thy with the internal squabbling at King's—especially when it might allow
Linus, of all people, to get the thrill of discovering the structure of still
another important molecule. Also aiding our cause was my work on tobacco mosaic virus. It had given Bragg the impression that I was on my own.
Thus he could fall asleep that night untroubled by the nightmare that he
had given Crick carte blanche for another foray into frenzied inconsiderate-
ness. I then dashed down the stairs to the machine shop to warn them that
I was about to draw up plans for models wanted within a week.

Shortly after I was back in our office, Francis strolled in to report that their last night's dinner party was a smashing success. Odile was positively enchanted with the French boy that my sister had brought along. A month previously Elizabeth had arrived for an indefinite stay on her way back to the States. Luckily I could both install her in Camille Prior's boarding house and arrange to take my evening meals there with Pop and her foreign girls. Thus in one blow Elizabeth had been saved from typical English digs, while I looked forward to a lessening of my stomach pains.

Also living at Pop's was Bertrand Fourcade, the most beautiful male, if not person, in Cambridge. Bertrand, then visiting for a few months to perfect his English, was not unconscious of his unusual beauty and so welcomed the companionship of a girl whose dress was not in shocking contrast with his well-cut clothes. As soon as I had mentioned that we knew the handsome foreigner, Odile expressed delight. She, like many Cambridge women, could not take her eyes off Bertrand whenever she spotted him walking down King's Parade or standing about looking very well-favored during the intermissions of plays at the amateur dramatic club. Elizabeth was thus given the task of seeing whether Bertrand would be free to join us for a meal with the Cricks at Portugal Place. The time finally arranged, however, had overlapped my visit to London. When I was watching Maurice meticulously finish all the food on his plate, Odile was admiring Bertrand's perfectly proportioned face as he spoke of his problems choosing among potential social engagements during his forthcoming summer on the Riviera.

This morning Francis saw that I did not have my usual interest in the French moneyed gentry. Instead, for a moment he feared that I was going to be unusually tiresome. Reporting that even a former birdwatcher could now solve DNA was not the way to greet a friend bearing a slight hangover. However, as soon as I revealed the B-pattern details, he knew I was not pulling his leg. Especially important was my insistence that the meridional reflection at 3.4 Å was much stronger than any other reflection. This could only mean that the 3.4 Å-thick purine and pyrimidine bases were stacked on top of each other in a direction perpendicular to the helical axis. In addition we could feel sure from both electron-microscope and X-ray evidence that the helix diameter was about 20 Å.

Francis, however, drew the line against accepting my assertion that the repeated finding of twoness in biological systems told us to build two-chain models. The way to get on, in his opinion, was to reject any argument which did not arise from the chemistry of nucleic-acid chains. Since the experimental evidence known to us could not yet distinguish between two- and three-chain models, he wanted to pay equal attention to both alternatives. Though I remained totally skeptical, I saw no reason to contest his words. I would of course start playing with two-chain models.

No serious models were built, however, for several days. Not only did we lack the purine and pyrimidine components, but we had never had the shop put together any phosphorus atoms. Since our machinist needed at least three days merely to turn out the more simple phosphorus atoms, I went back to Clare after lunch to hammer out the final draft of my genetics manuscript. Later, when I cycled over to Pop's for dinner, I found Bertrand and my sister talking to Peter Pauling, who the week before had charmed Pop into giving him dining rights. In contrast to Peter, who was complaining that the Perutzes had no right to keep Nina home on a Saturday night, Bertrand and Elizabeth looked pleased with themselves. They had just returned from motoring in a friend's Rolls to a celebrated country house near Bedford. Their host, an antiquarian architect, had never truckled under to modern civilization and kept his house free of gas and electricity. In all ways possible he maintained the life of an eighteenth-century squire, even to providing special walking sticks for his guests as they accompanied him around his grounds.

Dinner was hardly over before Bertrand whisked Elizabeth on to another party, leaving Peter and me at a loss for something to do. After first deciding to work on his hi-fi set, Peter came along with me to a film. This kept us in check until, as midnight approached, Peter held forth on how Lord Rothschild was avoiding his responsibility as a father by not inviting him to dinner with his daughter Sarah. I could not disagree, for if Peter moved into the fashionable world I might have a chance to escape acquiring a faculty-type wife.

Three days later the phosphorus atoms were ready, and I quickly strung together several short sections of the sugar-phosphate backbone. Then for a day and a half I tried to find a suitable two-chain model with the backbone in the center. All the possible models compatible with the B-form X-ray data, however, looked stereochemically even more unsatisfactory than our three-chained models of fifteen months before. So, seeing Francis absorbed by his thesis, I took off the afternoon to play tennis with Bertrand. After tea I returned to point out that it was lucky I found tennis more pleasing than model building. Francis, totally indifferent to the perfect spring day, immediately put down his pencil to point out that not only was DNA very important, but he could assure me that someday I would discover the unsatisfactory nature of outdoor games.

During dinner at Portugal Place I was back in a mood to worry about what was wrong. Though I kept insisting that we should keep the backbone in the center, I knew none of my reasons held water. Finally over coffee I admitted that my reluctance to place the bases inside partially arose from the suspicion that it would be possible to build an almost infinite number of models of this type. Then we would have the impossible task of deciding whether one was right. But the real stumbling block was the bases. As long as they were outside, we did not have to consider them. If they were pushed inside, the frightful problem existed of how to pack together two or more chains with irregular sequences of bases. Here Francis had to admit that he saw not the slightest ray of light. So when I walked up out of their basement dining room into the street, I left Francis with the impression that he would have to provide at least a semiplausible argument before I would seriously play about with base-centered models.

The next morning, however, as I took apart a particularly repulsive backbone-centered molecule, I decided that no harm could come from spending a few days building backbone-out models. This meant temporarily ignoring the bases, but in any case this had to happen since now another week was required before the shop could hand over the flat tin plates cut in the shapes of purines and pyrimidines.

There was no difficulty in twisting an externally situated backbone into a shape compatible with the X-ray evidence. In fact, both Francis and I had the impression that the most satisfactory angle of rotation between two adjacent bases was between 30 and 40 degrees. In contrast, an angle either twice as large or twice as small looked incompatible with the relevant bond angles. So if the backbone was on the outside, the crystallographic repeat of 34 Å had to represent the distance along the helical axis required for a complete rotation. At this stage Francis' interest began to perk up, and at increasing frequencies he would look up from his calculations to glance at the model. Nonetheless, neither of us had any hesitation in breaking off work for the weekend. There was a party at Trinity on Saturday night, and on Sunday Maurice was coming up to the Cricks' for a social visit arranged weeks before the arrival of the Pauling manuscript.

Maurice, however, was not allowed to forget DNA. Almost as soon as he arrived from the station, Francis started to probe him for fuller details of the B pattern. But by the end of lunch Francis knew no more than I had picked up the week before. Even the presence of Peter, saying he felt sure his father would soon spring into action, failed to ruffle Maurice's plans. Again he emphasized that he wanted to put off more model building until Rosy was gone, six weeks from then. Francis seized the occasion to ask Maurice whether he would mind if we started to play about with DNA models. When Maurice's slow answer emerged as no, he wouldn't mind, my pulse rate returned to normal. For even if the answer had been yes, our model building would have gone ahead.

The next few days saw Francis becoming increasingly agitated by my failure to stick close to the molecular models. It did not matter that before his tenish entrance I was usually in the lab. Almost every afternoon, knowing that I was on the tennis court, he would fretfully twist his head away from his work to see the polynucleotide backbone unattended. Moreover, after tea I would show up for only a few minutes of minor fiddling before dashing away to have sherry with the girls at Pop's. Francis' grumbles did not disturb me, however, because further refining of our latest backbone without a solution to the bases would not represent a real step forward.

I went ahead spending most evenings at the films, vaguely dreaming that any moment the answer would suddenly hit me. Occasionally my wild pursuit of the celluloid backfired, the worst occasion being an evening set aside for *Ecstasy*. Peter and I had both been too young to observe the origi-

nal showings of Hedy Lamarr's romps in the nude, and so on the long-
awaited night we collected Elizabeth and went up to the Rex. However, the
only swimming scene left intact by the English censor was an inverted
reflection from a pool of water. Before the film was half over we joined the
violent booing of the disgusted undergraduates as the dubbed voices
uttered words of uncontrolled passion.

Even during good films I found it almost impossible to forget the bases.
The fact that we had at last produced a stereochemically reasonable configuration for the backbone was always in the back of my head. Moreover,
there was no longer any fear that it would be incompatible with the experimental data. By then it had been checked out with Rosy's precise measurements. Rosy, of course, did not directly give us her data. For that matter, no
one at King's realized they were in our hands. We came upon them
because of Max's membership on a committee appointed by the Medical
Research Council to look into the research activities of Randall's lab to coordinate Biophysics research within its laboratories. Since Randall wished to
convince the outside committee that he had a productive research group,
he had instructed his people to draw up a comprehensive summary of their
accomplishments. In due time this was prepared in mimeograph form and
sent routinely to all the committee members. The report was not confiden-
tial and so Max saw no reason not to give it to Francis and me. Quickly
scanning its contents, Francis sensed with relief that following my return
from King's I had correctly reported to him the essential features of the B
pattern. Thus only minor modifications were necessary in our backbone
configuration.

Generally, it was late in the evening after I got back to my rooms that I
tried to puzzle out the mystery of the bases. Their formulas were written
out in J.N. Davidson's little book *The Biochemistry of Nucleic Acids*, a copy
of which I kept in Clare. So I could be sure that I had the correct structures
when I drew tiny pictures of the bases on sheets of Cavendish notepaper.
My aim was somehow to arrange the centrally located bases in such a way
that the backbones on the outside were completely regular—that is, giving
the sugar-phosphate groups of each nucleotide identical three-dimensional
configurations. But each time I tried to come up with a solution I ran into
the obstacle that the four bases each had a quite different shape. Moreover,
there were many reasons to believe that the sequences of the bases of a
given polynucleotide chain were very irregular. Thus, unless some very special trick existed, randomly twisting two polynucleotide chains around one
another should result in a mess. In some places the bigger bases must
touch each other, while in other regions, where the smaller bases would lie
opposite each other, there must exist a gap or else their backbone regions
must buckle in.

There was also the vexing problem of how the intertwined chains might
be held together by hydrogen bonds between the bases. Though for over a
year Francis and I had dismissed the possibility that bases formed regular

hydrogen bonds, it was now obvious to me that we had done so incorrectly. The observation that one or more hydrogen atoms on each of the bases
could move from one location to another (a tautomeric shift) had initially
led us to conclude that all the possible tautomeric forms of a given base
occurred in equal frequencies. But a recent rereading of J.M. Gulland's and
D.O. Jordan's papers on the acid and base titrations of DNA made me finally appreciate the strength of their conclusion that a large fraction, if not all,
of the bases formed hydrogen bonds to other bases. Even more important,
these hydrogen bonds were present at very low DNA concentrations,
strongly hinting that the bonds linked together bases in the same molecule.
There was in addition the X-ray crystallographic result that each pure base
so far examined formed as many irregular hydrogen bonds as stereochemically possible. Thus, conceivably the crux of the matter was a rule govern-
ing hydrogen bonding between bases.

My doodling of the bases on paper at first got nowhere, regardless of
whether or not I had been to a film. Even the necessity to expunge *Ecstasy*
from my mind did not lead to passable hydrogen bonds, and I fell asleep
hoping that an undergraduate party the next afternoon at Downing would
be full of pretty girls. But my expectations were dashed as soon as I arrived
to spot a group of healthy hockey players and several pallid debutantes.
Bertrand also instantly perceived he was out of place, and as we passed a
polite interval before scooting out, I explained how I was racing Peter's
father for the Nobel Prize.

Not until the middle of the next week, however, did a nontrivial idea
emerge. It came while I was drawing the fused rings of adenine on paper.
Suddenly I realized the potentially profound implications of a DNA structure
in which the adenine residue formed hydrogen bonds similar to those
found in crystals of pure adenine. If DNA was like this, each adenine
residue would form two hydrogen bonds to an adenine residue related to it
by a 180-degree rotation. Most important, two symmetrical hydrogen bonds
could also hold together pairs of guanine, cytosine, or thymine. I thus started wondering whether each DNA molecule consisted of two chains with
identical base sequences held together by hydrogen bonds between pairs of
identical bases. There was the complication, however, that such a structure
could not have a regular backbone, since the purines (adenine and guanine) and the pyrimidines (thymine and cytosine) have different shapes.
The resulting backbone would have to show minor in-and-out buckles
depending upon whether pairs of purines or pyrimidines were in the center.

Despite the messy backbone, my pulse began to race. If this was DNA, I
should create a bombshell by announcing its discovery. The existence of
two intertwined chains with identical base sequences could not be a chance
matter. Instead it would strongly suggest that one chain in each molecule
had at some earlier stage served as the template for the synthesis of the
other chain. Under this scheme, gene replication starts with the separation

of its two identical chains. Then two new daughter strands are made on the two parental templates, thereby forming two DNA molecules identical to the original molecule. Thus, the essential trick of gene replication could come from the requirement that each base in the newly synthesized chain always hydrogen-bonds to an identical base. That night, however, I could not see why the common tautomeric form of guanine would not hydrogen-bond to adenine. Likewise, several other pairing mistakes should also occur. But since there was no reason to rule out the participation of specific enzymes, I saw no need to be unduly disturbed. For example, there might exist an enzyme specific for adenine that caused adenine always to be inserted opposite an adenine residue on the template strands.

As the clock went past midnight I was becoming more and more pleased. There had been far too many days when Francis and I worried that the DNA structure might turn out to be superficially very dull, suggesting nothing about either its replication or its function in controlling cell biochemistry. But now, to my delight and amazement, the answer was turning out to be profoundly interesting. For over two hours I happily lay awake with pairs of adenine residues whirling in front of my closed eyes. Only for brief moments did the fear shoot through me that an idea this good could be wrong.

My scheme was torn to shreds by the following noon. Against me was the awkward chemical fact that I had chosen the wrong tautomeric forms of guanine and thymine. Before the disturbing truth came out, I had eaten a hurried breakfast at the Whim, then momentarily gone back to Clare to reply to a letter from Max Delbrück which reported that my manuscript on bacterial genetics looked unsound to the Cal Tech geneticists. Nevertheless, he would accede to my request that he send it to the *Proceedings of the National Academy*. In this way, I would still be young when I committed the folly of publishing a silly idea. Then I could sober up before my career was permanently fixed on a reckless course.

At first this message had its desired unsettling effect. But now, with my spirits soaring on the possibility that I had the self-duplicating structure, I reiterated my faith that I knew what happened when bacteria mated. Moreover, I could not refrain from adding a sentence saying that I had just devised a beautiful DNA structure which was completely different from Pauling's. For a few seconds I considered giving some details of what I was up to, but since I was in a rush I decided not to, quickly dropped the letter in the box, and dashed off to the lab.

The letter was not in the post for more than an hour before I knew that my claim was nonsense. I no sooner got to the office and began explaining my scheme than the American crystallographer Jerry Donohue protested that the idea would not work. The tautomeric forms I had copied out of Davidson's book were, in Jerry's opinion, incorrectly assigned. My immediate retort that several other texts also pictured guanine and thymine in the enol form cut no ice with Jerry. Happily he let out that for years organic chemists had been arbitrarily favoring particular tautomeric forms over their alternatives on only the flimsiest of grounds. In fact, organic-chemistry textbooks were littered with pictures of highly improbable tautomeric forms. The guanine picture I was thrusting toward his face was almost certainly bogus. All his chemical intuition told him that it would occur in the keto form. He was just as sure that thymine was also wrongly assigned an enol configuration. Again he strongly favored the keto alternative.

Jerry, however, did not give a foolproof reason for preferring the keto forms. He admitted that only one crystal structure bore on the problem. This was diketopiperazine, whose three-dimensional configuration had been carefully worked out in Pauling's lab several years before. Here there was no doubt that the keto form, not the enol, was present. Moreover, he felt sure that the quantum-mechanical arguments which showed why diketopiperazine has the keto form should also hold for guanine and thymine. I was thus firmly urged not to waste more time with my harebrained scheme.

Though my immediate reaction was to hope that Jerry was blowing hot air, I did not dismiss his criticism. Next to Linus himself, Jerry knew more about hydrogen bonds than anyone else in the world. Since for many years he had worked at Cal Tech on the crystal structures of small organic molecules, I couldn't kid myself that he did not grasp our problem. During the six months that he occupied a desk in our office, I had never heard him shooting off his mouth on subjects about which he knew nothing.

Thoroughly worried, I went back to my desk hoping that some gimmick might emerge to salvage the like-with-like idea. But it was obvious that the new assignments were its death blow. Shifting the hydrogen atoms to their keto locations made the size differences between the purines and pyrimidines even more important than would be the case if the enol forms existed. Only by the most special pleading could I imagine the polynucleotide backbone bending enough to accommodate irregular base sequences. Even this possibility vanished when Francis came in. He immediately realized that a like-with-like structure would give a 34 Å crystallographic repeat only if each chain had a complete rotation every 68 Å. But this would mean that the rotation angle between successive bases would be only 18 degrees, a value Francis believed was absolutely ruled out by his recent fiddling with the models. Also Francis did not like the fact that the structure gave no explanation for the Chargaff rules (adenine equals thymine, guanine equals cytosine). I, however, maintained my lukewarm response to Chargaff's data. So I welcomed the arrival of lunchtime, when Francis' cheerful prattle temporarily shifted my thoughts to why undergraduates could not satisfy *au pair* girls.

After lunch I was not anxious to return to work, for I was afraid that in trying to fit the keto forms into some new scheme I would run into a stone wall and have to face the fact that no regular hydrogen-bonding scheme was compatible with the X-ray evidence. As long as I remained outside gazing at the crocuses, hope could be maintained that some pretty base arrangement would fall out. Fortunately, when we walked upstairs, I found that I had an excuse to put off the crucial model-building step for at least several more hours. The metal purine and pyrimidine models, needed for systematically checking all the conceivable hydrogen-bonding possibilities, had not been finished on time. At least two more days were needed before they would be in our hands. This was much too long even for me to remain in limbo, so I spent the rest of the afternoon cutting accurate representations of the bases out of stiff cardboard. But by the time they were ready I realized that the answer must be put off till the next day. After dinner I was to join a group from Pop's at the theater.

When I got to our still empty office the following morning, I quickly cleared away the papers from my desk top so that I would have a large, flat surface on which to form pairs of bases held together by hydrogen bonds. Though I initially went back to my like-with-like prejudices, I saw all too well that they led nowhere. When Jerry came in I looked up, saw that it was not Francis, and began shifting the bases in and out of various other pairing possibilities. Suddenly I became aware that an adenine-thymine pair held together by two hydrogen bonds was identical in shape to a guanine-cytosine pair held together by at least two hydrogen bonds. All the hydrogen bonds seemed to form naturally; no fudging was required to make the two types of base pairs identical in shape. Quickly I called Jerry over to ask him whether this time he had any objection to my new base pairs.

When he said no, my morale skyrocketed, for I suspected that we now had the answer to the riddle of why the number of purine residues exactly equaled the number of pyrimidine residues. Two irregular sequences of bases could be regularly packed in the center of a helix if a purine always hydrogen-bonded to a pyrimidine. Furthermore, the hydrogen-bonding requirement meant that adenine would always pair with thymine, while guanine could pair only with cytosine. Chargaff's rules then suddenly stood out as a consequence of a double-helical structure for DNA. Even more exciting, this type of double helix suggested a replication scheme much more satisfactory than my briefly considered like-with-like pairing. Always pairing adenine with thymine and guanine with cytosine meant that the base sequences of the two intertwined chains were complementary to each other. Given the base sequence of one chain, that of its partner was automatically determined. Conceptually, it was thus very easy to visualize how a single chain could be the template for the synthesis of a chain with the complementary sequence.

Upon his arrival Francis did not get more than halfway through the door before I let loose that the answer to everything was in our hands. Though as a matter of principle he maintained skepticism for a few moments, the similarly shaped A-T and G-C pairs had their expected impact. His quickly pushing the bases together in a number of different ways did not reveal any other way to satisfy Chargaff's rules. A few minutes later he spotted the fact that the two glycosidic bonds (joining base and sugar) of each base pair were systematically related by a diad axis perpendicular to the helical axis. Thus, both pairs could be flipflopped over and still have their glycosidic bonds facing in the same direction. This had the important consequence that a given chain could contain both purines and pyrimidines. At the same time, it strongly suggested that the backbones of the two chains must run in opposite directions.

The question then became whether the A-T and G-C base pairs would easily fit the backbone configuration devised during the previous two weeks. At first glance this looked like a good bet, since I had left free in the center a large vacant area for the bases. However, we both knew that we would not be home until a complete model was built in which all the stereochemical contacts were satisfactory. There was also the obvious fact that the implications of its existence were far too important to risk crying wolf. Thus I felt slightly queasy when at lunch Francis winged into the Eagle to tell everyone within hearing distance that we had found the secret of life.

Understanding What You've Read

How did Watson determine how the bases fit together?

Why did Watson think that his base-pairing scheme made more sense than his older, incorrect scheme?

Write a short essay about Watson's claim that "we had found the secret of life." One approach might be that the secret of life goes beyond any biochemical explanation. An alternative approach could be that the better we understand the biochemistry of life, the better we will be able to appreciate it. Whatever approach you choose, suggest a way of interpreting what is meant by the "secret of life."

The Double Helix was written to be a best seller, and it was. How would you describe Watson's writing style? How does it differ from that of a biology textbook? Which style do you like best? Why?

Many people view scientists as bland, unemotional people. However, in *The Double Helix*, Watson shows that scientists can be fiercely competitive, sexist, abusive, and narrow-minded. Which is accurate? What are your stereotypes of scientists?

As mentioned in the Introduction, no scientist works in isolation. What scientists contributed to Watson's and Crick's success?

Lewis Thomas

Germs

The late Lewis Thomas (b. 1913) was a research physician who studied hypersensitivity, the pathogenicity of mycoplasmas, and infectious diseases. Most of his essays were originally published in *The New England Journal of Medicine*, and later in books. The first of these books was *Lives of a Cell: Notes of a Biology Watcher*, which won the National Book Award in 1975.

Watching television, you'd think we lived at bay, in total jeopardy, surrounded on all sides by human-seeking germs, shielded against infection and death only by a chemical technology that enables us to keep killing them off. We are instructed to spray disinfectants everywhere, into the air of our bedrooms and kitchens and with special energy into bathrooms, since it is our very own germs that seem the worst kind. We explode clouds of aerosol, mixed for good luck with deodorants, into our noses, mouths, underarms, privileged crannies—even into the intimate insides of our telephones. We apply potent antibiotics to minor scratches and seal them with plastic. Plastic is the new protector; we wrap the already plastic tumblers of hotels in more plastic, and seal the toilet seats like state secrets after irradiating them with ultraviolet light. We live in a world where the microbes are always trying to get at us, to tear us cell from cell, and we only stay alive and whole through diligence and fear.

We still think of human disease as the work of an organized, modernized kind of demonology, in which the bacteria are the most visible and centrally placed of our adversaries. We assume that they must somehow relish what they do. They come after us for profit, and there are so many of them that disease seems inevitable, a natural part of the human condition; if we succeed in eliminating one kind of disease there will always be a new one at hand, waiting to take its place.

These are paranoid delusions on a societal scale, explainable in part by our need for enemies, and in part by our memory of what things used to be like. Until a few decades ago, bacteria were a genuine household threat, and although most of us survived them, we were always aware of the nearness of death. We moved, with our families, in and out of death. We had lobar pneumonia, meningococcal meningitis, streptococcal infections, diphtheria, endocarditis, enteric fevers, various septicemias, syphilis, and, always, everywhere, tuberculosis. Most of these have now left most of us, thanks to antibiotics, plumbing, civilization, and money, but we remember.

In real life, however, even in our worst circumstances we have always been a relatively minor interest of the vast microbial world. Pathogenicity is not the rule. Indeed, it occurs so infrequently and involves such a relatively small number of species, considering the huge population of bacteria on the earth, that it has a freakish aspect. Disease usually results from inconclusive negotiations for symbiosis, an overstepping of the line by one side or the other, a biologic misinterpretation of borders.

Some bacteria are only harmful to us when they make exotoxins, and they only do this when they are, in a sense, diseased themselves. The toxins of diphtheria bacilli and streptococci are produced when the organisms have been infected by bacteriophage; it is the virus that provides the code for toxin. Uninfected bacteria are uninformed. When we catch diphtheria it is a virus infection, but not of us. Our involvement is not that of an adversary in a straightforward game, but more like blundering into someone else's accident.

I can think of a few microorganisms, possibly the tubercle bacillus, the syphilis spirochete, the malarial parasite, and a few others, that have a selective advantage in their ability to infect human beings, but there is nothing to be gained, in an evolutionary sense, by the capacity to cause illness or death. Pathogenicity may be something of a disadvantage for most microbes, carrying lethal risks more frightening to them than to us. The man who catches a meningococcus is in considerably less danger for his life, even without chemotherapy, than meningococci with the bad luck to catch a man. Most meningococci have the sense to stay out on the surface, in the rhinopharynx. During epidemics this is where they are to be found in the majority of the host population, and it generally goes well. It is only in the unaccountable minority, the "cases," that the line is crossed, and then there is the devil to pay on both sides, but most of all for the meningococci.

Staphylococci live all over us, and seem to have adapted to conditions in our skin that are uncongenial to most other bacteria. When you count them up, and us, it is remarkable how little trouble we have with the relation. Only a few of us are plagued by boils, and we can blame a large part of the destruction of tissues on the zeal of our own leukocytes. Hemolytic streptococci are among our closest intimates, even to the extent of sharing antigens with the membranes of our muscle cells; it is our reaction to their presence, in the form of rheumatic fever, that gets us into trouble. We can carry brucella for long periods in the cells of our reticuloendothelial system without any awareness of their existence; then cyclically, for reasons not understood but probably related to immunologic reactions on our part, we sense them, and the reaction of sensing is the clinical disease.

Most bacteria are totally preoccupied with browsing, altering the configurations of organic molecules so that they become usable for the energy
needs of other forms of life. They are, by and large, indispensable to each
other, living in interdependent communities in the soil or sea. Some have
become symbionts in more specialized, local relations, living as working
parts in the tissues of higher organisms. The root nodules of legumes would
have neither form nor function without the masses of rhizobial bacteria
swarming into root hairs, incorporating themselves with such intimacy that
only an electron microscope can detect which membranes are bacterial and
which plant. Insects have colonies of bacteria, the mycetocytes, living in
them like little glands, doing heaven knows what but being essential. The
microfloras of animal intestinal tracts are part of the nutritional system. And
then, of course, there are the mitochondria and chloroplasts, permanent res-
idents in everything.

The microorganisms that seem to have it in for us in the worst way—the
ones that really appear to wish us ill—turn out on close examination to be
rather more like bystanders, strays, strangers in from the cold. They will
invade and replicate if given the chance, and some of them will get into
our deepest tissues and set forth in the blood, but it is our response to their
presence that makes the disease. Our arsenals for fighting off bacteria are
so powerful, and involve so many different defense mechanisms, that we
are in more danger from them than from the invaders. We live in the midst
of explosive devices; we are mined.

It is the information carried by the bacteria that we cannot abide.

The gram-negative bacteria are the best examples of this. They display
lipopolysaccharide endotoxin in their walls, and these macromolecules are
read by our tissues as the very worst of bad news. When we sense
lipopolysaccharide, we are likely to turn on every defense at our disposal;
we will bomb, defoliate, blockade, seal off, and destroy all the tissues in
the area. Leukocytes become more actively phagocytic, release lysosomal
enzymes, turn sticky, and aggregate together in dense masses, occluding
capillaries and shutting off the blood supply. Complement is switched on at
the right point in its sequence to release chemotactic signals, calling in
leukocytes from everywhere. Vessels become hyperreactive to epinephrine
so that physiologic concentrations suddenly possess necrotizing properties.
Pyrogen is released from leukocytes, adding fever to hemorrhage, necrosis,
and shock. It is a shambles.

All of this seems unnecessary, panic-driven. There is nothing intrinsically
poisonous about endotoxin, but it must look awful, or feel awful when
sensed by cells. Cells believe that it signifies the presence of gram-negative
bacteria, and they will stop at nothing to avoid this threat.

I used to think that only the most highly developed, civilized animals
could be fooled in this way, but it is not so. The horseshoe crab is a primitive fossil of a beast, ancient and uncitified, but he is just as vulnerable to

disorganization by endotoxin as a rabbit or a man. Bang has shown that an injection of a very small dose into the body cavity will cause the aggregation of hemocytes in ponderous, immovable masses that block the vascular channels, and a gelatinous clot brings the circulation to a standstill. It is now known that a limulus clotting system, perhaps ancestral to ours, is centrally involved in the reaction. Extracts of the hemocytes can be made to jell by adding extremely small amounts of endotoxin. The self-disintegration of the whole animal that follows a systemic injection can be interpreted as a well-intentioned but lethal error. The mechanism is itself quite a good one, when used with precision and restraint, admirably designed for coping with intrusion by a single bacterium: the hemocyte would be attracted to the site, extrude the coagulable protein, the microorganism would be entrapped and immobilized, and the thing would be finished. It is when confronted by the overwhelming signal of free molecules of endotoxin, evoking memories of vibrios in great numbers, that the limulus flies into panic, launches all his defenses at once, and destroys himself.

It is, basically, a response to propaganda, something like the panic-producing pheromones that slave-taking ants release to disorganize the colonies of their prey.

I think it likely that many of our diseases work in this way. Sometimes, the mechanisms used for overkill are immunologic, but often, as in the limulus model, they are more primitive kinds of memory. We tear ourselves to pieces because of symbols, and we are more vulnerable to this than to any host or predators. We are, in effect, at the mercy of our own Pentagons, most of the time.

Understanding What You've Read

How does Thomas achieve such a graceful style?

Thomas has been described as a scientist with the mind of a humanist. Based on this essay, do you agree? Why or why not?

Why do you think that Thomas calls our fears of microbes "paranoid delusions on a societal scale"? Is this an accurate evaluation? Why or why not?

Thomas said, "Earth, the only truly closed ecosystem any of us knows, is an organism." What did he mean? Write a short essay describing your answer.

Exercises

1. In *The Double Helix*, James Watson writes about the relationship between meaning, style, and an author's personality:

By the time I was back in Copenhagen, the journal containing Linus' article had arrived from the states. I quickly read it and immediately reread it. Most of the language was above me, and so I could only get a general impression of his argument. I had no way of judging whether it made sense. The only thing I was sure of was that it was written with style. A few days later the next issue of the journal arrived, this time containing seven more Pauling articles. Again the language was dazzling and full of rhetorical tricks. One article started with the phrase, "Collagen is a very interesting protein." It inspired me to compose opening lines of the paper I would write about DNA, if I solved its structure. A sentence like "Genes are interesting to geneticists" would distinguish my way of thought from Pauling's.

Write a one-sentence summary of this paragraph. What are the implications of Watson's message?

2. Find a paragraph in a textbook that you find hard to read, not because of substance, but because of expression. Why is the paragraph hard to understand? Rewrite the paragraph to improve its readability.

3. Write an essay summarizing a scientific or socio-scientific controversy such as recombinant DNA, societal control of science, or cloning as a means of human reproduction. Your audience for this essay is the layperson. Do not tangle yourself in the argument. Rather, discuss the issues and facts on which the controversy rests. Where facts are contradictory, say so, but do not try to resolve the argument.

4. Russell Doolittle used these words to help readers understand the role of platelets in blood-clotting:

> Prick us and we bleed, but the bleeding stops; the blood clots. The sticky cell fragments called platelets clump at the site of the puncture, partially sealing the leak.

If Doolittle had written "Platelets help our blood to clot," he would have been accurate. Why, then, didn't he?

5. Select a science-related article from a recent issue of *Scientific American* or *Natural History*. Assume that you are the journal's editor and must decide whether to publish the article. Write an essay summarizing your evaluation of the article and stating your recommendation regarding its publication.

6. Discuss, support, or refute the ideas in these quotations:

> We, too, are silly enough to believe that all nature is intended for our benefit. — Bernard Le Bovier Sieur de Fontenelle

> Science when well digested is nothing but good sense and reason. — Stanislaw Leszczynski

> Equipped with his five senses, man explores the universe around him and calls the adventure Science. — Edwin Powell Hubble

> Science may be learned by rote, but Wisdom not. — Lawrence Sterne

> Truth comes out of error more readily than out of confusion. — Francis Bacon (1561-1626)

What is called science today consists of a haphazard heap of information, united by nothing, often unnecessary, and not only failing to present one unquestionable truth, but as often as not containing the grossest errors, today put forward as truths, and tomorrow overthrown. — Leo N. Tolstoy (1828-1910)

It is the weight, not numbers of experiments that is to be regarded. — Isaac Newton

Science is a way of thinking much more than it is a body of knowledge. — Carl Sagan

Imagination is more important than knowledge. — Albert Einstein (1879-1955)

The important thing in science is not so much to obtain new facts as to discover new ways of thinking about them. — [Sir] William Lawrence Bragg (1890-1971)

CHAPTER SEVEN

Writing for a

Professional Audience

Without publication, science is dead.
— Gerald Piel

To get to know, to discover, to publish—this is the
destiny of a scientist.
— Francois Arago

Only when he has published his ideas and findings
has the scientist made his scientific contribution, and
only when he has thus made it a part of the public
domain of science can he truly lay claim to it as his.
For his claim resides only in the recognition accorded by peers in the social system of science through
reference to his work.
— R.K. Merton

Doing an experiment is not more important than
writing.
— E.G. Boring

First, have something to say; second, say it; third,
stop when you have said it; and finally, give it an
accurate title.
— John Shaw Billings

Work, finish, publish.
— Michael Faraday

I have learned that when I write a research paper I
do far more than summarize conclusions already
neatly stored in my mind. Rather, the writing process
is where I carry out the final comprehension, analysis, and synthesis of my results.
— Sidney Perkowitz

Scientific research begins with a set of sentences
which point the way to certain observations and
experiments, the results of which do not become
fully scientific until they have been turned back into
language, yielding again a set of sentences which
then become the basis of further exploration into
the unknown . . .
— Benjamin Lee Whorf

What is a Research Paper?

A research paper is a written, published, and readily available report describing how the scientific method was used to study a problem. It is a repository of information—a permanent record describing repeatable experiments and original observations. It assesses observations, supports conclusions with data, evaluates conclusions, and describes the significance of the research. Research papers resemble a mystery with a solution documented by evidence, and form the foundation of science. Indeed, 80% of all references in the scientific literature are to journal articles, and 85% of biomedical scientists rank research papers in journals as their most frequently used source of information for staying up-to-date.[1] To be published in a journal, a paper must be first judged worthy of publication by experts in the field. This so-called "peer review" improves the quality of a published work and is a hallmark of quality journals.

The standard research paper was developed in the 17th century by Henry

---

[1]Levitan, Karen B. 1979. "Scientific Societies and Their Journals: Biomedical Scientists Assess the Relationship." *Social Studies of Science* 9: 393–400.

Oldenberg, secretary of the Royal Society of London.[2] Today, these papers are the common currency of all science. New journals appear every few months, and scientists' mailboxes regularly fill with announcements of new books and monographs. To stay up-to-date in their teaching and research, scientists must read a seemingly overwhelming number of books, papers, and journals.[3]

If you want a career involving scientific research, it's critical that you learn to write and publish research papers. This is largely an exercise in something that you studied in Chapter 2: Organization.

How to Write a Research Paper

Science is a collaborative enterprise based on the open sharing of information that can be tested and built upon. The vehicle for sharing this information is the research paper published in a peer-reviewed journal such as *Science* or *Nature*. Research is completed only when results are published and made available to others.

Scientists can't read all the scientific journals that are published. Even reading a few of these journals takes a tremendous amount of time and requires that scientists gather information efficiently. To do this, scientists have developed a clear, standard format for reporting their discoveries. This prescribed format distinguishes scientific writing from literature and helps readers quickly understand what is being reported in a paper. Everything in a research paper must fit into one of the following categories:

Title

Authors and Affiliations

Abstract

Introduction

Materials and Methods

Results

[2]*Philosophical Transactions*, the journal of the Royal Society of London, was founded by Oldenberg in 1665 and is the oldest scientific journal. Isaac Newton's discovery of the compound nature of sunlight, published in the February 19, 1672 issue of *Philosophical Transactions*, was the first major scientific discovery to be announced in a journal rather than in a book. Newton's paper was published 11 days after its receipt.

[3]A typical medical school library receives about 50 journals each day that require 61 cm (2 ft) of shelf space. Also, the number of scientific journals has doubled during the past decade. If this trend continues, in the year 2010 libraries will need 96,770 km (60,000 miles) of new shelf space each year for the new journals. Since 1960, the number of journals has increased by more than 2% per year, a bit slower than the growth in the number of scientists. This growth has occurred despite claims by scientists that few papers are innovative, that most papers are concerned only with details, and that the overall quality of research papers is poor. For more information, see Moran, Jeffrey B. 1989. "The Journal Glut: Scientific Publications Out of Control." *The Scientist* 10 July 1989, p. 11; Waksman, B.H. 1980. "Information Overload in Immunology: Possible Solutions to the Problem of Excessive Publication." *Journal of Immunology* 124: 1009–1015.

Discussion

Literature Cited

These categories are not just names. Rather, they are guidelines that define the strategy and direction of your paper: They tell you what to put in and what to leave out of a paper. This structure has not resulted merely from chance or tradition. Indeed, it reflects the logic and elegance of the scientific method by sequentially progressing from the nature of a problem to the problem's solution. This uniform organization also helps scientists quickly and efficiently determine the relevance of papers they read to their work or interests. To publish a paper, and therefore to advance science, you must understand how scientific papers are organized.

Title

The title of a paper is similar to that of a newspaper article. It is a short label (usually less than 12 words) that helps readers decide if they will read your article. The title is your first chance to catch the reader's attention. Consequently, the title should indicate what the paper is about and thus ensure that your article is read by its intended audience. A title must be informative, specific, concise, and represent the content of the paper.

Thousands of scientists may read the title of your paper, either in the journal or in indexing publications such as *Current Contents*. Therefore, choose the title of your paper carefully. The title of your paper should identify precisely the main topic of your article. Trying to write an informative title is an effective test of whether your research has led to a definite conclusion. If you can't write a specific and informative title, you may not be sure of what you have to say. A vague title will cause your paper to miss its intended audience, thereby reducing the impact of your work.

Here's how to write an effective title for your paper:

Use the fewest number of words possible to adequately describe the content of your paper. Delete superfluous phrases such as "Preliminary Studies of," "Aspects of," "Contributions to," and "Observations of" that tell the reader nothing. Similarly, avoid wordy and dull titles. "An Overview of the Structural and Conceptual Characteristics and Associated Features of Future Planning Systems for Scientific Research" strikes out on both counts. "Planning Scientific Research" might get you to first base.

Make the title specific and informative. Avoid vague titles such as, "Rock Formations in Wyoming." What kind of rock formations? Where in Wyoming?

Include taxonomic information where appropriate, and delete abbreviations, jargon, proprietary names, and chemical formulas.

Include the key words of the paper in the title.

If required to do so by the journal, provide a shortened title that will be used as a "running head" atop pages.

Not following these guidelines will lessen the impact of your paper. It may also produce a ridiculous title. For example, an article in the April 1991 issue of *BioScience* titled "Nematodes Win Prize at National Hardware Show" must have disappointed readers hoping that the prize would go to the raccoon's entry. Similarly, a paper published in *Clinical Research* was titled "Preliminary Canine and Clinical Evaluation of a New Antitumor Agent, Streptovitacin." When that pooch is through checking out streptovitacin, I hope it will drop by my office to help me with some of my research.

Authors and their Affiliations

The authors of a paper are the people who actively contributed to the design, execution, or analysis of the experiments and who take public intellectual responsibility for the paper. Authors are usually listed in the order of their importance to the work. The author listed first is the "senior author" and is the person who contributed most to the research. Similarly, the author listed last contributed least to the work.[4]

The "publish or perish" philosophy for promotion and tenure has increased the importance of authorship of scientific papers (see next page). The increased number of authors per paper reflects, in part, the increasing number of collaborations typical of modern science. This has increased the average number of authors of scientific papers from 1.8 per paper in 1955 to 3.5 today. Some papers have huge numbers of authors. For example, a paper published in *Physical Review Letters* lists 27 authors, yet is only 12 paragraphs long. Similarly, the first paper to describe a genetic linkage map had only one author,[5] whereas a more recent paper describing a linkage map of the human genome had 33 authors.[6] But if the authorship of these papers seems unusual, consider the papers that resulted from an exhausting 17-year search for the top quark. One paper describing the find, published in the April 3, 1995 issue of *Physical Review Letters*, devoted the first two of its six pages to the names of all 437 authors and their 35 institutions. A second paper, published in the same issue, listed 403 authors and their 42 institutions. In 1994 alone, there were 18 papers having more than 500 authors, and one paper had 972 authors. As one professor asked, "Which author is responsible for which word?" To combat this problem, some journals limit the number of authors of a paper; for example, the *New England Journal of Medicine* lists no more than 12 coauthors.

[4]Some journals have tried to avoid the ranking of authors by listing authors alphabetically. Interestingly, these journals received fewer papers from authors whose last names start with P–Z.

[5]Sturtevant, A.H. 1913. "The Linear Arrangement of Six Sex-Linked Factors in *Drosophila*, as Shown by Their Mode of Association." *Journal of Experimental Zoology* 14: 43-59.

[6]Donis-Keller, H. P. Green, C. Helms, et al. 1987. "A Genetic Linkage Map of the Human Genome." *Cell* 51: 319–337.

Publish or Perish

Scientists at most major universities are under tremendous pressure to regularly publish the results of their research in peer-reviewed journals. The motive for this "publish or perish" pressure is that more papers presumably mean more prestige for a researcher's university, and that prestige, administrators hope, will translate into more grant money for the university. This pressure to publish has many positive effects. For example, it encourages researchers to start projects that will generate publications that contribute to our understanding of nature. However, the pressure to publish also has negative effects. At the very least, publish or perish encourages scientists to delay more important work to prepare material for publication that otherwise might not be submitted. At its worst, the pressure to publish promotes disreputable, unethical, and even fraudulent practices.

Many universities evaluate a scientist's productivity by counting the papers he or she has published. Not surprisingly, some scientists have used different strategies to increase their number of publications. Some have published the same material more than once. This self-plagiarism indicates a lack of scientific objectivity, dilutes the literature, and pollutes the publication record of the scientist. A much more common strategy has been to publish several small papers rather than one large paper. This "salami science" of slicing results into increasingly smaller pieces helps explain why almost half of the peer-reviewed research papers that appeared in the top 4,500 (of 74,000) scientific journals between 1981 and 1985 were never cited in the following five years.[1] Most of these uncited papers amount to little more

[1]Institute of Scientific Information, as cited in Begley, Sharon. 1991. "Gridlock in the Labs." *Newsweek* 14 January 1991.

However, authors are sometimes added to papers as a courtesy, thus explaining why the authorship of some papers reads like a laundry list or routing slip of a department or building. Sometimes, lab directors declare themselves an author despite not having done any of the research—a practice similar to the custom of "the medieval seigneur, whose prerogative it was to spend the marriage night with each new bride in his domain."[7]

Decide the authorship of a paper before you start writing the paper, even if the decision is tentative. Authors should know why and how the observations were made, how the conclusions were made from the observations, and how to defend the work against criticism. Furthermore, each author should have participated in the design, observation, and interpretation of the work. People who provide space, money, and routine technical assistance should be acknowledged for their contributions (see below), but usually should not be listed as authors of the paper.

[7]Chernin, E. 1981. "First Do No Harm." In: Warren, K.S., ed. *Coping with the Biomedical Literature.* New York: Praeger.

than aggravating background noise in journals. Most of these papers are weak on objectivity, replicability, importance, competence, intelligibility, and efficiency.[2] These papers usually avoid important problems, do not challenge current beliefs, use complex methods, and are written obtusely to mask insignificance.

A final strategy to increase one's publication record has been to increase the number of authors on papers. When coauthors reciprocate the gratuity, they greatly increase their number of publications—all of the authors count the shared publications in their own "scores"—without increasing their workload. For example, the percentage of papers published with multiple authors in the *New England Journal of Medicine* has increased from 1.5% in 1886 (when the publication was called the *Boston Medical and Surgical Report*) to 96% in 1977. Some of this change is due to the shifting nature of scientific inquiry. However, much of the change is also due to the gratuitous inclusion of extra authors on publications (for a more comprehensive discussion of authorship, see pp. 263–265).

Fortunately, many administrators and granting agencies have realized the problems associated with using the number of publications to measure the scientist's worth. Agencies such as the National Science Foundation now require that scientists list no more than three publications per year (and 10 for a five-year period) in their research proposals. This stimulates researchers to publish more full-length, thorough papers rather than numerous smaller reports.

[2]Armstrong, J. Scott. 1982. "Research on Scientific Journals: Implications for Editors and Authors." *Journal of Forecasting* 1: 83–104. 1982.

According to one set of guidelines for authorship of scientific papers, participation solely in the collection of data or other evidence does not justify authorship.[8] Most papers will have fewer than four authors, all of whom must take intellectual responsibility for the contents of the paper.

Following the list of authors are the authors' affiliations. If the authors move to another institution before the paper is published, indicate their new address with a footnote.

Abstract

The abstract summarizes the major parts of a paper and is therefore a miniversion or skeleton of the paper. The abstract is important because it is a textual table of contents that maps your paper and helps readers assess the relevance of your paper

[8]Huth, E. J. 1986. "Guidelines on Authorship of Medical Papers." *Ann. Intern. Medicine* 104: 269–274.

to their interests. Abstracts are usually the first section of a paper read by reviewers (and readers of abstracting services such as *Scientific Abstracts*). Here are the features of an effective abstract:

> It is one paragraph (usually less than 250 words) that summarizes the objective or scope of the paper, methods used in the work, results, conclusions, and the significance of the research. It concisely states all of the paper's major results and conclusions. Notice how these authors stated their results in this abstract published in *Science*:

>> Homing pigeons that had never seen the sun before noon could not use the sun compass in the morning; nevertheless they were homeward oriented.[9]

> It is specific, concise, and must stand alone from the rest of the paper. Abstracts should include no abbreviations, acronyms, or citations of other papers.

> The abstract is not evaluative. Report only what's in your paper, and don't add your insights or thoughts.

Here's an example of a well written abstract:

> Primary roots of *Zea mays* cv. Tx 5855 treated with fluridone respond strongly to gravity but have undetectable levels of abscisic acid. Primary roots of the carotenoid-deficient w-3, vp-5, and vp-7 mutants of *Z. mays* also respond strongly to gravity and have undetectable amounts of abscisic acid. Graviresponsive roots of untreated and wild-type seedlings contain 286 to 317 ng abscisic acid per gram fresh weight, respectively. We conclude that abscisic acid is not necessary for root gravitropism.

Although abstracts are relatively short, they are important. Well written abstracts announce well written papers, and poorly written abstracts usually forecast poorly written papers and poor science.

Introduction: What Was the Problem and Why Was the Work Done?

The introduction of a paper concisely states why you did the research and puts your work into the context of previous research. Here's how to write an effective introduction:

> Keep it short (two to five paragraphs) and simple. Do not write a comprehensive review of the literature; instead, include only enough information to help readers appreciate your work and to establish the background and context of your research. Provide only enough background information to orient

---

[9]Wiltschko, R. D. Nohr, and W. Wiltschko. 1981. "Pigeons with a Deficient Sun Compass Use the Magnetic Compass." *Science* 214: 343.

and help readers determine how the research relates to what we know or don't know. Editors tire of reading comprehensive literature reviews and lengthy accounts of glorious accomplishments to which authors add their modest contribution in hopes of gaining glory by association. Long introductions waste time and space, and are unnecessary. Review papers have eliminated the need for an extensive literature review in the introduction of a paper. This doesn't mean that scientists can be ignorant of the history of the problem that they're studying. That knowledge will be revealed in your treatment of the data. There's no need to demonstrate this knowledge any other way.

Proceed from general to specific. Get quickly to the point to grab readers' attention. Do this with statements such as, "This paper is the first to show that calcium affects osmoregulation in birds."

Don't tell readers what they already know and don't cite elementary textbooks. Rather, cite only significant papers that, together, show that a problem or question exists. Use these references to establish what is being challenged or developed and to show why you did the work.

Explain the rationale, objective, purpose, or hypothesis of the work. Tell readers why you did the work with statements such as, "The objective of this research was to . . ." Define the problem, provide background information, and explain what your paper is about. Don't force readers to read other papers to understand why you did your work. Use the introduction to set the tone of the paper; do not use the introduction to try to convince readers of the importance of your work.

Do not confuse a chronological review of the literature with an introduction to your research. Chronological approaches are usually ineffective because most scientists who read your paper are interested in and already know something about the topic. Starting your introduction with a nebulous and extensive rehash of the history of your field usually bores readers and prepares them for nothing. Avoid starting the introduction with a sentence such as this:

> The last decade has seen a rapid development of new techniques for studying chemical reactions.

Here are two typical examples of Introductions taken from research papers. Their important parts are labeled to show their order. I've abbreviated examples and omitted details where they seemed unnecessary. The identification of the part is enclosed in brackets; it applies to all of the material between it and the preceding bracket.

> Although there have been various reports of multiple sclerosis (MS) in children (Low and Carter 1956, . . .), the onset of this disease occurs most frequently during adulthood and its manifestation during childhood is rare.

[literature] For this reason the occurrence of MS in children often is not even considered, which may mean that the correct diagnosis is not made for young patients with a relapsing neurological disease, although the clinical condition may be indicative. [problem]

A diagnosis of MS basically depends on clinical criteria, i.e., neurological signs suggesting multiple lesions of the nervous system and a relapsing course (Poser et al. 1983). Findings from laboratory investigations, such as delayed latencies of evoked responses and the presence of oligoclonal antibodies in the cerebrospinal fluid (CSF), may lend diagnostic support to clinically suspected MS. Moreover, the detection of cerebral plaques by means of computer tomography (CT) is of particular importance and is considered to be of pathognomonic value (Bye et. al. 1985). [discussion of past methods] However, there is no specific test to confirm this diagnosis. [problem] Since the introduction of magnetic resonance imaging (MRI), it has been expected that this method would increase diagnostic certainty in cases of suspected MS. Previous reports on MRI findings for adults suffering from MS confirm this expectation (Gebarski et al. 1985) . . . [discussion and literature]

In the present paper we report and contrast MRI and CT findings for three children who were diagnosed as definite MS cases according to the criteria of Poser et al. (1983) and who were treated in our hospital between 1980 and 1985. [objective, methods]

Heed the advice of an editor of *Nature*, a prestigious scientific journal: "If more authors . . . would attempt to make the first paragraph into a crystal clear description of what the paper is about rather than what other people's papers have been about, *Nature* would be an easier journal to read."

Materials and Methods: How Did I Study the Problem? What Did I Do?

The Materials and Methods section describes how, when, where, and what you did. The hallmark of the Materials and Methods section of a paper, and of the scientific method, is repeatability which, in turn, is determined by how well you write the Materials and Methods. If a competent scientist with a similar background can repeat your work and obtain similar results, the Materials and Methods section is well written. If the materials and methods are not repeatable, it represents poor science and the paper should be rejected, regardless of its results or conclusions. Write the Materials and Methods section while you're doing the work when your ideas are fresh.

Here's how to write an effective Materials and Methods section:

Materials

Describe the growth conditions, chemicals, lighting, temperature, diet, and apparatus used in the work. List sources of hard-to-find materials and avoid mentioning brand names unless they are critical for repeating the work. Provide diagrams or photographs of unusual set-ups or equipment. Be as specific as possible. For example, say "methanol" instead of "alcohol."

Include the genus and species (e.g., *Zea mays*, not corn), strain or cultivar, characteristics, age, and source of all organisms. In studies involving humans, indicate how the subjects were selected. If required by the journal, attach an "informed consent" statement to the manuscript (Appendix 6).

Italicize or underline the scientific name of the organism(s) that you studied. Capitalize the genus and write the species in lowercase letters.

human	*Homo sapiens* or <u>Homo</u> <u>sapiens</u>
corn	*Zea mays* or <u>Zea</u> <u>mays</u>

Capitalize larger divisions such as phyla, classes, orders, and families.

Chordata	Primates

Capitalize proper names that are part of a medical term.

Hodgkin's disease

Write the names of elements in lowercase letters. Capitalize the first letter of their symbols.

sodium	Na
iron	Fe

Describe field sites with photos or maps.

Methods

List methods chronologically with subheadings such as "Microscopy," "Sampling Techniques," and "Statistical Analyses." The Materials and Methods section is the first section of the paper with subheadings.

Describe all controls and variables that you tested. Describe features such as temperature, pH, photoperiod, and incubation conditions. Do not include information such as the day the experiments were done or the type of microscope you used unless this information is critical to repeating the experiment. State the sample sizes and describe statistical treatments (see Chapter 8).

Reference previously published methods only if they appear in widely available journals. Provide all information about new techniques, modified techniques, novel equipment, or those described in obscure journals. Avoid citing obscure journals such as *The Indonesian Journal of Zebra Nostril Hairs*. Chances are that most libraries won't have it.

Write the Materials and Methods section as soon as you've established your procedures and have overcome any initial problems. Keep the reader in mind. For example, ask yourself if the reader is familiar with and can reproduce the technique. If so, use references to provide details of what you did and to avoid long recipes. Remember that the issue is repeatability, and that evidence is invalid without repeatability.

Results: What Were the Findings?

The Results section of a paper reports new knowledge and is the heart of a scientific paper. The early parts of a paper describe how you got to the Results, and the later parts describe what the results mean. Thus, the entire paper stands on the Results section.

The Results section should summarize your findings and describe the evidence for your arguments. It is often the shortest section of a paper. Here's how to write an effective Results section:

Present relevant, representative data from your experiments, not all data from all experiments that you might have done. Summarize your data to show trends and patterns.

Make no comparisons with data in other publications, and do not discuss why your results agree or disagree with your predictions. Do that in the Discussion section.

Present repetitive determinations in tables or graphs (see Chapter 8). Use graphs and tables to support generalizations about your data.

Avoid redundancy. Present data in only one way—do not use text to describe data presented in a graph or table. Refer to all tables and figures in the text.

Guide the reader and point out trends. Avoid arrogant, opinionated statements such as, "Our data clearly show that . . ." Your readers, not you, will decide if your data show anything, much less if they show them clearly.

Make it simple and direct. For example, write that "Penicillin inhibited growth of *E. coli* (Table 1)."

Be sure that all methods used to obtain the results are described in the Materials and Methods section of the paper.

Support conclusions drawn from numerical data with brief statements of statistical criteria that you used.

Here's an example of a typical Results section published in the *Journal of Environmental Quality*:

In the first experiment, the highest acidity (pH 2.6) significantly reduced the mass of hypocotyls, but there were no significant effects of

Laboratory and Field Notebooks

When found, make a note of. — Charles Darwin

Laboratory and field notebooks are important tools of scientists. These notebooks are a cross between a journal and a class notebook, and contain a complete record of your work, observations, and ideas. Scientists use notebooks to: (1) record data about laboratory and field observations; (2) record background information relevant to their experiments; (3) evaluate, plan, and describe experiments, and (4) speculate about results. These notebooks are one of the few places where you can openly question your data.

All entries in notebooks should be complete, permanent, efficient, and systematic recordings of observations, and include the date, time, and conditions of the observations. Recording information and ideas as you go will reveal gaps in your knowledge and help you understand what you did. This, in turn, will help you better understand and prepare for future experiments. Notebooks also minimize your reliance on memory or random scraps of paper with cryptic notes. Document variations and details of your work, for this information could be important for later discoveries. Insert evidence such as photographs, sketches, and print-outs to help yourself formulate ideas.

Having a thorough, accurate, and well-organized notebook will greatly ease writing because the notebook will contain ideas and information that will be included in your report.

acidity on shoots (Table 2). There also were no significant effects of anions on the mass of either shoots or hypocotyls, nor was there an interaction between acidity and anions. Both linear and quadratic terms in the dose-response functions for effects of acidity on hypocotyls were significant.

In the second experiment, simulated rain at pH 3.0 and 3.4 reduced the mass of hypocotyls compared to pH 5.0, but there were no significant effects of acidity on shoots (Table 3). Anion composition of simulated rain did not significantly affect mass of shoots or hypocotyls nor was there a significant interaction between acidity and anions. The same results were found in harvests 1 and 2 although the effects of acidity on mass of shoots was close to being significant at $\alpha=0.05$ in harvest 1.

Generally, hypocotyls were more susceptible to effects of acidity than shoots (Tables 2 and 3) and effects tended to become less pronounced when plants were given a recovery period after treatment (harvest 2, Table 3). Daily exposures of simulated acidic rain during the period of rapid hypocotyl expansion appeared to have a slightly greater effect than three exposures per week beginning with the seedling stage (Table 2 vs. Table 3). Anions had no effect on dry mass, alone or in combination with acidity, either in the first or second experiment.

Discussion: Why are the Findings Important?

The Discussion section tells what the results mean and why they're important. It interprets the results relative to the objectives stated in the Introduction and answers the questions, "So what?" and "What does it mean?" The Discussion section is where the true nature of your paper comes to light. It is often the most difficult section of a paper to write.

Here's how to write an effective discussion section:

Tell the reader what the results mean. For example, do they support the hypothesis you tested? Why or why not? Don't leave your readers wishing that you would explain your explanation.

Clearly discuss generalizations, relationships, principles, and the significance of your results. Deliver on what you promised in the Introduction, and do not restate your results. Discuss and analyze your data, not the methods or statistics. Summarize evidence for each of your conclusions.

Use your data and those of others to argue for the most plausible interpretation for your results. Propose a simple, testable hypothesis to explain your results. Remember to keep it simple—piling hypotheses atop each other is bad for a reader's digestion and an author's reputation. Also remember that it's not enough to collect or report data—machines and other animals can do that. Rather, you must logically explain the *significance* of your results.

Focus on important discoveries and their underlying causes. Do not present all conceivable explanations of your data. It's OK to speculate if the speculation is reasonable, based on data, and testable. Use data to support your arguments.

Don't overstate your data, and remember that you can't show the "whole truth"—leave that to the loudmouths who proclaim it every day. Research is often incomplete, yet valuable because it points to new experiments or new ways of organizing information. Darwin recognized the incomplete nature of his work in the Introduction to *On the Origin of Species*:

No one ought to feel surprise at much remaining as yet unexplained in regard to the origin of species and varieties, if he make due allowance for our profound ignorance in regard to the mutual relations of the many beings which live around us. . . . Although much remains obscure, and will long remain obscure . . . I am convinced that natural selection has been the most important, but not the exclusive, means of modification.

Darwin left many findings obscure or unexplained. Nevertheless, his ideas are among the most important ideas ever proposed. Darwin's work redirected the ideas of other scientists and led to much important work, particularly in areas in which his theory was weak. Do not discount the importance of interim solutions.

Move from specifics to generalizations, and compare your findings with those of others. Point out exceptions, lack of correlation, and unexpected results. Seek explanations, not refutations, and never state the opinion of others or the majority as fact.

Discuss the implications and importance of your work. Remember that data do not suggest, research doesn't indicate, and results do not show—these are all actions of scientists, not data. Show your knowledge and take responsibility for your work by refusing to hedge, apologize, or retreat behind a wall of excuses and "hedge words." Take a stand and finish the section positively and forcefully with statements such as, "I conclude that . . ." Then stop.

Acknowledgments

> In most of mankind gratitude is merely a secret hope for greater favours.
> — Duc de la Rochefoucauld

> [Newton wrote to Halley . . . that he should not give Hooke any credit] That, alas, is vanity. You find it in so many scientists. You know, it has always hurt me to think that Galileo did not acknowledge the work of Kepler.
> — Albert Einstein 1879–1955

The Acknowledgments section of a paper is where authors acknowledge organizations, reviewers, and colleagues who helped with the paper or the research. The most common acknowledgments are for technical help, the use of equipment, and financial support. A typical acknowledgment would be, "I thank Randy Wayne for his help with the micromanipulators."

Acknowledgments are a professional courtesy and do not imply an endorsement of the work. Many scientists expect colleagues to acknowledge their help, while others do not want to be acknowledged. Therefore, check with your colleagues before acknowledging them in your paper.

Literature Cited

> . . . the function of a citation is no different from that of the paper itself: to supply the reader with information he doesn't already have.
> — E. Garfield

Verb Tense

You'll use present and past tense when you write a scientific paper. The work of others is part of an existing framework of knowledge. Therefore, use present tense when you discuss the published work of others. Since new data are not yet considered part of that framework of knowledge, use past tense when reporting your own present findings. If you follow these guidelines, most of the Abstract, Materials and Methods, and Results sections will be written in past tense because they cite or describe your present work. For example,

Abstract
We studied the role of penicillin in preventing bacterial infections.

Materials and Methods
We fed our rats a diet containing 4% (w/v) oat bran.

Results
The experimental drug increased the size of nuclei in the cancerous cells.

Similarly, the Introduction and Discussion sections will be written mostly in present tense because they include frequent references to published research. For example,

Introduction
The availability of nitrogen strongly influences plant growth (Baldridge, 1992).

Discussion
Bass dominate the fish population of Lake Waco (Vodopich 1992).

Here are the exceptions to these guidelines:

Use past tense when you refer directly to an author:
Baldridge (1992) studied leaf hoppers at Marlin Prairie.

Use present tense when referring to a table:
Table 2 shows that . . .

The man is most original who can adapt
from the greatest number of sources.
— Thomas Carlyle

The Literature Cited section of a paper provides complete bibliographic details of all work cited in the paper. As such, this section substantiates many claims made in the paper and is where you document evidence provided in other papers. Here's how to write an effective Literature Cited section:

Include all required details of all papers cited in the paper. Carefully check the accuracy of each citation. Interestingly, this section of a scientific paper is where most mistakes are made: Authors regularly include misspelled names, added or deleted names, incorrect dates, mangled titles (even of papers written by the author), and wrong journals.[10] The carelessness that produces these errors can damage your scientific credibility, for not checking the accuracy of supporting data is like making conclusions about a chemical reaction without checking the purity of the starting products. Examine all of the original papers yourself so that you won't perpetuate others' mistakes.

Cite only significant, published papers. Increasing the number of literature citations changes neither your data nor your conclusions; majority does not equal fact or truth. As William Roberts said, "Manuscripts containing innumerable references are more likely a sign of insecurity than a mark of scholarship."

Build the Literature Cited section as you write the paper. Cite the references in the format specified by the journal in its "Instructions to Authors." Although different journals require different formats for literature citations, three systems dominate the scientific literature: the Harvard System, the Alphabetical System, and the Citation Order System.

The **Harvard System** lists names of authors in the text: ". . . according to procedures described by Smith and Wesson (1990)." This system is easy to use when writing and allows readers to recognize authors of papers that you cite. Here's an example of a citation:

Much evidence suggests that dinosaurs were warm-blooded (Flintstone and Rubble, 1991).

The **Alphabetical System** refers to papers by number: "according to procedures described by Smith and Wesson (1)." Cited papers are listed alphabetically in the Literature Cited section of the paper. Here's an example of a citation:

Much evidence suggests that dinosaurs were warm-blooded (1).

The **Citation Order System** also refers to papers by number. References are arranged according to the order in which they appear in the text. *Science* uses the citation order system of literature citations. Here's an example of a citation from *Science*:

Much evidence suggests that dinosaurs were warm-blooded (8).

Unless required to do otherwise by the journal, cite the paper where the reference best applies in the sentence. For example, the sentence, "We used previously published methods to measure the protein content and respiratory

[10]Poyer, R. K. 1979. "Inaccurate References in Significant Journals of Sciences." *Bulletin of the Medical Librarians Association* 67: 396-398; Key, J. D. and C. G. Roland. 1977. "Reference Accuracy in Articles Accepted for Publication in the *Archives of Physical Medicine and Rehabilitation*." *Archives of Physical Medicine* 58: 136-137.

rate (Doe, 1986; Smith, 1990)" does not specify what reference matches each technique. The sentence is improved by placing each reference where it best applies: "We used previously published methods to measure the protein content (Doe, 1986) and respiratory rate (Smith, 1990) of the rats."

The "Instructions to Authors" will also include directions for other problems associated with citing literature, such as citing papers having more than one author and citing papers published by the same authors in the same year. Almost all journals use different formats to list cited papers. Check each journal to learn about the format that it uses for citing references.

Many journals require authors to list the entire name of a journal, while others require only the abbreviation of the journal's title. If you use abbreviations, don't abbreviate one-word titles of journals (e.g., *Science*). The word "Journal" is abbreviated "J." and "-ology" words are cut off after the "l" (e.g., *J. Bacteriol.* for *Journal of Bacteriology*).

Choosing a Journal

> In almost every scientific subfield there is a hierarchy of journals that reflects the relative quality of published papers. Although it does not exist overtly, this hierarchy is known to all sophisticated scientists within the field.
> — National Academy of Sciences,
> *The Life Sciences*

Decide to what journal you'll submit your paper before you begin writing. Use these criteria to decide where to submit your paper:

Scope of the Journal   All journals publish a statement of their purpose. For example, here's what *Evolution* tells its authors:

> Manuscripts submitted to *Evolution* should contain significant new results of empirical or theoretical investigations concerning facts, processes, mechanics, or concepts of evolutionary phenomena. Brief notes, or comments on previously published papers, should be submitted as "Notes and Comments." . . . Acceptance is based upon the significance of the article to the understanding of evolution; each paper must stand on its own merits and be a substantial contribution to the field.

Submit your paper to a journal whose purpose matches the objectives of your paper.

Prestige of the Journal   A journal's prestige results largely from the quality of papers that it publishes. Consequently, your paper will have its greatest impact if it is published in a prestigious multidisciplinary journal such as *Science* rather than in either a low-quality journal or a journal not read by your colleagues, in which case the significance of your work will probably be overlooked. However, submitting your paper to prestigious journals such as *Science, Nature,* and *The New England Journal of Medicine* increases your chances of having the paper rejected. Indeed, these journals reject about 85% of the papers they receive, whereas the *Journal of Scientific Chemistry,* a more specialized journal, rejects only about half of the papers that it receives. You can choose an appropriate journal for your paper by talking with colleagues and determining where the best papers in the discipline are published. Also examine *Journal Citation Reports,* a listing of the journal's impact based on the number of times scientists have cited the journal's papers.

Your choice of journals will help determine the impact of your paper. For example, Gregor Mendel's paper entitled "Experiments with Plant Hybrids" describing his plant-breeding experiments was published in 1866. Although Mendel mailed copies of the paper to 120 scientific societies and universities, it remained unknown and was ignored for almost 40 years. This was no accident: Mendel published the paper in an obscure journal—*Transactions of the Brünn Natural History Society.* Not until DeVries, Correns, and Tschermak all thought that they had independently reached the same conclusion in about 1990 did they discover Mendel's paper and realize Mendel's genius.[11]

Quality of Reproduction   If your paper includes photographs such as electron micrographs, choose a journal that will reproduce your micrographs well.

Circulation   Journals with a large circulation are usually read by more scientists than those with a smaller circulation. Journals report their circulation in their "Statement of Ownership" that's published each year, usually in the November or December issue. Here are the 1990 circulation figures for a few journals:

Scientific American	606,826
Natural History	546,345
Science Digest	230,000
New England Journal of Medicine	226,000
Science	153,000
Skeptical Inquirer	38,000
Nature	31,000
Science Teacher	25,000
Journal of Irreproducible Results	12,000
BioScience	12,000
Proceedings of the National Academy of Science	10,000
American Science Teacher	9,413

[11]Gasking, E. B. 1959. "Why Was Mendel's Work Ignored?" *Journal of Historical Ideas* 20: 60–84.

Ecology	8,434
Journal of Scientific Chemistry	7,200
Biochemistry	6,011
Evolution	4,223
Quarterly Review of Science	2,725
Scientific Bulletin	2,600
Lipids	2,600
Human Science	1,654
Biophysics	1,000
Journal of the Alabama Academy of Science	800
Newsweek	3,057,081

Speed of Publication   Most scientists are in a rush to publish their work. This rush to publish dates to the 17th century when, in an effort to force scientists to divulge their data, a secretary of the Royal Society of London invented the rule that priority goes to whoever publishes first, not to who discovers first. Scientists have been vying to publish first ever since. For example, Watson's and Crick's seminal paper in the April 1953 issue of *Nature* was published "as is" just three weeks after it was submitted for publication. Such haste to publish is also driven by journal editors wanting to publish "hot" papers. Indeed, the competition between journals such as *Science* and *Cell* is "like an outright war." Many worry that such rapid publication may prompt premature publication—that in their rush to publish, scientists may cut corners and the review process may be compromised, leading to incorrect or incomplete work. Stanley Pons's and Martin Fleischmann's paper about cold fusion, published by the *Journal of Electroanalytical Chemistry* in just four weeks, is a precedent that no one wants to repeat.

If you work in a fast moving field or are in a rush to publish your work, consider the time required by the journal to publish your paper. Journals such as *American Journal of Botany* take more than a year to publish a paper, while others such as *Science* and *Cell* take an average of about 4.5 months. To hasten publication, some journals receive manuscripts on disk, thereby easing editing and typesetting. However, relatively few journals receive manuscripts via electronic mail.

Formatting, Typing, Packaging, and Mailing Your Manuscript

> It's a damn poor mind that can think of
> only one way to spell a word.
>   — Andrew Jackson

Formatting

Papers published in scientific journals must be submitted in a format specified by the journal. This format is detailed in the journal's "Instructions to Authors." For example, here are the instructions for how to prepare a manuscript for publication in *BioScience*:

Information for *BioScience* Contributors

How to Prepare a Manuscript

The Bioscience Staff

The editors welcome manuscripts written for a broad audience of professional biologists and advanced students. *BioScience* publishes peer-reviewed Articles, summarizing recent advances in important areas of biological research. *BioScience* also publishes short opinion pieces in the Viewpoint section, longer opinion pieces and essays on policy issues important to biologists in the Roundtable section, essays on the teaching of biology to students and to the general public in the Education section, and Letters pertaining to material previously published in *BioScience*. In addition, the Professional Biologist discusses issues in the practice of the biological profession, and Thinking of Biology contains essays on biology history and philosphy. Finally, the special book issues of *BioScience* include an article on some aspect of book writing, publishing, or reading.

The editors reserve the right to edit all manuscripts for style and clarity. Contributions are accepted for review and publication on the condition that they are submitted solely to *BioScience,* and they will not be reprinted or translated without the publisher's permission. All of the authors must transfer certain copyrights to the publisher.

Articles

Articles should review significant scientific findings in an area of interest to a broad range of biologists. They should include background for biologists in disparate fields. The writing should be as free of jargon. All articles, whether invited or independently submitted, undergo peer review for content and writing style. Articles must be no longer than 20–25 double-spaced typed pages, including all figures, tables, and references. No more than 50 references should be cited.

Viewpoint

The Viewpoint page may cover any topic of interest to biologists, from science policy to technical controversy. Viewpoint manuscripts must not exceed two and a half double-spaced pages; Roundtable contributions may be up to 15 double-spaced pages. Viewpoint submissions may not have references, footnotes or a reference list.

Other Departments

Manuscripts for Roundtable, Education, The Professional Biologist, Thinking of Biology, and special book articles must not exceed 15 double-spaced pages and may cite no more than 25 references. They may include a few photographs, drawings, figures, or tables. These manuscripts will undergo review. The book reviews are generally solicited. If you are interested in writing a book review, contact the Assistant Editor.

Manuscript Preparation

Submission Submit an original and four copies of all manuscripts along with a cover letter to the

Editor, *BioScience*, 1444 Eye Street, NW, Suite 200, Washington, DC 20005. Be sure to include your telephone and FAX numbers and your e-mail address. Authors must obtain written permission to use in their articles any material copyrighted by another author or publisher. Include with your manuscript photocopies of letters granting permission; be sure credit to the source is complete. List in your cover letter names of colleagues who have reviewed your paper plus the names, addresses, and telephone numbers of four potential referees from outside your institution but within North America.

Typing   Use double-spacing throughout all text, tables, references, and figure captions. Type on one side only of $8\frac{1}{2} \times 11$-inch white paper. Type all tables, figure captions, and footnotes on sheets separate from the text. Provide a separate title page with authors' names, titles, affiliations, and addresses; include a sentence or two of relevant biographical information and research interests.

Style   Follow *Scientific Style and Format*, 6th edition (CBE 1994), for conventions in biology. For general style and spelling, consult *The Chicago Manual of Style*, 14th edition (Chicago 1982) and a dictionary such as *Webster's Third International Dictionary* (Gove 1968).

Abstract   Manuscripts for Articles and Roundtable, Education, The Professional Biologist and Thinking of Biology sections should include an informative abstract no longer than 50 words. The abstract is useful to the editors and reviewers. It is not published. Do not include a summary in the article.

Symbols, Acronyms, and Measurement   Define all symbols and spell out all acronyms the first time they are used. All weights and measures must be in the metric system, SI units. In Articles and Toolbox contributions, abbreviations may be used for units of weight or measurement that describe data.

References Cited and Footnotes   No more than 50 references should be cited in Articles and no more than 25 in the department. Personal communications, unpublished data, and manuscripts in preparation should be cited in footnotes containing the date and source's name and affiliation.[1] Keep other footnotes to a minimum. Number text footnotes with consecutive superscript numerals. For footnotes in tables, use symbols (page 419, Chicago 1993). In-text citations must take the form: (Author date). Multiple citations should be listed in alphabetical order: (Author date, Author date). Use the first author's name and "et al." for in-text citation of works with more than two authors or editors; list every author or editor in the "References cited" list. All works cited in the text must be listed alphabetically in References cited; works not cited in the text should not be listed. Provide the full names of all journals. Underline the titles of all books and journals. Refer to recent issues of *BioScience* for additional formatting; some examples:

- A journal article: Bryant, PJ, and Simpson P. 1984. Intrinsic and extrinsic control of growth in developing organs. *Quarterly Review of Biology* 59: 387–415.

- A book: Ling, GN 1984. In search of the physical basis of life. New York: Plenum Press.

- Chapter in book: Southwood TRE. 1981. Bionomic strategies and population parameters. Pages 30–52 in May RM, ed. Theoretical ecology. Sunderland (MA): Sinauer Associates.

- Technical report: Lassiter RR and Cooley JL. 1985. Prediction of ecological effects of toxic chemicals, overall strategy and theoretical basis for the ecosystem model. Washington (DC): Government Printing Office. Reporter nr 83-261-685. Available from: National Technical Information Service, Springfield, VA.

[1]Footnote format: H. J. Smurd, 1989, personal communication, University or other affiliation, city, state.

- Meeting paper: Kleiman RLP, Hedin RS, Edenbom HM. 1991. Biological treatment of minewater—and overview. Paper presented at the Second International Conference on Abatement of Acid Drainage; 16–18 Sep 1991; Montreal, Canada.

Disk Submission

Along with the hard copies of the manuscript, enclose a high-density floppy disk ($3\frac{1}{2}$ or $5\frac{1}{4}$ inch IBM compatible or Macintosh) containing an ASCII text version (i.e., no formatting or control characters) of the article.

Illustrations

Black-and-white Photographs, maps, line drawings, and graphs must be camera-ready, glossy black-and white prints, photostats, or original art. On the reverse side of the artwork, number and identify figures and indicate "top" of photographs. All photographs must be untrimmed and unmounted, 4×5 to 8×10 inches in size, and clearly focused; photomicrographs should have a scale bar. Line drawings and graphs should be done by professional artists or scientific illustrators. Figures will be reduced to fit one column (2.25") or two columns (4.75"). Therefore, when labeling figures, make sure to use large enough letters so that they will reduce to no smaller than 8 points. In drawings and graphs, please capitalize only the first letter of the first word of each label; please use helvetica or a similar sans serif typeface, without bold or italics (unless genus, species, or gene names). Do not place keys within a graph. Letters labeling the sections of a figure should be lowercase, placed in the upper left hand area of the section, and not followed by a period.

Color Authors must pay the cost of printing color art within an article. Contact the Editor for details.

Cover Authors are encouraged to submit color transparencies related to the manuscript for consideration as cover art. Cover photographs must be in a vertical format, sharply focused, and colorful. They must have a light or dark area at the top for the logo.

References Cited

[CBE] CBE Style Manual Committee 1994. Scientific style and format: the CBE manual for authors, editor, and publishers. Chicago (IL): Council of Biology Editors.

[Chicago] The Chicago manual of style. 1993. 14th ed. Chicago (IL): University of Chicago Press.

Gove PB, ed. 1968. Webster's third new international dictionary of the English language unabridged. Springfield (MA): G. & E. Merriam Co.

CORRESPONDENCE: Direct all correspondence to *BioScience* Editor, American Institute of Biological Sciences, 730 11th Street, NW, Washington, DC 20001-4585. Tel: 202/628-1500; FAX 202/628-1509.

Spelling

Long ago, people paid little attention to spelling. Indeed, Chaucer often changed the spelling of words to fit his rhymes. Today we have one correct way to spell each word (words having two spellings usually have one spelling that is preferred).

We often accept different pronunciations, but not different spellings. Therefore, type your paper carefully, being sure to check the spelling of every word. Papers loaded with typos usually indicate that the writing and science are poor. For example, when people read *Noble laureate* for *Nobel laureate*, they usually question the writer's intelligence, education, and everything else that he or she has to say. This may not seem logical or fair, but it's the way it is. Moreover, typos can have important consequences—for example, spinach may have got its reputation as a dietary supplement because of a misplaced decimal point in which the iron content was reported as ten times higher than it was.[12] Other reports claim that typos have resulted in deaths.[13]

You'll probably not know how to spell every word that you use. However, don't be overly concerned—everyone has words that trouble them. For example, John Irving looked up *strictly* 14 times during one five-year period; during the same time, he also looked up the word *ubiquitous* 20 times. Not only can he still not spell these words, he can barely remember what they mean. However, that's no excuse to misspell words. To avoid problems, always follow spelling's only rule:

IF YOU DON'T KNOW, LOOK IT UP.

Double-space your paper and type on only one side of each sheet. Use margins at least 3 cm wide and begin the Abstract on a new page. Place tables and figures at the end of the paper. Consult the journal's "Instructions to Authors" for information such as how many copies to submit.

Packaging and Mailing

Package your manuscript carefully. Editors are continually perplexed by authors who mail elaborate artwork and photographs in flimsy envelopes that are damaged or destroyed by the postal service. To help ensure that your paper arrives in good condition, mail the paper in a heavy-duty, padded envelope whose ends and seams are reinforced with tape. Use no staples—they can tear or scratch artwork. Keep a copy of your paper for your files.

Include a cover letter with the paper that states (1) the work reported in the paper is original and has not been submitted elsewhere for publication; (2) the name and address of a person to contact about the paper; and (3) that all authors

---

[12]Hamblin, T. J. 1981. "Fake." *British Medical Journal* 283: 1671–1674

[13]An article entitled "Mortal Consequences of a Typographical Error" described how a German physician had read in a journal that cases of pruritis ani had been treated successfully with a 1% solution of percaine. When he gave this solution to a patient, the patient died. The physician was unaware that the journal's editor, in a subsequent issue of the same journal, had published a correction stating that the solution should have been one per thousand, not one per hundred. Nevertheless, a German court held the physician guilty of negligence on the grounds that, considering the novelty of the treatment, he should have followed the literature more closely and sought advice from competent authorities. (See "Consequences mortelle d'une erreur typographique." *Med dans le Monde* [Suppl. Semaine Med] 26: 537, 1950.)

Copyright and Ownership

You may wish to include quotations, illustrations, and photographs from other sources, including your own, in your papers. If your paper is used for educational purposes, involves no profits, and does not significantly impact the original paper, you can probably use the material without worrying about copyright problems. For example, including short quotations from other papers does not require that you obtain any permissions because such use is considered "fair use" by the government. However, if you want to use large quotations, photographs, or other illustrations from other papers in your paper, you must know something about copyright.

Copyright is a legal protection against unauthorized use of your work. Publication or registration with the U.S. Copyright Office is not required for a work to be protected by copyright: The copyright belongs to a writer or artist as soon as they finish writing a paper, book, or essay. This copyright provides exclusive rights to reproduce copies, derive works from the work, rent or sell the work, and the right to transfer the copyright to someone else.

Most publishers require that you transfer the copyright of your work to them to protect their interests and to prevent unauthorized use of your paper. As dictated by the U.S. Copyright Act of 1976, this transfer must be in writing on a copyright transfer form. This transfer helps ensure that your paper is original, published only once, and that everyone, including yourself, must obtain written permission to use data or photographs included in your paper. If you want to republish anyone's work, seek their permission early. Request from the holder of the copyright exactly what's wanted. For example, send a photocopy of the table or figure that you want to re-publish. It is the job of the author, not the editor, to obtain permission to re-publish copyrighted information. Do this by sending a "permission request letter" (Appendix 3) to the copyright holder and author. The copyright holder will stipulate what to include in the credit line and if a fee is required.

Transfer of copyright gives to someone else the right to use the article, not the article itself. Such a transfer is necessary to maintain the integrity of scientific publications because it allows publishers to protect your work from unauthorized use. It also means that you'll need written permission from the publisher to republish tables, figures, or substantial parts of the text of your paper. Publishers will honor all reasonable requests to publish your work in review articles, collections of reprints, or institutional collections of papers. When you republish this material, indicate that you've obtained written permission to reprint material with a legend such as "Reprinted with permission from (source); copyright (year) by (owner of copyright)." There's no need for you to obtain permission to use material in the "public domain," which includes works whose copyright has expired and works prepared by officers or employees of the U.S. government as part of their official duties.

If you have questions about copyright, write to the U.S. Copyright Office, Library of Congress, Washington, D.C. 20559.

participated in the research. Don't bias the editor against your paper with a long or poorly written cover letter. For example, one author closed a cover letter with "I hope you will find this paper exceptable." The editor did.

What Happens Next

Within a week or two the journal's editor will send you a letter acknowledging receipt of your paper. The acknowledgment will probably also tell you when you can expect a decision about the fate of your paper. After you receive the acknowledgment letter you'll hear nothing more about your manuscript for one or two months. During this time, the editor will send your paper to other experts (usually two) for their opinion about its publication in the journal. These reviewers recommend to the editor that the paper either be published or rejected. Based on these reviews and his or her opinion of the paper, the editor will then decide the fate of your paper.[14] This decision will usually be one of the following:

Publish the paper "as is." If this happens, congratulate yourself. Such a decision is rare—less than 5% of papers are published "as is."

Publish the paper after it's revised according to suggestions of the editor and reviewers. About 45% of papers fit this category. In these cases, you must enclose with your revised manuscript a cover letter describing how you've met the criticisms made by the reviewers. If you've ignored any of the reviewers' suggestions, you must defend your actions. Such explanations are extremely important. Indeed, the fastest way to have your paper rejected is to return it without giving reasons for ignoring the reviewers' suggestions. Few editors will tolerate such arrogance and disregard for the criticism of your peers. If you make the suggested changes or present a good reason for not changing your paper, most editors will accept your paper for publication.

Reject the paper. About half of all papers submitted to quality journals are rejected, meaning that the editor has decided that your paper will not be

---

[14]A negative review will not necessarily cause a good editor to reject a paper for publication. Editors ignore reviews that include no evidence for the recommendation. Indeed, reviews such as "This paper shouldn't be published because it is not worthy of publication" tell editors and authors nothing. Good editors demand that reviewers document their recommendation with evidence.

The history of science is filled with examples of editors ignoring reviewers' recommendations to publish what became important papers. For example, shortly before Christmas 1671, Newton's reflecting telescope (the first ever built) was greeted with acclaim by the Royal Society. Soon thereafter, Newton told the Society that he would submit a paper that he considered "the oddest if not the most considerable detection which hath hitherto beene made in the operation of Nature." Robert Hooke, at the time England's leading authority on optics, was highly critical of Newton's paper, but his criticisms neither halted nor delayed publication. However, the matter did not end there. Hooke and Newton continued their bitter dispute for many more years. Scientists today often misjudge papers they review. For example, four of the six most-cited papers from *The Lancet* and *British Medical Journal* from 1955–1988 contained ideas that were initially rejected or disbelieved (see Dixon, Bernard. 1989. "Disbelief Greeted Classics in Top U.K. Medical Journals." *The Scientist* 17 April 1989).

published in the journal. If you get a rejection letter, read it carefully. Perhaps your paper was rejected because it describes research that is irrelevant to the scope of the journal. Although these decisions are pointless to challenge, remember that such a decision is not a criticism of your data or conclusions: You've merely sent your paper to the wrong journal. Other papers are rejected because of defects in the research, such as no data for a critical control experiment. In these cases, do the needed experiment and resubmit the paper. Finally, other papers are rejected because they represent poor science—poor experimental design and unfounded conclusions. In this case, reconsider the validity of your paper. You may need to make wholesale revisions in the experiments underlying the work.

Despite your best attempts, you'll probably receive a rejection letter someday. Do not despair—it's happened to just about everyone. For example, F. Scott Fitzgerald got 122 rejection slips before he published his first book, and Richard Bach's *Jonathan Livingston Seagull* was rejected 18 times before Macmillan agreed to print 7500 copies; within five years, more than seven million copies of the book had been sold in the United States alone. Great scientists have also had papers rejected. For example, Hans Krebs's first paper on what came to be known as the Krebs cycle was published in *Enzymologia* two months after being rejected by *Nature*.

A few months before your paper is published you'll receive page proofs of the paper. These are typeset pages of your paper for you to examine and correct. The editor or printer will give you detailed instructions about how to mark the page proofs and when they must be returned. Here are some suggestions for handling the page proofs:

Do not just read the page proofs. If you do that, you'll have no way of catching mistakes such as "61" substituted for "16." Rather, *study* the page proofs carefully —they're the last you'll see of your paper before it's published. Any mistakes that you don't catch will appear in your paper when it is published.

Have someone read the manuscript aloud—word by word, line by line— while you check the accuracy of the proofs. After spending eight months and $300,000 to gather data, don't sleep through a typesetter's change of your paper.

Read the corrected page proofs aloud to yourself. Speaking makes you think about what you wrote and helps highlight mistakes.

Pay special attention to all numbers (dates, numbers, and formulas), scientific and proper names, punctuation, and figure placements.

Keep a copy of the corrected proofs and return one copy to the editor.

Near the time you get the page proofs you'll also receive a form for ordering reprints of your article. You'll probably want to order reprints, especially if your paper contains photographs that won't photocopy well.

A Final Word About Dealing With Editors

Editors of journals are the gatekeepers of scientific literature: They decide which papers will be published and which papers will be rejected. Although their decisions are usually guided by reviewers, it is the editor who has the final word on the fate of a paper. If you deal with editors courteously, you'll learn that they are reasonable scientists and are on your side: They *want* to publish quality papers in their journals. You'll get nowhere by conceitedly insisting that science can't advance unless your ideas are enshrined in print.

Other Kinds of Writing For Professional Audiences

Although you must learn to write research papers if you want a career in scientific research, your writing tasks won't stop there. You'll probably also need to write review articles, abstracts, poster presentations, a thesis and dissertation, grants, and conference reports.

Review Articles

Reviews are papers that describe someone else's interpretation and evaluation of the primary literature. Reviews summarize, analyze, and critically evaluate literature already published. They are similar to term papers in that they review the literature and put it into a new perspective. Reviews do not report original research and are usually 10 to 50 pages long. The audience for a review paper is more general than that for a research paper. Although review papers have no prescribed format, most concentrate on reviewing and discussing the literature. Useful reviews should try to answer important questions. Review papers may not provide a definite answer, but they should indicate directions of future research.

Review articles seldom contain discussions of materials and methods, and often are longer than research papers because they cover much material derived from many sources about a relatively broad subject. Reading review articles will immerse you in the scientific literature, thereby showing you how scientists think. Reading these papers is also the best way to stay up-to-date on developments in fields other than your own. The following books and journals publish review articles for scientists and nonscientists:

American Scientist. Review articles covering a variety of sciences.

Annual Reviews. This series of review articles began in 1950 with the publication of *Annual Review of Medicine*. There are now more than 30 versions of *Annual Reviews* (for example, *Annual Review of Plant Physiology*).

Scientific Reviews. Review articles on scientific topics.

Quarterly Review of Science. Excellent review papers about science.

Scientific American. Readable articles about many scientific topics.

Many scientists write review articles in response to an invitation from a journal's editor. If you've not received such an invitation, ask the journal's editor if she or he is interested in publishing such a review. *Science Citation Index* codes reviews with an "R."

Theses and Dissertations

A thesis is a statement, proposition, or position that a person advances and is prepared to defend. It describes original research and therefore is a type of research paper. Thesis requirements vary at different universities, and usually only the format of the thesis (for example, where to place page numbers and the sizes of margins) is prescribed. Theses usually take several months to write, so students should start writing them long before their anticipated date of graduation.

Many scientists question the validity of a traditional thesis. Since a thesis is a published report of original research, it should be written with the same rigor as a research paper. However, this is seldom the case. Many scientists think that theses differ from primary publications. Consequently, most theses are not subjected to the same criticism as a primary publication. This produces excessively long theses containing irrelevant and trivial data—50 pages of science crammed into a 250-page thesis. They're then filed away in the basement of the library, never to be opened again except perhaps by a bored janitor. J. Frank Dobie said it well: "The average Ph.D. thesis is nothing but a transference of bones from one graveyard to another."

Many theses are written poorly because many scientists do not understand writing. Indeed, many fear a thin thesis because they equate an article's length with its significance. Others defend a verbose thesis as critical to measuring characteristics such as mastery of the subject, investigation, and critical thinking. However, these characteristics are features of the student determined long before the thesis is written. A thesis should measure scientific preparation, not activity. Consequently, verbiage in a thesis indicates either laziness or an attempt to hide one's scientific shortcomings. A 250-page thesis containing trivial data, poor analyses, and a "comprehensive" literature review and bibliography shows only that a student has learned to read, write, and stuff their papers with "filler" rather than to think or write effectively. There is no correlation between the length of a scientific paper and its significance. Similarly, including "everything" in a thesis merely indicates that the student hasn't learned to discriminate what's significant from what's trivial and unnecessary. To quote an old saying, "Only fools collect facts; wise people use them."

Don't accept the argument that a thesis (or anything else that you write) must be long to be significant. Get help from your major professor, and approach a thesis with the same rigor as you would a research paper. Meet the local requirements for the thesis, but remember that professional scientists publish research papers, not theses. Your success as a scientist will depend largely on your ability to write a research paper. Once you graduate, you'll never write a thesis again. Therefore, concentrate on learning to write quality papers for publication in quality, peer-reviewed journals. One journal article contributes much more to science than does a pile of poorly written theses filed in a dusty library.

Grants

Before the nineteenth century, science was a hobby of the rich. Members of the nobility and the literate bourgeoisie set aside rooms of their houses as scientific labs. People without the leisure afforded by wealth had to raise their own money to pursue science. For example,

> Johannes Kepler moonlighted as a court astrologer. His three laws of planetary motion, notably his deduction that planets move in elliptical orbits around the sun, vindicated Copernicus and served as the basis for Isaac Newton's research.

> Francis Bacon was a lawyer who became Attorney General and Lord Chancellor of England. His legal service ended when he was charged with taking a bribe. However, neither Bacon's public service nor his dismissal slowed his studies showing that nature could be confirmed and understood only by systematic experimentation.

> Antoine Lavoisier was a tax-farmer (a tax-collector empowered to make a personal profit) in France. He named oxygen and hydrogen, wrote the first textbook of modern chemistry, and is regarded as the father of modern chemistry. His profits as a tax-farmer supported his research, but also were his downfall: Soon after the French Revolution, Lavoisier's career ended at the guillotine.

> Leonardo da Vinci's pathetic letter to Lodovico Sforza seeking a job emphasized his ability to contrive instruments of war as a means of supporting his scientific studies.

During the nineteenth century, scientists went out of their way to show the relevance of their work to the public as a means of obtaining money to support their work. For example, Thomas Edison sponsored an "electric breakfast" at which he cooked all of the food with electricity. He used the profits of the demonstration to support his research. In the half century before World War I, the increasing professionalism of science burdened science with much jargon that isolated scientists from the public. Scientists began to speak only to their peers, causing much of the public to lose its understanding of science. Many scientists began to respect only "pure" research, and viewed any popularization of science as vulgar. Soon thereafter, scientists' work began to be funded primarily by governmental grants, a trend that continues today. To get this governmental money, a scientist must write a grant proposal.

Grant proposals are written by scientists requesting money to support a research or teaching project. They're read and evaluated by a small audience—usually only a few people. They have many forms, ranging from letters to long, involved proposals. The format and length of a proposal are determined by funding agencies such as the National Science Foundation and the National Institutes of Health.

Obtaining grants is one of the most important and stressful aspects of being a research scientist. Although many books and seminars claim that following a simple set of rules will ensure getting a grant, scientists know better: No one set of instructions can guarantee that you'll get a grant. However, all scientists agree that you will increase your chances of getting a grant if you do the following things:

> Propose an original, defined idea with a measurable conclusion. Be sure the experiments that you propose can be done within the time frame of the grant.

> Understand the problem and the proposed work. Your proposal should demonstrate your knowledge of the field and the relevance of the proposed work.

> Include a detailed plan and timetable for doing the work. Include preliminary data, if they are available.

> Include a curriculum vitae documenting your credentials for doing the work.

> Include a detailed budget for the work.

Don't waste your time by throwing together a poorly written proposal that outlines trivial work or work that you can't do. Research funds are hard to get, even when a proposal is written well by a competent scientist. Poorly written proposals have no chance of succeeding. If you want money for your work, learn to write well.

Abstracts

"Reviewed" abstracts are abstracts written as part of a research paper. Scientists also write "nonreviewed" abstracts, which are summaries of research papers written to obtain a place on the program at a scientific meeting. Nonreviewed abstracts are published "as is" and without review, while reviewed abstracts are published only after the research paper is reviewed by experts. Nonreviewed abstracts are written like reviewed abstracts (see above).

Unlike conference reports, abstracts often contain original data, and are therefore considered a type of primary literature. Most organizations and professional societies provide forms on which you must type the abstract.

Poster Presentations

Many scientists use posters to communicate their findings to colleagues at professional meetings. Indeed, poster sessions are the primary way that information is exchanged between scientists at many meetings. Posters are displayed for several hours or days, thereby allowing scientists to exchange information rather than merely present data.

Posters are written summaries of papers and are organized like a research paper. They announce new results, contain few details, and should be visually interesting to help attract your target audience. Posters are usually allotted about 1.2 m × 2.5 m of space. Pattern your poster after the most effective posters that you see at a poster session at a professional meeting. Print the title in letters about 3 cm high, the authors' names about 2 cm high, and the text in letters about 4 mm high. Use simple writing to highlight your most important data and conclusions.

Conference Proceedings

Each year there are about 10,000 scientific meetings, two-thirds of which publish their proceedings. These reports are usually published in response to an invitation from an editor, and often describe the proceedings of a symposium, congress, or workshop. Conference reports are usually 1,000 to 2,000 words long and seldom contain many details or historical perspectives. Furthermore, they are seldom reviewed prior to publication and often are published in "proceedings volumes" that are ignored by most other scientists. However, conference reports are good places for speculation, tentative conclusions, and preliminary data. Since many of these data are published later in research papers, authors must beware of copyright problems (see "Copyright and Ownership").

Some Research Papers You Might Have Missed

In Chapter One you read some humorous titles to newspaper articles about science. Here are some unusual titles that have appeared on research papers:

Impact of wet underwear on thermoregulatroy responses and thermal comfort in the cold. *Ergonomics* 37: 1375–89, 1994.

Seasonal changes of circadian pattern in human rectal temperature rhythm under semi-natural conditions. *Experimentia* 43: 294–6, 1987.

A limited mathematical model for the fermentation of dry sausage. *Journal of Food Engineering* 21: 17, 1994.

I hope that you enjoy reading them.

Pranks, Hoaxes, and Faked Data

> Science does not select or mold especially honest people: it simply places them in a situation where cheating does not pay. . . . For all I know, scientists may lie to the IRS or to their spouses just as frequently or as infrequently as everybody else. — S.E. Luria

William Osler was a great scientist and writer—indeed, his celebrated textbook *Principles and Practice of Medicine* is a masterpiece of English literature and a monument to Osler as a scientist, clinician, and teacher. However, Osler was addicted to pranks. He published at least 18 pranks under the pen name of Egerton Y. Davis. Most of these pranks had a Rabelaisian bent, such as this description of "Peyronie's disease" or "*strabismus du penis*" in which "when erect it curved to one side in such a way as to form a semicircle, hopeless and useless for any practical purpose." Osler submitted this paper under the name of a prominent Philadelphia urologist, who responded by confirming the observation and signing it EYD Jr.[1] Some of Osler's pranks were cited in other papers. For example, one writer cited the "Peyronie's disease" paper as being by "a medical man called Davis, not otherwise identified."[2]

Unfortunately, many scientists have published faked data. Among the most notorious stories of faked data is the story of the "midwife toad" reported by Austrian zoologist Paul Kammerer in the 1920s. Kammerer used midwife toads which breed on dry land and therefore never develop "nuptial pads," which are characteristic of related species that breed in water. When Kammerer bred the dry-land toads in water, he claimed to have transmitted the ability to develop nuptial pads from one generation to another. This induced transmission of acquired characters provoked many evolutionary scientists. Other scientists were skeptical, and one visitor to Kammerer's lab saw where India ink was injected where pads were said to exist. Kammerer was accused of fraud, but the charges were never proved. Kammerer later committed suicide.[3]

A famous case of faked data centered on James Shearer. In 1916 while British troops were dying in trenches of Europe, James Shearer published a paper in the *British Medical Journal* describing a method he devised for depicting the path of gunshot wounds. British military officials quickly implemented Shearer's methods. However, it quickly became obvious that the data were faked, and the journal published a retraction. Shearer, who had been in His Majesty's Army, was sentenced to death by firing squad. The sentence was later commuted, and Shearer died in prison a year later of

---

[1]Nation, E. F. 1973. "William Osler on Penis Captivus and Other Urological Topics." *Urology* 2: 468–470; Teigen, P.M. and E.H. Bensley. 1981. "An Egerton Y. Davis Checklist." *Osler Library Newsletter* 38: 1–5; Taylor, F. K. 1979. "Penis Captivus—Did it Occur?" *British Medical Journal* 2: 977–978.

[2]Taylor, F. K. 1979. "Penis captivus—Did it Occur?" *British Medical Journal* 2: 977–978.

[3]Koestler, A. 1972. *The Case of the Midwife Toad*. New York: Random House.

tuberculosis.[4]

Others have faked data with less serious consequences. For example, German zoologist Ernest Haeckel altered illustrations by labelling three copies of the same plate as human, dog, and rabbit to "prove" their developmental similarities. He later defended himself against these accusations by saying ". . . hundreds of the best observers and scientists lie under the same charges."[5] Similarly, Pasteur may have been guilty of "creative reporting"—data recorded in his laboratory notebook differ from those in his papers about anthrax and rabies vaccines.[6] Other examples of faked research include the following:[7]

A Yale researcher plagiarized others' work and falsified work in a dozen papers he coauthored with his department head.[8]

A Stanford professor published an article containing citations to papers not published.[9]

A researcher in 1975 used a black felt-tipped marker to fake a successful skin graft on a white mouse.[10]

A more recent case of fraud involved a paper published in the April, 1986 issue of *Cell*. That paper came from Nobel laureate David Baltimore's laboratory at the Massachusetts Institute of Technology (MIT). The paper claimed that a foreign gene inserted into mice had influenced the way the mice produced antibodies, a finding that was described by many scientists as "surprising" and "important."

What began as a laboratory dispute between Tufts University immunologist Thereza Imanishi-Kari and her then postdoc, Margot O'Toole, escalated into one of the most celebrated cases of scientific fraud. The trouble started when O'Toole, unable to repeat the results and puzzled by data purporting to support them, charged that the paper was based on data faked by Imanishi-Kari, one of the six authors of the paper. Rather than refute these claims by repeating the experiments, the scientists, led by Baltimore, closed ranks. O'Toole was asked to relinquish her place in the lab. Unable to get

[4]Lock, S. 1988. "Misconduct in Medical Research: Does it Exist in Britain?" *British Medical Journal* 297: 1531.

[5]Hamblin, T. J. 1981. "Fake." *British Medical Journal* 283: 1671–1674.

[6]Le Fanu, J. 1983. "Pasteur's Notes Tells a Different Story." *Medical News* 7: 9.

[7]In "Pathological Science" (Colloquium at the Knolls Research Lab, December 18, 1953), Irving Langmuir provides an amusing and tragic account of a series of "discoveries" (primarily by physicists) in which they found things that were never there. Some of these "discoveries" generated hundreds of supporting studies.

[8]Broad, W. J. 1980. "Imbroglio at Yale." *Science* 210: 38–41, 171–173.

[9]"Stanford Denies Cover-Up of Research Fraud." *NY Times* 23 August 1981, p. 31.

[10]Broad, W. and N. Wade. 1982. *Betrayers of the Truth*. New York: Simon and Schuster.

recommendations for another job, O'Toole took a job in her brother's moving company. Conversely, Imanishi-Kari was hired at Tufts University. Baltimore claimed that, "the errors that have been identified in the *Cell* paper were inconsequential to the conclusions." Later investigations by two different panels at the National Institutes of Health (NIH) and by three Congressional hearings prompted many scientists to rally behind Baltimore and call the hearings a "witch hunt." However, the investigation continued and produced remarkable evidence. For example, the radiation-counter tapes attached to Imanishi-Kari's notebooks were green. When the Secret Service studied more than 60 notebooks from other scientists in the same lab, they found no similar tapes dated later than January 1984. A more detailed study of tape color, type font, and ink produced what the Secret Service called a "full match" between Imanishi-Kari's tapes and those produced by Charles Maplethorpe, then a graduate student in her lab, between November 26, 1981 and April 20, 1982. This suggested to investigators that the tapes were not generated as part of the research for the *Cell* paper at all. The 121-page minutely detailed draft-report of the NIH's Office of Scientific Integrity claimed that Imanishi-Kari committed "serious scientific misconduct" by "repeatedly present[ing] false and misleading information" to federal investigators. The report described O'Toole's actions as "heroic" and praised her "dedication to the belief that truth In science matters." The report criticized Baltimore who, even after an earlier NIH panel found "significant errors of misstatement and omission" in the paper, called for "all scientists" to support Imanishi-Kari in the face of mounting evidence of fraud. It also claimed that Imanishi-Kari "fabricated" and "falsified" data.

The statements of Baltimore's that the Office of Scientific Integrity found "most deeply troubling" were made to the NIH in 1990. These statements concerned data not included in the original paper, but which Imanishi-Kari gave to an NIH panel to support her published claims against O'Toole's charges. When the panel asked for the new data to be published as a correction to the original paper, Baltimore said, "In my mind you can make up anything you want in your notebooks, but you can't call it fraud if it wasn't published. Now, you managed to trick us into publishing—sort of tricked Thereza—into publishing a few numbers and now you're going to go back and see if you can produce those as fraud."

Throughout the investigation Baltimore had publicly battled all assaults on the integrity of his colleagues and on the paper itself. He had tried to divert attention from the controversy by depicting the controversy as a political incident—a political attack on scientific freedom. Baltimore even claimed NIH was somehow responsible for the fraud when he said that, "If those data were not real, then she (Dr. Imanishi-Kari) was driven by the process of investigation into an unseemly act. . . ." However, in late March of 1991 when drafts of the NIH report began to circulate, Baltimore suddenly tried to distance himself from the paper, saying that it raised "very serious questions." Moreover, on March 20, 1991 Baltimore finally asked that the *Cell* paper—the paper he had so adamantly defended—be retracted.[11] Despite his claims that he would not abandon a colleague in distress, he also then severed ties with Imanishi-Kari.

Late in 1994, the Office of Research Integrity concluded that Imanishi-Kari committed scientific misconduct when she "fabricated and falsified" data. That report found Imanishi-Kari guilty of 19 charges of misconduct.

Although most scientists claim that such incidents are rare, other evidence suggests that fraud may be more common than is believed.[12] For example, a 1993 report by the Department of Health and Human Services found Robert C. Gallo—a scientist who hogged credit for finding the cause of AIDS—guilty of misconduct for misrepresenting his work. In 1994, it was reported that at least 11 institutions falsified data in a national breast cancer research project. Incredibly, the leader of the study defended the study's findings: "Yes, there was falsified data. That does not mean that the study was flawed."

Interestingly, scientific fraud is a growth industry—each month there are conferences and symposia on the topic. As you might guess, most retractions regarding faked data are ambiguous and loaded with doublespeak.

Scientists who publish faked data break the strong contract they have with their peers to honestly obtain, record, and publish their observations. Since everyone's work depends, in some way, on the work of others, scientists who consciously deceive other scientists by providing too few details to repeat their work, by reporting information that could lead others astray, and by publishing faked data should be pariahs.

[11]Elmer-DeWitt, P. 1991. "Thin Skins and Fraud at MIT." *Time* 1 April 1991; Anonymous. 1991. "The Baltimore Affair: Ignoble." *The Economist* 30 March 1991; Hamilton, David P. 1991. "NIH Panel Finds Fraud in *Cell* Paper." *Science* 251: 1552–1554; Hamilton, David P. 1991. "Verdict in Sight in the Baltimore Case." *Science* 251: 1168–1172.

[12]St. James-Robert, I. "Cheating in Science." 1976. *New Scientist* 72: 466–469; Altman, L. and L. Melcher. 1983. "Fraud in Science." *British Medical Journal* 286: 2003–2006.

James Watson and Francis Crick

A Structure for Deoxyribose Nucleic Acid

In the previous chapter you read part of *The Double Helix,* a best-selling book written by James Watson about his search for the structure of DNA. Here is the *Nature* paper that came from Watson's and Crick's research:

We wish to suggest a structure for the salt of deoxyribose nucleic acid (D.N.A.). This structure has novel features which are of considerable biological interest.

A structure for nucleic acid has already been proposed by Pauling and Corey.[1] They kindly made their manuscript available to us in advance of publication. Their model consists of three intertwined chains, with the phosphates near the fibre axis, and the bases on the outside. In our opinion, this structure is unsatisfactory for two reasons: (1) We believe that the material which gives the X-ray diagrams is the salt, not the free acid. Without the acidic hydrogen atoms it is not clear what forces would hold the structure together, especially as the negatively charged phosphates near the axis will repeal each other. (2) Some of the van der Waals distances appear to be too small.

Another three-chain structure has also been suggested by Fraser (in the press). In his model the phosphates are on the outside and the bases on the inside, linked together by hydrogen bonds. This structure as described is rather ill-defined, and for this reason we shall not comment on it.

We wish to put forward a radically different structure for the salt of deoxyribose nucleic acid. This structure has two helical chains each coiled round the same axis (see diagram). We have made the usual chemical

[1] Pauling, L., and Corey, R. B., *Nature,* 171, 346 (1953); *Proc. U.S. Nat. Acad. Sci.,* 39, 84 (1953).

This figure is purely diagrammatic. The two ribbons symbolize the two phosphate-sugar chains, and the horizontal rods the pairs of bases holding the chains together. The vertical line marks the fibre axis.

assumptions, namely, that each chain consists of phosphate diester groups joining β-D-deoxyribofuranose residues with $3',5'$ linkages. The two chains (but not their bases) are related by a dyad perpendicular to the fibre axis. Both chains follow right-handed helices, but owing to the dyad the sequences of the atoms in the two chains run in opposite directions. Each chain loosely resembles Furberg's[2] model No. 1; that is, the bases are on the inside of the helix and the phosphates on the outside. The configuration of the sugar and the atoms near it is close to Furberg's "standard configuration," the sugar being roughly perpendicular to the attached base. There is a residue on each chain every 3–4 A. In the z-direction. We have assumed an angle of 36° between adjacent residues in the same chain, so that the structure repeats after 10 residues on each chain, that is, after 34 A. The distance of a phosphorus atom from the fibre axis is 10 A. As the phosphates are on the outside, cations have easy access to them.

The structure is an open one, and its water content is rather high. At lower water contents we would expect the bases to tilt so that the structure could become more compact.

The novel feature of the structure is the manner in which the two chains are held together by the purine and pyrimidine bases. The planes of the bases are perpendicular to the fibre axis. They are joined together in pairs, a single base from one chain being hydrogen-bonded to a single base from

[2]Furberg, S., *Acta Chem. Scand.*, 6, 634 (1952).

the other chain, so that the two lie side by side with identical z-co-ordinates. One of the pair must be a purine and the other a pyrimidine for bonding to occur. The hydrogen bonds are made as follows: purine position 1 to pyrimidine position 1; purine position 6 to pyrimidine position 6.

If it is assumed that the bases only occur in the structure in the most plausible tautometric forms (that is, with the keto rather than the enol configurations) it is found that only specific pairs of bases can bond together. These pairs are: adenine (purine) with thymine (pyrimidine), and guanine (purine) and cytosine (pyrimidine).

In other words, if an adenine forms one member of a pair, on either chain, then on these assumptions the other member must be thymine; similarly for guanine and cytosine. The sequence of bases on a single chain does not appear to be restricted in any way. However, if only specific pairs of bases can be formed, it follows that if the sequence of bases on one chain is given, then the sequence on the other chain is automatically determined.

It has been found experimentally[3,4] that the ratio of the amounts of adenine to thymine, and the ratio of guanine to cytosine, are always very close to unity for deoxyribose nucleic acid.

It is probably impossible to build this structure with a ribose sugar in place of the deoxyribose, as the extra oxygen atom would make too close a van der Waals contact.

The previously published X-ray data[5,6] on deoxyribose nucleic acid are insufficient for a rigorous test of our structure. So far as we can tell, it is roughly compatible with the experimental data, but it must be regarded as unproved until it has been checked against more exact results. Some of these are given in the following communications. We were not aware of the details of the results presented there when we devised our structure, which rests mainly though not entirely on published experimental data and stereochemical arguments.

It has not escaped our notice that the specific pairing we have postulated immediately suggests a possible copying mechanism for the genetic material.

Full details of the structure, including the conditions assumed in building it, together with a set of co-ordinates for the atoms, will be published elsewhere.

We are much indebted to Dr. Jerry Donohue for constant advice and criticism, especially on interatomic distances. We have also been stimulated by a knowing of the general nature of the unpublished experimental results

---

[3]Chargaff, E., for references see Zamenhof, S., Brawerman, G., and Chargaff, E., *Biochim. et Biophys. Acta.* 9, 402 (1952).

[4]Wyatt, G. R., *J. Gen. Physiol.*, 36, 201 (1952).

[5]Ashbury, W. T., *Symp. Soc. Exp. Biol.* 1, Nucleic Acid, 66 (Camb. Univ. Press, 1947)..

[6]Wilkins, M. H. F., and Randall, J. T., *Biochim. et Biophys. Acta*, 10, 192 (1953).

and ideas of Dr. M. H. F. Wilkins, Dr. R. E. Franklin and their co-workers at King's College, London. One of us (J.D.W.) has been aided by a fellowship from the National Foundation for Infantile Paralysis.

Understanding What You've Read

Compare the writing style in this article with that of *The Double Helix*. How do the writing styles differ? For example, how does the terminology, sentence length, and title differ in the two articles? What does the *Nature* article conceal? How does each style of writing achieve its purpose?

Exercises

1. Examine a paper published in a scientific journal of your choice. What hypotheses were the scientists testing? What were their assumptions? Do you agree with their conclusions? How would you rewrite the paper?

2. Read the "Letters" section of an issue of *Science*. What is the purpose of each letter? To provide information? To persuade? To sway opinion?

3. Go to a library and examine a thesis, research paper, review article, book, and an article written for a nonscientist. How does the writing differ? How is it similar?

4. Write an abstract for an article of your choice in *Scientific American*.

5. Discuss, support, or refute the ideas in these quotations:

    > The total energy of the universe is constant and the total entropy is continually increasing. — Isaac Newton

    > In the year 1657 I discovered very small living creatures in rain water. — Antonie van Leeuwenhoek

    > You can observe a lot just by watching. — Yogi Berra

    > Chance favors the trained mind. — Louis Pasteur

    > I am now convinced that theoretical physics is actual philosophy. — Max Born (1882–1970)

    > Physics, beware of metaphysics. — [Sir] Isaac Newton (1642–1727)

    > It behooves us always to remember that in physics it has taken great men to discover simple things. They are very great names indeed which we couple with the explanation of the path of a stone, the droop of a chain, the tints of a bubble, and the shadows in a cup. — d'Arcy Wentworth Thompson (1860–1948)

UNIT THREE

PRESENTING
INFORMATION

A picture shows me at a glance what it takes dozens
of pages of a book to expound.
— Turgenev

A tabular presentation of data is often the heart of,
better, the brain, of a scientific paper.
— Peter Morgan

CHAPTER EIGHT

Numbers, Tables, and Figures

All science as it grows toward perfection becomes mathematical in its ideas.
> — Alfred North Whitehead

Mathematics is both the door and the key to the sciences.
> — Roger Bacon

And ye who wish to represent by words the form of man and all the aspects of his membranification, get away from the idea. For the more minutely you describe, the more you will confuse the mind of the reader, and the more you will prevent him from a knowledge of the thing described. And so it is necessary to draw and describe.
> — Leonardo da Vinci

The use of diagrams is a particular instance of that method of symbols which is so powerful an aid in the advancement of science.
> — James Clerk Maxwell

Every picture tells a story, don't it?
> — Rod Stewart

Numbers

> For he who knows not mathematics
> cannot know any other science; what is
> more, he cannot discover his own igno-
> rance, or find its proper remedy.
> — Roger Bacon

> Numerical precision is the very soul of
> science.
> — Sir D'Arcy Thompson

> Through and through the world is infest-
> ed with quantity: to talk sense is to talk
> quantities. It is no use saying the nation
> is large—how large? It is no use saying
> that radium is scarce—how scarce? You
> cannot evade quantity.
> — Alfred North Whitehead

> If you can measure that of which you
> speak, and can express it by a number,
> you know something of your subject;
> but if you cannot measure it, your
> knowledge is meager and unsatisfactory.
> — William Thomson (Lord Kelvin)

> Even in the valley of the shadow of
> death, two and two do not make six.
> — Leo Tolstoy

Science removes the fuzz and mystery from the world by using experiments and precise language. The most precise language is math. For example, there was no doubt that Galileo reported four—not three, not five—moons around Jupiter.

Scientists rely on numbers to describe many aspects of science, such as sizes of animals, heights of seedlings, and numbers of base-pairs in genes. Similarly, words such as *larger* and *smaller* are merely verbal expressions of numbers. Consequently, your ability to understand and write about numbers will affect your success as a scientist. To show this, consider the work of Gregor Mendel. Mendel's experiments with pea plants weren't original; several other scientists had done the same crosses and noted the same kinds of offspring seen by Mendel. Mendel's work differed from that of others in only one regard: Mendel *counted* his results. This allowed Mendel to determine the ratios that helped him explain some kinds of inheritance.

Scientists must know how to write effectively about numbers. Writing effectively about numerical data involves knowing (1) what measurements to make, (2) how to manipulate numbers statistically, and (3) how to write about numbers.

Measurements in Science

Improved technology has greatly improved the accuracy of many of our measurements. For example, consider the speed of light, the one measure that stands as the most absolute in all of science. In late 1972, the accuracy of the measurement of that speed was increased 100-fold by Ken Evenson and his colleagues at the National Bureau of Standards when they estimated the speed of light to an accuracy of 0.5 meter per second. Such accuracy seems remarkable, but we still do not know exactly how fast light travels. All we have is a more accurate estimate. Nevertheless, such improved accuracy can have many effects. For example, before 1983 we determined the length of a meter independently, and the speed of light was specified by the length of a meter. Now, because the speed of light is considered constant, it defines the length of a meter: A meter is defined as the distance light travels in 1/299,792,458 of a second. Any straightedge that long is a meter long. Obviously, we typically use indirect and inexact methods to make a measuring stick one meter long.

Another constant often used by scientists is Avogadro's number, which is the number of molecules in one mole (gram-molecular weight) of a substance. Avogadro's number is given as 6.02486×10^{23} molecules, but there is a 0.0027% error in this estimate. Although that error is relatively small, it comes to about 16,000,000,000,000,000,000 molecules. Stated another way, Avogadro's number is accurate to about 16 quintillion molecules.

Do not be alarmed by this lack of complete accuracy and certainty in measurements. It simply means that many of our measurements can be improved. All of our measurements are imperfect.

The International System of Units

Scientists report measurements with a type of metric system called The International System of Units. This system is abbreviated SI and is the standard system of measurement used by scientists around the world. The system has several advantages, including that each quantity in SI has only one associated unit.

Measurements used most often by scientists include distance, mass, amount of substance, time, temperature, electric current, and luminous intensity. Prefixes to convert these so-called "base units" to other units—for example, to change meters to kilometers—are listed in Appendix 5.

Distance

The SI unit to measure distance is the **meter** (m). Area is measured as square meters (m²), and volume is measured as cubed meters (m³). Although the hectare and liter are used often by scientists, they are not SI units.

Here are equations to convert some English units to SI units:

1 m = 39.4 inches = 1.1 yd

1 km = 1,000 m = 0.62 mi

1 ha = 10,000 m² = 2.47 acres

1 in = 2.54 cm

1 ft = 30.5 cm

1 yd = 0.91 m

1 mi = 1.61 km

1 L = 2.1 pt = 1.06 quart = 0.26 gal = 1,000 ml = 0.001 m³

1 ml = 0.03 fl oz

Mass

The SI unit to measure mass is the **kilogram** (kg). The kilogram is the only SI unit containing a prefix (kilo). Units derived from kilogram measure force (newton, N), pressure (pascal, Pa), work and energy (joule, J), and power (watt, W). Calorie is not an SI unit because it can't be defined without using an experimentally derived factor.

Remember that mass and weight are not equivalent. Mass measures an object's potential to interact with gravity and cannot be measured directly. Weight is force exerted by gravity on an object. An object weightless in outer space has the same mass as it has on earth. For more on misused words, see Appendix 2.

Amount of a Substance

The SI unit to measure the amount of a substance is the **mole** (mol). This mass is numerically equal to the molecular weight of a substance. The elementary entities (for example, atoms or ions) must be specified in the measurement. Concentration is expressed as moles per cubic meter (mol $\cdot$ m^{-3}).

Time

The SI unit to measure time is the **second** (s). Units derived from the second are used to measure frequency (hertz, Hz), speed (m $\cdot$ s^{-1}), and acceleration (m $\cdot$ s^{-2}).

> ## Water Boiling at Zero Degrees Celsius?
>
> The Celsius temperature scale was developed in 1742 by Anders Celsius, a Swedish astronomer. Celsius originally set the boiling point of water at 0°C and the freezing point of water at 100°C. J. P. Christine later revised the scale to its present-day form, with water boiling at 100°C and freezing at 0°C. Therefore, °C is appropriate in any case, but "degrees Christine" is technically more accurate than "degrees Celsius."

Temperature

The SI unit to measure temperature is the **kelvin** (K). Zero kelvin (0 K) is absolute zero, the hypothetical temperature characterized by the absence of heat and molecular motion. There are no prefixes for kelvin. Since temperature measures the intensity of molecular motion, all Kelvin temperatures exceed zero. A Kelvin unit is the same as a degree Celsius, and to get from Celsius to Kelvin add 273 (K = °C + 273). Thus, water freezes at 273 K, room temperature is about 293 K, and body temperature is about 310 K. Note that kelvin measurements lack degree signs (°).

Most scientists use the Celsius scale to measure temperature. This scale uses 100 units to separate the temperatures at which water freezes and boils. Convert Fahrenheit temperatures to Celsius temperatures with this formula:

$$5(°F) = 9(°C) + 160$$

Electric Current

The SI unit to measure electric current is the **ampere** (A). Units derived from the ampere measure force or potential (volt, V) and electrical resistance (ohm).

Luminous Intensity

The SI unit to measure luminous intensity is the **candela** (cd). Units derived from the candela measure flux (lumen, lm). An incandescent light emits 10 to 20 lm for each watt of energy consumed, while a fluorescent light emits about 60 or more lumens for each watt.

Writing About SI Measurements

Follow these conventions when working with SI units and measurements:

SI uses symbols, not abbreviations. Use a period after a symbol only at the end of a sentence. SI symbols are always singular.

1.5 kg, not 1.5 kg.

10 km, not 10 kms

Symbols for prefixes are lowercase letters except for symbols denoting a million or more.

M (mega; 10^6) P (peta; 10^{15}) G (giga; 10^9)

E (exa; 10^{18}) T (tera; 10^{12})

Write symbols immediately after their prefix.

kg, not k g

State units completely. Do not use a prefix alone and do not use compound prefixes.

km, not kilo

nm, not mmm

Insert a space between the number and unit. However, do not space between the symbol and its prefix.

10 km, not 10km

Write symbols in lowercase letters except for unit names derived from a person's name.

A (ampere) J (joule)

Hz (hertz) V (volt)

K (kelvin) W (watt)

Pa (Pascal) N (newton)

Use a centered dot for a compound unit formed by multiplying two or more units.

kg · sec⁻¹, not kg × sec⁻¹

Express measurements with units requiring the fewest number of decimals.

1 km, not 1,000 m or 10^3 m

Do not mix symbols or units.

6.2 m, not 6 m 200 mm

Statistics

> Statistics are no substitute for judgment.
> — Henry Clay

> If your experiment needs statistics, you
> ought to have done a better experiment.
> — [Lord] Ernest Rutherford

> There are three kinds of lies—lies,
> damned lies and statistics.
> — Mark Twain

> One-third of the mice used in this experi-
> ment were cured by the test drug; one-
> third of the test population were unaf-
> fected by the drug and remained in a
> moribund condition; the third mouse
> got away.
> — related by Erwin Neter, Editor-in-
> Chief, *Infection and Immunity*

Statistics is a tool used to answer general questions on the basis of only a limited amount of specific information. Since statistics involves basing a judgment on a sample of a population (rather than the entire population), it helps us make a decision with incomplete knowledge of the population. Although this sounds unscientific, we do it all the time. For example, we diagnose disease with only a drop of blood, and we judge a watermelon with only a few thumps on its rind. Statistical studies of populations are necessary when it is impossible or unrealistic to test the entire population.

Scientists use **statistics** to reduce a population to a few characteristic numbers such as average height or percent protein. Without statistics, scientists would have to report raw data about each individual. This would be unwieldy as well as frustrating. For example, knowing the protein content of a particular grain of corn (*Zea mays*) is usually trivial; what we want to know is the protein content of a *typical* corn grain. To know that, we must study groups of corn grains.

The following is not meant to be a comprehensive description of statistical techniques. Rather, I present only the basic measures you'll need to include in most lab reports. If you want to know more, read a good introductory statistics book.

Let's start our discussion with the mean and median.

The **mean** is the arithmetic average value of the variables.

The **median** is the middle value of a group of measurements.

The mean is more sensitive to extreme values than is the median. To appreciate this, consider a group of 14 plants having the following heights: 80 cm, 69 cm, 62 cm, 74 cm, 69 cm, 50 cm, 45 cm, 40 cm, 9 cm, 64 cm, 65 cm, 64 cm, 61 cm, and 67 cm. The mean height is 58.6 cm. However, none of the plants are that height, and most of the plants are taller than 60 cm. Does the mean describe the "typical" plant?

To determine the median, first arrange the measurements in numerical order. Our sample would look like this: 9 cm, 40 cm, 45 cm, 51 cm, 61 cm, 63 cm, 64 cm, 64 cm, 65 cm, 67 cm, 69 cm, 69 cm, 73 cm, 80 cm. The median is between the seventh and eighth measurement; that is, the median is 64 cm. Note that in this example the mean differs from the median. What causes this difference? How would the mean change if the shortest plant were not in the sample?

The mean of two samples could be the same, yet the samples could differ significantly. For example, consider these samples:

- <u>Group 1</u>: 25 cm, 35 cm, 28 cm, 32 cm   Mean = 30 cm
- <u>Group 2</u>: 10 cm, 15 cm, 20 cm, 75 cm   Mean = 30 cm

Thus, reporting only the mean does not give readers a good description of the sample. To better describe the sample, you must provide some measure of the spread and variability of the data. You can do that by reporting the range and standard deviation.

The **range** is the difference between the largest and smallest values of the sample. For example, the range of Group 1 is 10 (35–25), whereas that of Group 2 is 65 (75–10). Although these measures are informative, they ignore all other data between the extremes.

The **standard deviation** measures how much each value differs from the mean. The standard deviation is easy to calculate: calculate the mean, measure the deviation of each sample from the mean, square each deviation, and then sum the deviations. For example, consider a group of trees that are 22 years, 19 years, 21 years, and 18 years old. The mean age is 20 years.

Individual	Mean	Deviation	(Deviation)^2
22	20	2	4
19	20	−1	1
21	20	1	1
18	20	−2	4

The sum of the squared deviations is 4 + 1 + 1 + 4 = 10. Divide this number by the number of observations minus one (4 − 1 = 3). This produces a value of 10/3 = 3.3 years^2, which is the **variance** (note that the units are years squared). The standard deviation is the square root of the variance: 1.8 years. The standard deviation is usually reported with the mean. For example, "The mean age of the trees was 20 ± 1.8 years."

The standard deviation (SD) is important for understanding the spread of a sample. For many distributions of values for a variable, the mean ± 1 SD encompasses 68% of the observations, whereas the mean ± 2 SD encompasses 95% of the observations. These distributions are the basis for determining what values of some variable are considered normal; those that are not considered normal (in terms of these distributions) may be scientificly significant and worthy of more study.

Most numerical data must be subjected to appropriate statistical tests to determine if differences are substantial enough to be convincing. Indeed, many journals require that you use statistical tests to verify your results. Here are things to remember when using statistical tests to compare numerical data:

Most measurements will produce different means. Are the differences between means due to real differences between the samples or due only to a small sample size? Consider again the corn plants mentioned above. The silo of seeds from which you picked your experimental plants may have contained thousands of other seeds. How do you know that you didn't pick the ones that are most different from the "average" seed in the bag? One way to test this would be to increase your sample size—to repeat the experiment with, say, 1,000 seedlings. Better yet, measure 100,000 seedlings. Better yet, measure 1,000,000,000 seedlings. Better yet, . . .

The only way to know "the truth" and be certain of your conclusions would be to measure all of the corn seedlings in the world. Since this is impossible, we must measure seedlings that *represent* all of the corn seedlings in the world. Such representative samples only estimate the truth—they only tell us how convincing or wacky our results are. Similarly, statistics tell us how representative our samples are.

Define a hypothesis, which is the issue that you'll test. This hypothesis is termed the null hypothesis and is designated H_0. The null hypothesis assumes nothing unusual has happened. Although your data may be convincing, they cannot *prove* a hypothesis. Rather, they can only discredit or support the hypothesis.

Statistics reduces a population to a few characteristic numbers. Statistics are not data, and you lose information as you reduce your population to its characteristic numbers. The farther you reduce your data, the farther you remove yourself from the data. Therefore, keep the statistics as simple as you can.

Decide on a reasonable degree of risk (usually 5%) of incorrectly rejecting the null hypothesis. No matter what statistical test you use, there is always a chance that you will make the wrong decision. All that a statistical test will tell you is the *chance* of making that wrong decision. You cannot prove a hypothesis with statistics.

Common statistical tests include Student's *t*, Chi-square, and analysis of variance (ANOVA). The type of test that you'll use will depend on the amount and type of your data and the nature of your null hypothesis. Each test manipulates data in standard formulas and produces a test number. Test numbers near zero suggest that your data are consistent with the null hypothesis, while test numbers far from zero suggest that the null hypothesis is incorrect.

Look up the test number in a standard table to see if you're in the expected range of values. If your test number falls within this range, your data support the null hypothesis. If your test number is outside this range, your data support the alternative.

If you must do a statistical analysis, be sure to do the *correct* statistical analysis.[1] Consult a statistics book or a statistician to determine which test you must use in your experiment. When comparing two means, you'll probably need only to use Student's *t* test, a simple calculation.

Remember that oddball data are rare, but possible.

Always report the number of observations (n), the arithmetic mean (x), and the standard deviation or standard error to give readers a sense of the data's variability (see below). Such information may be reported in the text or a table—for example, as 496 ± 60 (9), in which the first two numbers show the mean plus or minus the standard deviation or standard error (indicate which), and the third number (in parentheses) representing the number of data points.

The standard deviation summarizes the spread of data about the mean, while the standard error estimates the accuracy of the mean, not the spread of the data. A small standard deviation or standard error indicates a uniform population.

The standard deviation tells you if an individual is close to the mean. About 67% of a normal population is within one standard deviation, 95% within two standard deviations, and 99% within 2.6 standard deviations.

State the test of significance that you used with your data (a significance test summarizes the comparisons of your data with other populations). Also state the probability (P) values obtained with the test; these values indicate the confidence of your conclusion. Do this with statements such as $P < 0.001$, which means that the probability of the results being due to chance are less than one in a thousand. Traditionally, a result is not significant if the likelihood of obtaining the result by chance alone exceeds 5% ($P > 0.05$). Use the word *significant* only to denote statistical significance. *Significant* is not synonymous with *important* or *meaningful*.

If you do no statistical analyses, be especially cautious when making conclusions about your data. Do not say that any differences between groups of measurements are significant or insignificant.

---

[1] In medical journals, about half of the articles that use statistics use them incorrectly. See Glantz, S.A. 1980. "Biostatistics: How to Detect, Correct and Prevent Error in the Medical Literature." *Circulation* 61: 1–7; Gore, S.M., I.G. Jones, and E.C. Rytter. 1977. "Misuse of Statistical Methods: Critical Assessment of Articles in BMJ from January to March 1976." *British Medical Journal* 1: 85–87.

Writing About Numbers

Follow these conventions when you include numbers in text:

Spell out one-digit numbers. Write all other numbers as the numeral except at the beginning of a sentence and as a measurement. Spell out numbers at the beginning of a sentence.

Fourteen grams of . . .

. . . was 14 times larger than . . .

Use numerals with standard units of measurement.

I added 104 g of NaCl to the water.

Express large numbers in figures or in mixed figure–word form.

$9,000,000 or $9 million

5,700,000 or 5.7 million

In America and Britain, scientists use commas to separate long numbers into groups of three digits. However, Europeans use periods or spaces to group these digits.

AMERICAN AND BRITISH USAGE: 1,354,000,000

EUROPEAN USAGE: 1 354 000 000

If two or more numbers appear in the same sentence and if one of the numbers exceeds 10, use numerals for all of the numbers.

I added 7 g of NaCl and 19 g of $CaCl_2$ to the media.

Express related numbers or amounts within a sentence entirely in figures or entirely in words unless doing so would confuse the reader.

Avoid placing next to each other two numbers referring to different things.

NO: In 1981 15 states ratified the law.

YES: Fifteen states ratified the law in 1981.

Hyphenate cardinal numbers (e.g., twenty-one to ninety-nine) and ordinal numbers (twenty-first).

Don't present two numbers together.

two 100-watt lamps, *not* 2 100-watt lamps

Use a zero before a decimal point when the number is less than one.

0.21, *not* .21

Use numerals to designate negative numbers, numbers with decimals, percentages, amounts of money, and parts of a book.

-40 °C, -11, page 102, Figure 3, 4%, $8.60, Chapter 4, Table 8

Use the correct number of significant figures. Do not express the result of a calculation in more decimal places than in the least accurate part of the calculation. The accuracy of a calculation can't compensate for the lack of accuracy in collecting or recording data. A measurement reported as "5" is not synonymous with one reported as "5.000."

Be careful how you compare numbers. For example, "three times more" equals "four times as much as." Similarly, "1:5" equals "1 in 6."

Don't write "approximately 123.5 ml." Just write "123.5 ml."

Remember that "not statistically significant" is not synonymous with "insignificant." Improbable does not equal never.

Avoid using the words representing numbers greater than a billion. These words have different meanings in other countries. For example, in the United States a billion equals a thousand million (that is, giga = 10^9); in some other countries, a billion equals a million million (i.e., tera = 10^{12}).

Centering and numbering equations make a document neater and more readable. Include short equations in the text. For example:

On Ludwig Boltzmann's tombstone in the Zentralfriedhof in Vienna is carved his formula for entropy, $S = K \log \Omega$. Boltzmann committed suicide in 1906 at the age of 62.

Tables and Figures

Everything should be as simple as it can be, yet no simpler.
— Albert Einstein

Good visuals can make or break a science paper because people understand and remember images and pictures more readily than they do abstract words. Visuals can clarify images too complex to be described with words.

Scientists use two types of visuals in their writing: tables and figures. Tables are arrangements of numbers or words in rows and columns. Figures are everything else, including photographs, drawings, diagrams, and graphs. Several principles hold true for tables and figures:

Do not include a table or figure unless you need it and will discuss it.

Do not present all of the data that you gathered. Rather, include only *representative* data.

Place the table or figure near where it is referenced in the text. Use words to draw attention to the table or figure.

Refer in the text to every figure, and use a consistent style for each figure. If possible, put only one table or figure on a page.

Label all axes and columns. Use large letters and numbers so that the information will be readable when the table or figure is reduced.

Include an informative title that explains the figure without referring to the text.

Be precise, clear, and concise. This doesn't mean that you should make the figures small; rather, ensure that each figure quickly informs readers of your message.

Present a realistic depiction of a process, organism, or object. Base this realism on detail included in the illustration.

Strive for clarity and simplicity. If your figures contain too much information, the reader will learn nothing. If your figures are too complex, either delete the figure or split it into two or more parts.

Plan the figures before you write the text so that they are an integral part of the article rather than an appendage tacked on as an afterthought. Figures must mesh with the text, like gears that drive a machine. If they don't, you're likely to end up with text and photos that resemble this advertisement that appeared in a magazine:

Figure 8–1
Why you need to coordinate your illustrations with your text.

Tables

Tables are an excellent way to summarize and emphasize material, reveal comparisons, show significance, and present absolute values where precision is important. However, tables must do more than merely "show the data." Well designed tables summarize and point out the meaning and significance of data without interrupting the text. Use a table only if you must present repetitive data. Purposeless tables mystify readers, as does purposeless writing.

Here's how to design an effective table:

Include a clear, intelligible title and legend for each table. Number the tables consecutively.

Make the tables understandable and independent of the text. Don't force readers to refer to the text to understand a table. Be sure that data in the table agree with those in the text.

Align the decimal points of numbers in the table.

Ensure that like elements in the table read down. Include units in column headings and avoid using exponents in headings, for they often confuse readers. Specifically, readers don't know if the data are to be, or have been, multiplied by these numbers. For example, "cpm $\times$ 10^3" refers to thousands in the *Journal of Bacteriology*, whereas "cpm $\times$ 10^{-3}" refers to thousands in the *Journal of Scientific Chemistry*.

Check each table for internal accuracy. For example, do all of the percentages add to 100? Place columns to be compared next to each other.

Explain data with footnotes at the end of the table.

Indicate the position of each table by adding a note in the margin of the text. Be sure that all tables are cited and in the correct order.

Do not use a table to describe only one or two tests. Describe such results in the text of the paper.

Include only significant, representative results in a table. If a variable has no effect, delete it from the table. Use tables to present variables and data, not standard conditions.

Provide camera-ready copies of complicated tables. This will help you avoid laborious proofreading and help ensure accuracy.

Table 8–1 is an example of an effective table:

Table 8–1 Use of Contraceptives in the United States.*

Method	Estimated % Use	% Accidental Pregnancy in One Year of Use†
Male sterilization	15	0.15
Female sterilization	19	0.4
Oral contraceptive pill	32	3
Condom	17	12
Diaphram + spermicide	5	18
"Rhythm" (periodic abstinence)	4	20
IUD	3	6
Contraceptive sponge	3	18
Vaginal foams, jellies	2	21

* Data from *Developing New Contraceptives: Obstacles and Opportunities.* Washington, DC: National Academy Press, 1990.
† About 89% of women using no contraceptives become pregnant within one year.
Source: Arms and Camp, A Journey Into Life, 2nd ed., 1991.

Figures

Although tables and figures have the same goals as language—namely, to communicate quickly, precisely, and clearly—the kind of figure that you use must be appropriate to the concept and your purpose:

Figure	Best Illustrates
bar graph ("histogram")	comparisons and relative quantities; emphasize high-low comparisons; excellent for displaying tabular data when no mathematical relationship exists between the variables; meaningless for small numbers or small differences

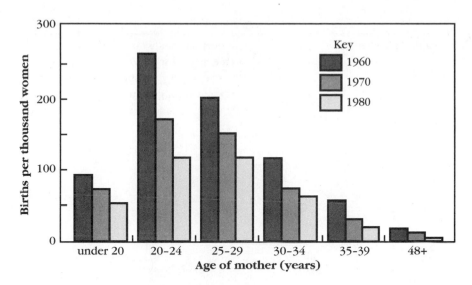

Figure 8–2
America's falling birth rate. This graph shows the number of births per thousand women in various age groups during 1960, 1970, and 1980. The number of births in all age groups fell from 1960 to 1980, while the average age of the mother increased. Both of these trends slow population growth.

Figure	Best Illustrates
line graph	trends and relationships between variables to be compared over time

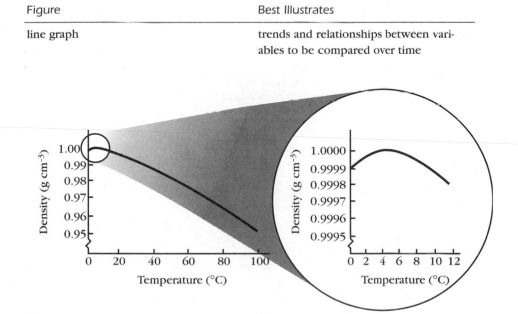

Figure 8–3
How the density of water changes with temperature. The inset at the right shows that water is most dense at 4°C.

Figure	Best Illustrates
circle graph ("pie chart")	proportions of a whole; make abstract percentages appear visually as parts of the whole; poor for showing differences between small percentages; if you use a pie chart, include no more than eight slices

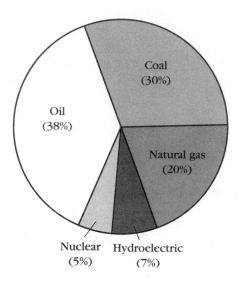

Figure 8–4
Sources of energy for industrial use in the United States. Most of our energy comes from fossil fuels.

Figure	Best Illustrates
map	locations

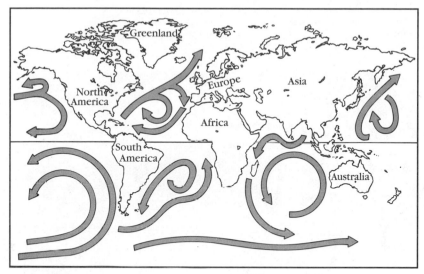

Figure 8–5
The basic patterns of surface currents in oceans are primarily due to winds. The main flow of ocean currents—that is, clockwise in the Northern Hemisphere and counter-clockwise in the Southern Hemisphere—results partly from the Earth's rotation.

Figure	Best Illustrates
line drawing	simple part of a process or an enlarge-ment to show detail

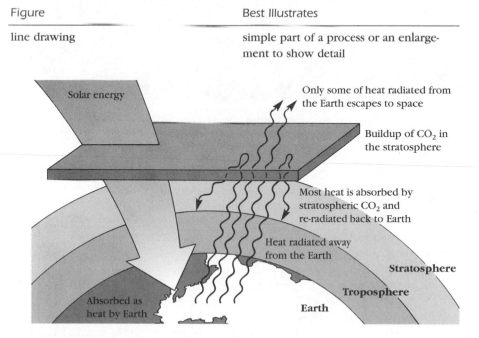

Figure 8–6
How the greenhouse effect and carbon dioxide promote global warming. Greenhouse gases in the atmosphere function like a glass roof on a greenhouse, trapping heat near the surface of the Earth.

Figure	Best Illustrates
photograph	how something looks

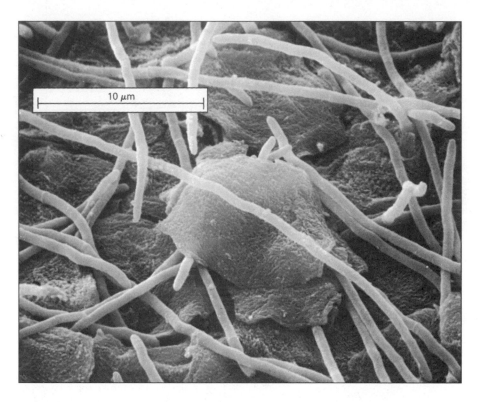

Figure 8–7
Fungal infection of human skin. Slender, hair-like hyphae grow over and under flat epidermal cells of the skin. × 4,400. (Biophoto Associates)

Graph

A graph is a pictorial table that allows fast comparisons of relationships or trends. As such, an effective graph illustrates trends, patterns, or relationships, promotes understanding, and suggests interpretations of data. Graphs do not add any importance or truth to data; they merely present data in a pictorial form. Use a graph only if it is difficult to discuss your data in the text. Include no more data than needed to make your point.

The horizontal axis (abscissa) of a graph is called the x-axis, and the vertical axis (ordinate) is called the y-axis. The x-axis describes the *independent variable*, which is the variable that's selected or controlled. Examples of independent variables include time, temperature, and protein content of a diet. The y-axis describes the *dependent variable*, which is what was measured in response to the independent variable. Examples of dependent variables include growth rate, weight, and curvature.

Here's how to design an effective graph:

Label both axes and point the scribes (the marks on the axis) inward. Be sure that numbers on the *y*-axis are upright, and indicate breaks in the axis with separated lines.

Title the graph. An example of an effective title of a graph would be "The influence of temperature on root elongation."

Use the entire area of the graph. Do not bunch data at one end of the graph.

Use appropriate, evenly spaced markings on axes. Avoid empty spaces on the graph by not extending the axes.

If the independent variable is continuous (for example, time or temperature), use a line graph. If the independent variable is not continuous, such as when you're comparing the heights of seedlings, use a bar graph.

Use common symbols such as ■, □, ●, and ○ for data points. Include a key for each symbol in the graph. Use symbols consistently— that is, use the same symbols in each graph.

If you plot more than one set of data on a graph, use different lines for each set of data. Label each line clearly. Plot no more than five lines per graph.

Do not extrapolate beyond the data points.

Box the graph. This makes it easier for readers to estimate values on the right side of the graph.

Unless you're writing a term paper or lab report, print the legend for the graph on a separate page.

Indicate the standard-error bars for each mean with vertical lines extending up and down from each point. These "error bars" tell readers how reliably each mean estimates the population mean. You can also communicate this information by plotting 95% confidence intervals at each mean. To show variability within each sample, plot the standard deviation or standard error. Specify in the legend what statistic you've plotted.

How you should connect the data points on the graph depends on your data and how they were obtained, analyzed, and interpreted.

To show a trend suggested by the data, use a smooth line. Most of the data points will touch or be close to the line. Sampling error accounts for why all points do not lie on the line.

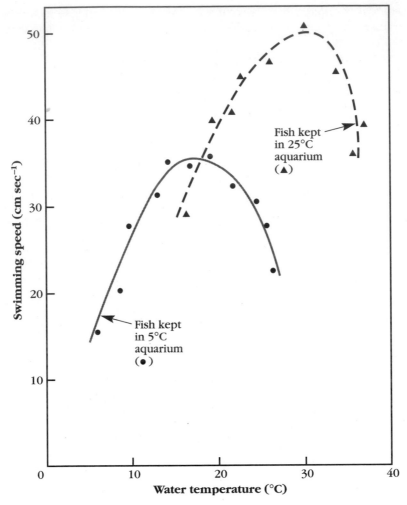

Figure 8–8
How water temperature affects swimming speed in goldfish. There is an optimal water temperature for swimming and a range of temperatures the goldfish can tolerate. By gradually changing the water temperature, a goldfish can tolerate higher or lower temperatures than it could had it been moved to the new environment abruptly.

To show high-low patterns, connect the data points with straight lines. The gaps between these points will not be due to sampling error, but rather to meaningful fluctuations. For example, Figure 8–9 shows how the size of a population of fruit flies changes when kept in a closed container and fed a constant diet of protein. The changes in the population are scientificly important.

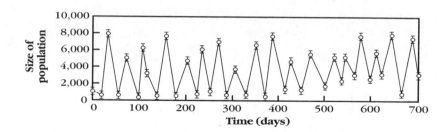

Figure 8–9

Oscillation in size (± standard deviation) of a population of *Drosophila melanogaster*. Flies were kept in a closed container and fed a constant diet of protein.

To show a mathematical relationship between the two variables, include a statistically-derived "best fit" line. In the legend, define the regression coefficient for the line.

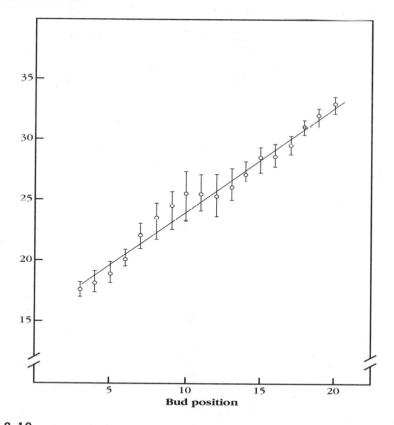

Figure 8–10

Position-dependent growth of axillary buds on *Nicotiana tabacum*. I've plotted the average number of nodes ± standard deviation produced by four buds at each position. Buds were numbered basipetally with bud number 1 being the first axillary bud immediately below the inflorescence. The regression coefficient (R^2) for the line equals 0.92.

Many computer programs such as *Excel* can generate a variety of graphs from data. Most of these graphs, when printed on a laser printer, are acceptable by journals for publication.

Line Drawings

A line drawing is an artist's concept of an object or concept constructed with lines, as shown in Figure 8-11.

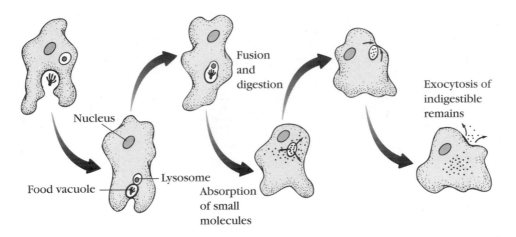

Figure 8–11
Phagocytosis, a type of endocytosis. An amoeba engulfs its prey, and part of the plasma membrane pinches off to form a food vacuole in the cell. This food vacuole then fuses with a lysosome, a membrane-bound organelle full of digestive enzymes. The small food molecules formed by digestion are absorbed into the cytoplasm before the amoeba expels the indigestible remains of the prey by exocytosis.

Line drawings are usually drawn with black ink. They should not show everything; if you want to show everything, use a photograph. Use drawings to omit extraneous details such as with a cutaway depicting the inner workings of a device. Make the drawing as simple as possible so that it won't hide the key ideas. Label the most important parts of the drawing.

Make drawings to scale and delete excess information. Make the drawings twice their final size to ensure accuracy and detail. Lines should be at least 0.3 mm thick.

Photographs

When it comes to showing precisely how an object appears, nothing is better than the realism of a photograph. Photographs show exact details, but can't speak for

Figure 8–12
Different eyes see different things. Human eyes see a marsh marigold with
no markings (*left*). Insects see large patches on the same flower (*right*, taken
with ultravoilet-sensitive film) because its eyes react to ultraviolet light, whereas our eyes do not. (Biophoto Associates)

themselves. Similarly, photographs can't lie, but they can mislead. Therefore, be
sure that each photograph represents the data you are trying to communicate. Like
illustrations, photographs are not proof; the burden of proof lies with the author.
However, photographs are more credible than drawings because photographs
show that the object actually exists and was not constructed from an artist's imagination.

Contrary to what many scientists think, not all pictures are worth 1,000 words.
You can help make them so by keeping these these things in mind:

Take photographs so that they highlight the purpose of the photograph. Be
sure that each photograph is in excellent focus, has good contrast, and has a
background that doesn't detract from the subject. Mount the photographs
into a montage (also called a photographic "plate") according to the journal's
instructions to authors.

When taking photographs for a paper, fill the frame of the photograph with
the subject. Remember what many consider Ansel Adams's best piece of
advice: "If you want to take a good picture, take a lot of pictures."

Photographs should be glossy black-and-white prints. These photographs are referred to as "halftones" because, unlike line drawings, they include shades of gray. Submit color photographs only if the journal accepts color photographs and you are willing to pay for printing the color photographs. Ask the managing editor of the journal about these costs.

Attach an adhesive sticker to the back of each photograph to indicate the top and figure number of the photograph. Similarly, indicate the approximate location of each photograph with a note in the margin of the text.

Use rub-on letters and arrows to highlight important features of the photograph, and letters to label other important parts. Include a scale marker on each photograph.

Include a legend describing the important aspects of each photograph. Type the figure legend on a separate page of paper. The reason for doing this is that the figure and legend are dealt with differently by publishers—legends go to typesetting, while photographs are handled by photographers.

A Final Check

Just as effective illustrations and photographs can improve the clarity and impact of your paper, so too can poor illustrations and photographs damage your credibility and discredit your work. Therefore, before turning in a paper or submitting a paper for publication, check the following:

Check every illustration against your original artwork. Did the artist introduce any mistakes?

Do the figure numbers correspond with those cited in the text?

Are the words, letters, and abbreviations on the figures consistent with those used in the text?

Is the figure drawn to scale? Is this scale marked on the illustration?

Does the figure contain too much information?

Are the units of measurement clearly marked on all axes and scales?

Should any photograph be replaced by a line drawing?

Is a concise, understandable legend included with each figure?

Do the photographs show what I want them to show? Are they in good focus?

Do all of the illustrations mesh with the text?

Do not sacrifice truth for appeal. However, make sure that each photograph or illustration is aesthetically appealing and professionally done. Editors are puzzled by authors who, after spending $100,000 and several months to gather data, then refuse to spend $10 to have a draftsperson prepare a professionally rendered graph.

Marie Curie

The Energy and Nature of Radiation

Marie Sklodowska Curie (1867–1934) was the most outstanding scientist to come out of Poland since Copernicus, who in 1543, claimed that the earth is a satellite of the sun. In an age when women were discouraged—and often barred—from many professions (especially chemistry and physics), Marie Curie became famous. She earned degrees in physics and math, and in 1903 became one of the first women in France to earn a doctorate. That same year, she and her husband Pierre Curie (1859–1906) shared the Nobel Prize for Physics with Henri Becquerel. Marie Curie's subsequent work led to a second Nobel Prize for Chemistry in 1911. Incredibly, the Curies' daughter Irène Joliot-Curie also won a Nobel Prize in 1935 for her discovery of artificial isotopes. Irène bombarded aluminum with alpha particles (i.e., helium nuclei), thereby converting the aluminum to a radioactive form of phosphorous.

Marie Curie studied radiation. She named polonium to honor her homeland, Poland, and in 1898 used a ton of pitchblende (a major ore of uranium) to produce about 0.1 g of pure radium, a luminous, whitish metal. This radium—the first radioactive element to be discovered—was pure enough to assign to atomic weight 225.93 (it is now calculated at 226.025), immediately below uranium and thorium. Curie's feats made her immensely famous; after Einstein, Curie was the most famous scientist of her time. Interestingly, she and her husband refused to patent their techniques for extracting radium and polonium, which would have made them very rich.

Marie Curie spent her last years studying the

Reprinted from Bowen and Schneller, *Writing About Science*, 2d ed. 1991. Gilford University Press.

applications of radioactivity to medicine. Marie Curie died at the age of 66 in 1934 of leukemia, almost certainly the result of her prolonged exposure to radiation. The notebooks of the Curies are still dangerously radioactive.

In this article, Curie describes three types of radiation.

Energy of Radiation

Whatever be the method of research employed, the energy of radiation of the new radioactive substances is always found to be considerably greater than that of uranium and thorium. Thus it is that, at a short distance, they act instantaneously upon a photographic plate, whereas an exposure of twenty-four hours is necessary when operating with uranium and thorium. A fluorescent screen is vividly illuminated by contact with the new radioactive bodies, whilst no trace of luminosity is visible with uranium and thorium. Finally, the ionizing action upon air is considerably stronger in the ratio of 10^6 approximately. But it is, strictly speaking, not possible to estimate the *total intensity of the radiation*, as in the case of uranium, by the electrical method described at the beginning. . . . With uranium, for example, the radiation is almost completely absorbed by the layer of air between the plates, and the limiting current is reached at a tension of 100 volts. But the case is different for strongly radioactive bodies. One portion of the radiation of radium consists of very penetrating rays, which penetrate the condenser and the metallic plates, and are not utilized in ionizing the air between the plates. Further, the limiting current cannot always be obtained for the tensions supplied; for example, with very active polonium the current remains proportional to the tension between 100 and 500 volts. Therefore the experimental conditions which give a simple interpretation are not realized, and, consequently, the numbers obtained cannot be taken as representing the measurement of the total radiation; they merely point to a rough approximation.

Complex Nature of the Radiation

The researchers of various physicists (M.M. Becquerel, Meyer and von Schweidler, Giesel, Villard, Rutherford, M. and Mme Curie) have proved the complex nature of the radiation of radioactive bodies. It will be convenient to specify three kinds of rays, which I shall denote, according to the notation adopted by Mr. Rutherford, by the letters α, β, γ.

I. The α are very slightly penetrating, and appear to constitute the principal part of the radiation. These rays are characterized by the laws by which they are absorbed by matter. The magnetic field acts very slightly upon them, and they were formerly thought to be quite unaffected by the action of this field. However, in a strong magnetic field, the α-rays are slightly deflected; the deflection is caused in the same manner as with cathode rays, but the direction of the deflection is reversed; it is the same as for the canal rays of the Crookes tubes.

II. The β-rays are less absorbable as a whole than the preceding ones. They are deflected by a magnetic field in the same manner and direction as cathode rays.

III. The γ-rays are penetrating rays, unaffected by the magnetic field, and comparable to Röntgen rays.

Consider the following imaginary experiment: Some radium, R, is placed at the bottom of a small deep cavity, hollowed in a block of lead, P [see Figure]. A sheaf of rays, rectilinear and slightly expanded, streams from the receptacle. Let us suppose that a strong uniform magnetic field is established in the neighborhood of the receptacle, normal to the plane of the figure and directed toward the back. The three groups of rays, α, β, γ, will now be separated. Then rather faint γ-rays continue in their straight path without a trace of deviation. The β-rays are deflected in the manner of cathode rays, and describe circular paths in the plane of the figure. If the receptacle is placed on a photographic plate, A C, the portion, B C, of the plate which receives the β-rays is acted upon. Lastly, the α-rays form a very intense shaft which is slightly deflected, and which is soon absorbed by the air. These rays describe in the plane of the figure a path of great curvature, the direction of the deflection being the reverse of that with the β-rays.

If the receptacle is covered with a thin sheet of aluminum (0.1 mm. thick), the α-rays are suppressed almost entirely, the β-rays are lessened, and the γ-rays do not appear to be absorbed to any great extent.

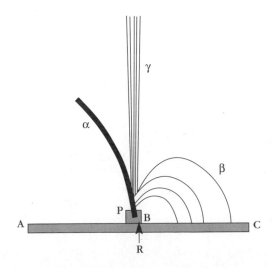

Understanding What You've Read

Marie Curie worked with several other scientists, including her husband, Rutherford, and Becquerel. Why are such collaborations so common and important?

How does radiation emitted from radium differ from that from uranium?

For whom did Curie write this article? How can you tell?

Marie Curie said that, "There are sadistic scientists who hurry to hunt down error instead of establishing truth." What did she mean by this?

Would you describe Curie's writing style as personal or impersonal? Why?

How does Curie determine the level of intensity of a radioactive substance?

How do alpha, beta, and gamma radiation differ?

Exercises

1. Examine the data shown in each of the figures and tables in this chapter. Write an essay about each set of data. Do not repeat the data; instead, discuss what they mean.

2. Examine the data shown in question 6 of Chapter 2. What is the most effective way of displaying these data? Why? In the space below, plot the data.

3.   Discuss, support, or refute the ideas of these scientists:

Biology

 Population, when unchecked, increases in geometrical ratio. Subsistence increases only in an arithmetical ratio.
— Thomas Robert Malthus

The most intensively social animals can only adapt to group behavior. Bees and ants have no option when isolated, except to die. There is really no such creature as a single individual; he has no more life of his own than a cast-off cell marooned from the surface of your skin.
— Lewis Thomas

Geology

 The present state of the earth and of the organisms now inhabiting, is but the last stage of a long and uninterrupted series of changes which it has undergone, and consequently, that to endeavor to explain and account for its present condition without any reference to those changes (as has frequently been done) must lead to very imperfect and erroneous conclusions.
— Alfred Russel Wallace (1823–1913)

We can hardly pick up a copy of a newspaper or magazine nowadays without being informed exactly how many million years ago some remarkable event in the history of the earth occurred.
— Adolph Knopf

Physics

 The electron is not as simple as it looks.
— [Sir] William Lawrence Bragg (1890–1971)

There is no democracy in physics. We can't say that some second-rate guy has as much right to opinion as Fermi.
— D.S. Greenburg

Every statement in physics has to state relations between observable quantities.
— Ernst Mach (1836-1916; this is a statement of Mach's principle)

CHAPTER NINE

Oral Presentations

A speech is a solemn responsibility. The man who makes a bad thirty-minute speech to two hundred people wastes only a half hour of his own time. But he wastes one hundred hours of the audience's time —more than four days—which should be a hanging offense.
— Jenkin Lloyd Jones

The art of writing is combined with the art of speaking.
— Quintilian

Listen up, because I've got nothing to say and I'm only going to say it once.
— Yogi Berra

Image is the great instrument of instruction.
— John Dewey

Speech is the representation of the mind, and writing is the representation of speech.
— Aristotle

First learn the meaning of what you say, and then
speak.
> — Epictetus

Make sure you have finished speaking before your
audience has finished listening.
> — Dorothy Sarnoff

Be sincere; be brief; be seated.
> — Franklin D. Roosevelt

You're probably puzzled by a chapter entitled "Oral Presentations" in a book entitled *Writing to Learn Science*. I've included this chapter because talking effectively about science, like writing effectively about science, is important to a scientist's professional development. Most scientific meetings include "paper sessions" in which scientists make oral presentations of their work.

The objective of such a talk is to communicate your discovery to your audience. To do this, you must understand the nature of your audience and capture their interest with logic, effective graphics, well-organized ideas, and simplicity. Since your audience listens to and looks at you, you must also concentrate on what you say and do in your talk.

Here's how to present an effective talk:

Know your audience. Gear your talk to their needs and interests.

Unlike writing, oral presentations are designed for immediate consumption. Therefore, present less data and stress generalizations. Define all terms in the introduction of your talk and don't hesitate to restate your conclusions.

Mix your discussions of results and discussion. Since listeners, unlike readers, have no way to re-examine what came before, having distinct "Results" and "Discussion" sections is usually not effective.

Capture the group's attention by politely saying "Good morning" or "Good afternoon" at the beginning of your talk. Don't start your talk by saying that you can't compress a year's worth of work into the 15 minutes you've been allocated. You've not been asked to compress any information—you're expected only to communicate well.

Project restrained self-confidence by projecting your voice, punctuating your words with animation, pronouncing each word crisply, and including no fillers such as "okay," "you know," and "uhhh."

Practice your talk several times to ease your nerves about giving it before a group. Face your audience and don't be a sleeping pill; show your enthusiasm for your work.

State the objectives of your work clearly, concentrate on concepts, and eliminate details. People can learn the details of your work by reading your papers or asking questions after your talk.

Focus on your results and conclusions. Discuss the purpose, rationale, and conclusions of your work as you go. Emphasize these points by repeating them in different ways. Lead readers to your conclusion.

Use large-print note cards, but don't read your talk. Rather, maintain eye contact with your audience to convey sincerity and conviction. To help avert a case of stage fright, memorize the first few sentences of your talk.

Obey the time limit. Most presenters are given 15 minutes for a talk. Use about 2 minutes to introduce your talk, 11 minutes for the body of your talk, and the remaining 3 minutes for conclusions and questions. There are few things more disrespectful to your peers than exceeding the time allocated for your presentation. Practice your talk so that you do not exceed your allotted time.

Maintain good posture. Avoid distracting mannerisms such as rocking back and forth, picking your nose, or grabbing your crotch as you talk.

Speak slowly and clearly at about 100 words per minute. At that rate, you'll cover a double spaced page in about 2.5 minutes. Use short, forceful sentences. If a microphone is available, stay about 15 cm from the microphone. If a microphone isn't available, speak loudly and project your voice as if you're talking to someone sitting in the back row.

Don't turn your back on the audience. This is impolite and reduces the projection of your voice.

At the end of your talk be prepared to answer questions about your work. When asked questions, first repeat the question to the group. Then answer the question directly. If you don't know the answer, just say "I don't know."

Announce the end of your talk by saying something like "Thank you" or "I'll be glad to answer questions." Don't just abruptly stop talking; people won't know if they should applaud or if you've just gone blank.

Photographic slides, handouts, and overhead transparencies are critical for most presentations, especially those given at meetings. Be sure that they're done professionally and that data aren't crowded onto any of the slides. When practicing your talk, sit at the back of a large hall while your slides are projected so that you can check them for accuracy and readability.

> Prepare slides, transparencies, and handouts expressly for your talk. If you use those made from tables and figures of a manuscript, you'll probably end up telling the audience things such as, "Ignore everything on this slide except this column." Design your talk so that the audience does not have to ignore anything you say or show.

Allow about one minute per slide or overhead. Give the audience time to look at each slide and then help them interpret each slide. You and your audience should be able to understand your slide in less than five seconds. Title each slide and do not read each word on a slide to the audience. Rather, use slides to supplement what you say. Cramming too many slides into your talk will only frustrate your audience.

Do not show a slide if it includes too much detail or anything that is irrelevant to your talk. Use one slide to convey one idea, not detailed data. Make that idea brief, clear, and simple so that it can be understood quickly. Be sure that all slides have less than eight lines and four columns.

Do not show a table if it has too many numbers or if the numbers are so small that some people cannot read them. If the slide is not readable when held at arm's length, it will not be readable when projected from the back of a conference hall.

Remove each slide after it has served its purpose so that it is not shown while you are trying to interest your audience in something else.

Make sure that all slides are in their proper orientation so that they will be projected right-side-up.

Use labels to focus the attention of the audience to the important parts of the slide.

Locate the pointer before you speak. Don't wait for the lights to go off before you start fumbling for the pointer.

Use slide carousels that hold 80 rather than 140 slides. The 140-capacity carousels jam more frequently than do 80-capacity carousels.

If the projector jams during your talk, keep talking. If you stop talking, you quickly lose the attention of the audience.

Plan slides so that their longest dimension is horizontal when projected. It's hard to see vertically oriented slides in rooms having low ceilings.

Use uppercase and lowercase letters, especially if your slide has more than about five words.

Start and end your talk with an opaque slide. Don't blind your audience with an empty frame.

Carry your slides, transparencies, and handouts with you if you travel to give your talk. Don't pack your slides in your checked baggage unless you want to risk having the airline lose or destroy them.

Stephen Jay Gould

False Premise, Good Science

Stephen Jay Gould's (b. 1941) award-winning essays and books describe riddles of evolution. His essay "False Premise, Good Science" from *The Flamingo's Smile: Reflections on Natural History* is reprinted here.

My vote for the most arrogant of all scientific titles goes without hesitation to a famous paper written in 1866 by Lord Kelvin, "The 'Doctrine of Uniformity' in Geology Briefly Reputed." In it, Britain's greatest physicist claimed that he had destroyed the foundation of an entire profession not his own. Kelvin wrote:

> The "Doctrine of Uniformity" in Geology, as held by many of the most eminent of British geologists, assumes that the earth's surface and upper crust have been nearly as they are at present in temperature and other physical qualities during millions of millions of years. But the heat which we know, by observation, to be now conducted out of the earth yearly is so great, that if *this* action had been going on with any approach to uniformity for 20,000 million years, the amount of heat lost out of the earth would have been about as much as would heat, by 100° Cent., a quantity of ordinary surface rock of 100 times the earth's bulk. (See calculation appended.) This would be more than enough to melt a mass of surface rock equal in bulk to the *whole earth*. No hypothesis as to chemical action, internal fluidity, effects of pressure at great depth, or possible character of substances in the interior of the earth, possessing the smallest vestige of probability, can justify the supposition that the earth's crust has remained nearly as it is, while from the whole, or from any part, of the earth, so great a quantity of heat has been lost.

I apologize for inflicting so long a quote so early in the essay, but this is not an extract from Kelvin's paper. It is the whole thing (minus the appended calculation). In a mere paragraph, Kelvin felt he had thoroughly undermined the very basis of his sister discipline.

Kelvin's arrogance was so extreme, and his later comeuppance so spectacular, that the tale of his 1866 paper, and his entire, relentless forty-year campaign for a young earth, has become the classical moral homily of our geological textbooks. But beware of conventional moral homilies. Their probability of accuracy is about equal to the chance that George Washington really scaled that silver dollar clear across the Rappahannock.

The story, as usually told, goes something like this. Geology, for several centuries, had languished under the thrall of Archbishop Ussher and his biblical chronology of but a few thousand years for the earth's age. This restriction of time led to the unscientific doctrine of catastrophism—the notion that miraculous upheavals and paroxysms must characterize our earth's history if its entire geological story must be compressed into the Mosaic chronology. After long struggle, Hutton and Lyell won the day for science with their alternative idea of uniformitarianism—the claim that current rates of change, extrapolated over limitless time, can explain all our history from a scientific standpoint by direct observation of present processes and their results. Uniformity, so the story goes, rests on two propositions: essentially unlimited time (so that slow processes can achieve their accumulated effect), and an earth that does not alter its basic form and style of change throughout this vast time. Uniformity in geology led to evolution in biology and the scientific revolution spread. If we deny uniformity, the homily continues, we undermine science itself and plunge geology back into its own dark ages.

Yet Kelvin, perhaps unaware, attempted to undo this triumph of scientific geology. Arguing that the earth began as a molten body, and basing his calculation upon loss of heat from the earth's interior (as measured, for example, in mines), Kelvin recognized that the earth's solid surface could not be very old—probably 100 million years, and 400 million at most (although he later revised the estimate downward, possibly to only 20 million years). With so little time to harbor all of evolution—not to mention the physical history of solid rocks—what recourse did geology have except to its discredited idea of catastrophes? Kelvin had plunged geology into an inextricable dilemma while clothing it with all the prestige of quantitative physics, queen of the sciences. One popular geological textbook writes (C. W. Barnes), for example:

> Geologic time, freed from the constraints of literal biblical interpretation, had become unlimited; the concepts of uniform change first suggested by Hutton now embraced the concept of the origin and evolution of life. Kelvin single-handedly destroyed, for a time, uniformitarian and evolutionary thought. Geologic time was still restricted because the laws of physics bound as tightly as biblical literalism ever had.

Fortunately for a scientific geology, Kelvin's argument rested on a false premise—the assumption that the earth's current heat is a residue of its original molten state and not a quantity constantly renewed. For if the earth continues to generate heat, then the current rate of loss cannot be used to infer an ancient condition. In fact, unbeknown to Kelvin, most of the earth's internal heat is newly generated by the process of radioactive decay. However elegant his calculations, they were based on a false premise, and Kelvin's argument collapsed with the discovery of radioactivity early in our century. Geologists should have trusted their own intuitions from the start and not bowed before the false lure of physics. In any case, uniformity finally won and scientific geology was restored. This transient episode teaches us that we must trust the careful empirical data of a profession and not rely too heavily on theoretical interventions from outside, whatever their apparent credentials.

So much for the heroic mythology. The actual story is by no means so simple or as easily given as evident moral interpretation. First of all, Kelvin's arguments, although fatally flawed as outlined above, were neither so coarse nor as unacceptable to geologists as the usual story goes. Most geologists were inclined to treat them as a genuine reform of their profession until Kelvin got carried away with further restrictions upon his original estimate of 100 million years. Darwin's strong opposition was a personal campaign based on his own extreme gradualism, not a consensus. Both Wallace and Huxley accepted Kelvin's age and pronounced it consonant with evolution. Secondly, Kelvin's reform did not plunge geology into an unscientific past, but presented instead a different *scientific* account based on another concept of history that may be more valid than the strict uniformitarianism preached by Lyell. Uniformitarianism, as advocated by Lyell, was a specific and restrictive theory of history, not (as often misunderstood) a general account of how science must operate. Kelvin had attacked a legitimate target.

Kelvin's Arguments and the Reaction of Geologists

As codiscoverer of the second law of thermodynamics, Lord Kelvin based his arguments for the earth's minimum age on the dissipation of the solar system's original energy as heat. He advanced three distinct claims and tried to form a single quantitative estimate for the earth's age by seeking agreement among them (see Joe D. Burchfield's *Lord Kelvin and the Age of the Earth*, the source for most of the technical information reported here).

Kelvin based his first argument on the age of the sun. He imagined that the sun had formed through the falling together of smaller meteoric masses. As these meteors were drawn together by their mutual gravitational attraction, their potential energy was transformed into kinetic energy, which, upon collision, was finally converted into heat, causing the sun to shine. Kelvin felt that he could calculate the total potential energy in a mass of meteors equal to the sun's bulk and, from this, obtain an estimate of the sun's original heat. From this estimate, he could calculate a minimum age for the sun, assuming that it has been shining at its present intensity since the beginning. But this calculation was crucially dependent on a set of factors that Kelvin could not really estimate—including the original number of meteors and their original distance from each other—and he never ventured a precise figure for the sun's age. He settled on a number between 100 and 500 million years as a best estimate, probably closer to the younger age.

Kelvin based his second argument on the probable age of the earth's solid crust. He assumed that the earth had cooled from an originally molten state and that the heat now issuing from its mines recorded the same process of cooling that had caused the crust to solidify. If he could measure the rate of heat loss from the earth's interior, he could reason back to a time when the earth must have contained enough heat to keep its globe entirely molten—assuming that this rate of dissipation had not changed through time. (This is the argument for his "brief" refutation of uniformity, cited at the beginning of this essay.) This argument sounds more "solid" than the first claim based on a hypothesis about how the sun formed. At least one can hope to measure directly its primary ingredient—the earth's current loss of heat. But Kelvin's second argument still depends upon several crucial and unprovable assumptions about the earth's composition. To make his calculation work, Kelvin had to treat the earth as a body of virtually uniform composition that had solidified from the center outward and had been, at the time its crust formed, a solid sphere of similar temperature throughout. These restrictions also prevented Kelvin from assigning a definite age for the solidification of the earth's crust. He ventured between 100 and 400 million years, again with a stated preference for the smaller figure.

Kelvin based his third argument on the earth's shape as a spheroid flattened at the poles. He felt that he could relate this degree of polar shortening to the speed of the earth's rotation when it formed in a molten state amenable to flattening. Now we know—and Kelvin knew also—that the earth's rotation has been slowing down continually as a result of tidal friction. The earth rotated more rapidly when it first formed. Its current shape should therefore indicate its age. If the earth formed a long time ago, when rotation was quite rapid, it should now be very flat. If the earth is not so ancient, then it formed at a rate of rotation not so different from its current pace, and flattening should be less. Kelvin felt that the small degree of actual flattening indicated a relatively young age for the earth. Again, and for the third time, Kelvin based his argument upon so many unprovable assumptions (about the earth's uniform composition, for example) that he could not calculate a precise figure for the earth's age.

Thus, although all three arguments had a quantitative patina, none was precise. All depended upon simplifying assumptions that Kelvin could not justify. All therefore yielded only vague estimates with large margins of error. During most of Kelvin's forty-year campaign, he usually cited a figure of 100 million years for the earth's age—plenty of time, as it turned out, to satisfy nearly all geologists and biologists.

Darwin's strenuous opposition to Kelvin is well recorded, and later commentators have assumed that he spoke for a troubled consensus. In fact, Darwin's antipathy to Kelvin was idiosyncratic and based on the strong personal commitment to gradualism so characteristic of his world view. So wedded was Darwin to the virtual necessity of unlimited time as a prerequisite for evolution by natural selection that he invited readers to abandon *On The Origin of Species* if they could not accept this premise: "He who can read Sir Charles Lyell's grand work on the *Principles of Geology*, and yet does not admit how incomprehensively vast have been the past periods of time, may at once close this volume." Here Darwin commits a fallacy of reasoning—the confusion of gradualism with natural selection—that characterized all his work and that inspired Huxley's major criticism of the *Origin*: "You load yourself with an unnecessary difficulty in adopting *Natura non facit saltum* [Nature does not proceed by leaps] so unreservedly." Still, Darwin cannot be entirely blamed, for Kelvin made the same error in arguing explicitly that his young age for the earth cast grave doubt upon natural selection as an evolutionary mechanism (while not arguing against evolution itself). Kelvin wrote:

> The limitations of geological periods, imposed by physical science, cannot, of course, disprove the hypothesis of transmutation of species; but it does seem sufficient to disprove the doctrine that transmutation has taken place through "descent with modification by natural selection."

Thus, Darwin continued to regard Kelvin's calculation of the earth's age as perhaps the gravest objection to his theory. He wrote to Wallace in 1869 that "Thomson's [Lord Kelvin's] views on the recent age of the world have been for some time one of my sorest troubles." And, in 1871, in striking metaphor, "But then comes Sir W. Thomson like an odious spectre." Although Darwin generally stuck to his guns and felt in his heart of hearts that something must be wrong with Kelvin's calculations, he did finally compromise in the last edition of the *Origin* (1872), writing that more rapid changes on the early earth would have accelerated the pace of evolution, perhaps permitting all the changes we observe in Kelvin's limited time:

> It is, however, probable, as Sir William Thompson [sic] insists, that the world at a very early period was subjected to more rapid and violent changes in its physical conditions than those now occurring; and such changes would have tended to induce changes at a corresponding rate in the organisms which then existed.

Darwin's distress was not shared by his two leading supporters in England, Wallace and Huxley. Wallace did not tie the action of natural selection to Darwin's glacially slow time scale; he simply argued that, if Kelvin limited the earth to 100 million years, then natural selection must operate at generally higher rates than we had previously imagined. "It is within that time [Kelvin's 100 million years], therefore, that the whole series of geological changes, the origin and development of all forms of life, must be compressed." In 1870, Wallace even proclaimed his happiness with a time scale of but 24 million years since the inception of our fossil record in the Cambrian explosion.

Huxley was even less troubled, especially since he had long argued that evolution might occur by saltation, as well as by slow natural selection. Huxley maintained that our conviction about the slothfulness of evolutionary change had been based on false and circular logic in the first place. We have no independent evidence for regarding evolution as slow; this impression was only an inference based on the assumed vast duration of fossil strata. If Kelvin now tells us that these strata were deposited in far less time, then our estimate of evolutionary rate must be revised correspondingly.

> Biology takes her time from geology. The only reason we have for believing in the slow rate of the change in living forms is the fact that they persist through a series of deposits which, geology informs us, have taken a long while to make. If the geological clock is wrong, all the naturalist will have to do is to modify his notions of the rapidity of change accordingly.

Britain's leading geologists tended to follow Wallace and Huxley rather than Darwin. They stated that Kelvin had performed a service for geology in challenging the virtual eternity of Lyell's world and in "restraining the reckless drafts" that geologists so rashly make on the "bank of time," in T. C. Chamberlin's apt metaphor. Only late in his campaign, when Kelvin began to restrict his estimate from a vague and comfortable 100 million years (or perhaps a good deal more) to a more rigidly circumscribed 20 million years or so did geologists finally rebel. A. Geikie, who had been a staunch supporter of Kelvin, then wrote:

> Geologists have not been slow to admit that they were in error in assuming that they had an eternity of past time for the evolution of the earth's history. They have frankly acknowledged the validity of the physical arguments which go to place more or less definite limits to the antiquity of the earth. They were, on the whole, disposed to acquiesce in the allowance of 100 millions of years granted them by Lord Kelvin, for the transaction of the long cycles of geological history. But the physicists have been insatiable and inexorable. As remorseless as Lear's daughters, they have cut down their grant of years by successive slices, until some of them have brought the number to something less than ten millions. In vain have geologists protested that there must be somewhere a flaw in a line of argument which tends to results so entirely at variance with the strong evidence for a higher antiquity.

Kelvin's Scientific Challenge and the Multiple Meanings of Uniformity

As a master of rhetoric, Charles Lyell did charge that anyone who challenged his uniformity might herald a reaction that would send geology back to its prescientific age of catastrophes. One meaning of uniformity did uphold the integrity of science in this sense—the claim that nature's laws are constant in space and time, and that miraculous intervention to suspend these laws cannot be permitted as an agent of geological change. But uniformity, in this methodological meaning, was no longer an issue in Kelvin's time, or even (at least in scientific circles) when Lyell first published his *Principles of Geology* in 1830. The scientific catastrophists were not miracle mongers, but men who fully accepted the uniformity of natural law and sought to render earth history as a tale of *natural* calamities occurring infrequently on an ancient earth.

But uniformity also had a more restricted, substantive meaning for Lyell. He also used the term for a particular theory of earth history based on two questionable postulates: first, that rates of change did not vary much throughout time and that slow and current processes could therefore account for all geological phenomena in their accumulated impact; second, that the earth had always been about the same, and that its history had no direction, but represented a steady state of dynamically constant conditions.

Lyell, probably unconsciously, then performed a clever and invalid trick of argument. Uniformity had two distinct meanings—a methodological postulate about uniform laws, which all scientists had to accept in order to practice their profession, and a substantive claim of dubious validity about the actual history of the earth. By calling them both uniformity, and by showing that all scientists were uniformitarians in the first sense, Lyell also cleverly implied that, to be a scientist, one had to accept uniformity in its substantive meaning as well. Thus, the myth developed that any opposition to uniformity could only be a rearguard action against science itself—and the impression arose that if Kelvin was attacking the "doctrine of uniformity" in geology, he must represent the forces of reaction.

In fact, Kelvin fully accepted the uniformity of law and even based his calculations about heat loss upon it. He directed his attack against uniformity only upon the substantive (and dubious) side of Lyell's vision. Kelvin advanced two complaints about this substantive meaning of uniformity. First, on the question of rates. If the earth were substantially younger than Lyell and the strict uniformitarians believed, then modern, slow rates of change would not be sufficient to render its history. Early in its history, when the earth was hotter, causes must have been more energetic and intense. (This is the "compromise" position that Darwin finally adopted to explain faster rates of change early in the history of life.) Second, on the question of direction. If the earth began as a molten sphere and lost heat continually through time, then its history had a definite pattern and path of change. The earth had not been perennially the same, merely changing the position of its lands and seas in a never-ending dance leading nowhere. Its history followed a definite road, from a hot, energetic sphere to a cold, listless world that, eventually, would sustain life no longer. Kelvin fought, within a scientific context, for a *short-term*, *directional* history against Lyell's vision of an essentially eternal steady-state. Our current view represents the triumph of neither vision, but a creative synthesis of both. Kelvin was both as right and as wrong as Lyell.

Radioactivity and Kelvin's Downfall

Kelvin was surely correct in labeling as extreme Lyell's vision of an earth in steady-state, going nowhere over untold ages. Yet, our modern time scale stands closer to Lyell's concept of no appreciable limit than to Kelvin's 100 million years and its consequent constraint on rates of change. The earth is 4.5 billion years old.

Lyell won this round of a complicated battle because Kelvin's argument contained a fatal flaw. In this respect, the story as conventionally told has validity. Kelvin's argument was not an inevitable and mathematically necessary set of claims. It rested upon a crucial and untested assumption that underlay all Kelvin's calculations. Kelvin's figures for heat loss could measure the earth's age only if that heat represented an original quantity gradually dissipated through time—a clock ticking at a steady rate from its initial reservoir until its final exhaustion. But suppose that new heat is constantly created and that its current radiation from the earth reflects no original quantity, but a modern process of generation. Heat then ceases to be a gauge of age.

Kelvin recognized the contingent nature of his calculations, but the physics of his day included no force capable of generating new heat, and he therefore felt secure in his assumption. Early in his campaign, in calculating the sun's age, he had admitted his crucial dependence upon no new source of energy, for he had declared his results valid "unless new sources now unknown to us are prepared in the great storehouse of creation."

Then, in 1903, Pierre Curie announced that radium salts constantly release newly generated heat. The unknown source had been discovered. Early students of radioactivity quickly recognized that most of the earth's heat must be continually generated by radioactive decay, not merely dissipating from an originally molten condition—and they realized that Kelvin's argument had collapsed. In 1904, Ernest Rutherford gave this account of a lecture, given in Lord Kelvin's presence, and heralding the downfall of Kelvin's forty-year campaign for a young earth:

> I came into the room, which was half dark, and presently spotted Lord
> Kelvin in the audience and realized that I was in for trouble at the last

part of the speech dealing with the age of the earth, where my views conflicted with his. To my relief, Kelvin fell fast asleep, but as I came to the important point, I saw the old bird sit up, open an eye and cock a baleful glance at me! Then a sudden inspiration came, and I said Lord Kelvin had limited the age of the earth, provided no new source of heat was discovered. That prophetic utterance refers to what we are now considering tonight, radium!

Thus, Kelvin lived into the new age of radioactivity. He never admitted his error or published any retraction, but he privately conceded that the discovery of radium had invalidated some of his assumptions.

The discovery of radioactivity highlights a delicious double irony. Not only did radioactivity supply a new source of heat that destroyed Kelvin's argument; it also provided the clock that could then measure the earth's age and proclaim it ancient after all! For radioactive atoms decay at a constant rate, and their dissipation does measure the duration of time. Less than ten years after the discovery of radium's newly generated heat, the first calculations for radioactive decay were already giving ages in billions of years for some of the earth's oldest rocks.

We sometimes suppose that the history of science is a simple story of progress, proceeding inexorably by objective accumulation of better and better data. Such a view underlies the moral homilies that build our usual account of the advance of science—for Kelvin, in this context, clearly impeded progress with a false assumption. We should not be beguiled by such comforting and inadequate stories. Kelvin proceeded by using the best science of his day, and colleagues accepted his calculations. We cannot blame him for not knowing that a new source of heat would be discovered. The framework of his time included no such force. Just as Maupertuis lacked a proper metaphor for recognizing that embryos might contain coded instructions rather than preformed parts, Kelvin's physics contained no context for a new source of heat.

The progress of science requires more than new data; it needs novel frameworks and contexts. And where do these fundamentally new views of the world arise? They are not simply discovered by pure observation; they require new modes of thought. And where can we find them, if old modes do not even include the right metaphors? The nature of true genius must lie in the elusive capacity to construct these new modes from apparent darkness. The basic chanciness and unpredictability of science must also reside in the inherent difficulty of such a task.

Understanding What You've Read

What stylistic techniques does Gould use to engage his audience? What is "progress" in science? Can false premise lead to good science? Why or why not?

Exercises

1. Prepare a 10-minute talk about any of the subjects that you've written about in this course. Be prepared to give the talk to your class.

2. Prepare an oral and written essay about the same subject. How are they different? How are they similar?

3. Prepare a short essay and talk about how this *The Far Side* cartoon relates to science:

Great moments in evolution.

Figure 9–1
The Far Side. Gary Larson (© 1982 Universal Press Syndicate).

EPILOG

Using What You've Learned

We are all apprentices in a craft where no one ever
becomes a master.
　　　Ernest Hemingway

What can you learn in college that will help in being
an employee?. . . There you learn the one thing that
is perhaps the most valuable for the future employee
to know. This one basic skill is the ability to organize
and express ideas in writing and speaking.
　　　— Peter Drucker, *Fortune*

It's difficult to think of any skill that is more essential
for career success than being able to write well.
Most often it's the written word that conveys the
personality, the ideas, and beliefs of the individual.
It's worth all the effort necessary to do it well.
　　　— Burnell R. Roberts, Chairman and Chief
　　　Executive Officer, *Mead Corporation*

Good writing skills are an invaluable, but often
understated, necessity for career and business. The
person who develops these skills will certainly be
recognized early . . .
　　　— Terry C. Carder, Chairman of the Board,
　　　Reynolds & Reynolds Information Systems

Practice, practice. Put your hope in that.
— M.S. Merwin

Science does not contradict art. I cannot accept the opinion of those who think that scientific positivism must be fatal to inspiration. Quite the contrary: the artist will find in science a more stable formulation, and the scientist will draw from art a more certain intuition.
— Claude Bernard, *La Science Experimentale*

More men become good through practice than by nature.
— Democritus of Abdera

If you've studied the principles discussed in this book, you're now a better writer than before: You know how to use writing to discover your ideas, how to organize those ideas into a first draft, and how to shape those ideas into a paper or laboratory report that communicates your ideas effectively. In short, you have a better understanding of the writing process and therefore can learn science by writing about science. Your ability to write well will benefit you throughout your career. Scientists and others who insist on writing poorly will envy your ability to communicate well; they'll also continue to cloak their ideas in secrecy by preferring jargon to clear writing, by sacrificing the beauty of active voice to seem objective, and by remaining blind to imagination despite their claims of seeing "the big picture." Don't let these peoples' attitudes bother you. Insist on writing clearly. If you do, people will be impressed by your ideas rather than confused by your writing.

If you can now write clearly, coherently, and emphatically, you'll be tempted to stop. Most would be satisfied to have accomplished so much. However, like other tools of scientists, writing is a craft that can be improved by practice and study. To show this, consider the following essays of William Osler. In 1889 Osler wrote like this:

> In a true and perfect form, imperturbability is indissolubly associated with wide experience and an intimate knowledge of the varied aspects of disease. With such advantages he is so equipped that no eventuality can disturb the mental equilibrium of the physician; the possibilities are always manifest, and the course of action clear. From its very nature this precious quality is liable to be misinterpreted, and the general accusation of hardness, so often brought against the profession, has here its foundation.[1]

Although this essay has a worthwhile message, Osler uses too many big words and

---

[1] Osler, W. 1932. "Aequanimitas." In *Aequanimitas* (3rd. ed.). Philadelphia: Blakiston. p. 315.

phrases such as *imperturbability is indissolubly associated* that hinder readers. Three years later Osler wrote like this about one of his favorite topics, the need for medical students to work with patients:

> I would fain dwell upon many other points in the relation of the hospital to the medical school—on the necessity of ample, full and prolonged clinical instruction, and on the importance of bringing the student and the patient into close contact, not through the cloudy knowledge of the wards; on the propriety of encouraging the younger men as instructors and helpers in ward work; and on the duty of hospital physicians and surgeons to contribute to the advance of their art . . .[2]

Although this essay is simpler and uses fewer big words than did his earlier essay, it still does not say things simply. Moreover, it lacks grace, has no rhythm, and sounds pompous. Compare this with what Osler wrote more than a decade later:

> Ask any physician of twenty years' standing how he has become proficient in his art, and he will reply, by constant contact with disease; and he will add that the medicine he learned in the schools was totally different from the medicine he learned at the bedside. The graduate of a quarter of a century ago went out with little practical knowledge, which increased only as his practice increased. In what may be called the natural method of teaching the student begins with the patient, continues with the patient, and ends with the patient, using books and lectures as tools, as means to an end. The student starts, in fact, as a practitioner. . . .[3]

What a difference! Unlike in the previous essays, here Osler relied on strong verbs and simple words (most have only one or two syllables), avoided useless repetition, and used few adjectives. Consequently, this essay is forceful, clear, and easy to understand.

Another excellent way to improve your writing is to read good writing. Just as a child isolated by deafness has problems learning to speak, so too does a writer who does not read have problems in writing. Therefore, take time to read the works of classical and modern writers. It's always a good idea to watch an expert. Therefore, don't be afraid to model your writing after your favorite writers; after all, Bach and Picasso used models, and you'll also benefit by studying the works of great writers.

Reading great science books will help you enjoy and appreciate how to use words effectively. Here are some good scientists to start with:

Rachel Carson's *Silent Spring* heralded the age of ecological awareness by describing the hidden effects of human dominance of nature.

Lewis Thomas's *Lives of a Cell: Notes of a Science Watcher* and *The Medusa*

---

[2]Osler, W. 1932. "Teacher and Student." In *Aequanimitas* (3rd ed.). Philadelphia: Blakiston. p. 31.

[3]Osler, W. 1932. "The Hospital as a College." In *Aequanimitas* (3rd ed.). Philadelphia: Blakiston. p. 315.

and the Snail are compilations of his essays from *The New England Journal of Medicine.* Thomas writes about the common aspects of science so readers can follow his ideas into the world of their experiences. *Lives of a Cell* won the National Book Award in 1975.

James Watson's best seller *The Double Helix* describes an exhilarating race to solve a scientific puzzle. His book, along with C.P. Snow's *The Search,* will show you that scientists have their share of vanity, greed, lust, sloth, and indecisiveness.

Jane Goodall's *In The Shadow of Man* describes Goodall's classic research on primate behavior and shows the power of science to enthrall.

Stephen Jay Gould's books and *Natural History* essays are popular because Gould realizes that everyone likes a good story. Gould views science as a man trying to understand forces much greater than he is. His *The Mismeasure of Man* won the National Book Critics' Award in 1981, and his *The Panda's Thumb* won the American Book Award for Science in 1981.

Although you'll like these books, don't restrict your reading to only books written by scientists. Also read the works of writers such as E.B. White, Ernest Hemingway, Ralph Waldo Emerson, Rebecca West, William James, Mark Twain, Henry David Thoreau, Robert Louis Stevenson, Thornton Wilder, John Steinbeck, Ernest Hemingway, James Thurber, Barbara Tuchman, Art Buchwald, and H.L. Mencken. Reading these writers' works will get the shapes and rhythms of good writing into your head. They'll also help you more than reading about good writing, and are much more entertaining. Moreover, you'll be surprised by the scientific insights of many of these writers. For example, Oliver Wendell Holmes became famous for his poem entitled "Old Ironsides." However, Holmes was also a physician who coined the term *anesthetic* and wrote extensively about scientific topics such as the contagiousness of puerperal fever. He was known as so astute a scientific reasoner that his British colleague, Arthur Conan Doyle, named the incomparable and eccentric detective Sherlock Holmes after him.

Finally, remember that what you write will be used by other scientists and will comprise much of your legacy as a scientist. Give it great care.

APPENDIX ONE

Useful References

Figures and Tables

Allen, A. 1977. *Steps Toward Better Scientific Illustrations*. 2nd edition. Lawrence, KS: Allen Press, Inc. An excellent discussion of illustrations and photographs.

Cleveland, W.S. 1985. *The Elements of Graphing Data*. Monterey, CA: Wadsworth Advanced Books and Software. This book for advanced readers includes many good and bad examples from the scientific literature.

Selby, P.H. 1976. *Interpreting Graphs and Tables*. New York: John Wiley and Sons. A self-teaching manual aimed at beginners.

Writing, Grammar, and Punctuation

Barnet, S., and M. Stubbs. 1986. *Practical Guide to Writing*. 5th edition. Boston: Little, Brown and Co.

Barzun, J. 1985. *Simple & Direct: A Rhetoric for Writers*. New York: Harper & Row. Includes 20 principles of clear writing.

Brittain, Robert. 1981. *A Pocket Guide to Correct Punctuation*. Woodbury, NY: Barron's Educational Series.

Chicago Manual of Style, The, 14th edition. Chicago: University of Chicago Press, 1993. The standard reference and source book for writers. Covers a wide range of scientific writing and is an excellent supplement to guides devoted to scientific writing.

Collinson, Diane, et al. *Plain English*. 1977. Milton Keynes, England: The Open University Press. A self-teaching text that contains many exercises, tests, and clear explanations.

Cook, Claire K. 1985. *Line by Line: How to Improve Your Own Writing*. Boston: Houghton Mifflin.

Corbett, Edward P. J. 1982. *The Little Rhetoric and Handbook*. Glenview, IL: Scott, Foresman, and Co. One of the few "writing books" that relates writing to reasoning.

Delton, Lucy. 1985. *The 29 Most Common Writing Mistakes and How to Avoid Them*. Cincinnati: Writer's Digest Books.

Flower, L. 1985. *Problem-Solving Strategies for Writing*. 2nd edition. New York: Harcourt Brace Jovanovich.

Fowler, H. W. 1965. *A Dictionary of Modern English Usage*. 2nd edition. Oxford: Oxford University Press. The most reliable reference for matters of usage.

Gordon, Karen E. 1983. *The Well-tempered Sentence*. New York: Ticknor & Fields. A short, funny survey of the rules of punctuation.

Gordon, Karen E. 1984. *The Transitive Vampire*. New York: Times Books. A funny, readable introduction to grammar.

Hairston, M. C. 1981. *Successful Writing*. 2nd edition. New York: W. W. Norton. Includes a thorough discussion of revising papers.

Hall, D. 1985. *Writing Well*. 5th edition. Boston: Little, Brown and Co.

Hooper, Vincent R., Cedric Gale, Ronald C. Foote, and Benjamin W. Griffith. 1982. *Essentials of English*. 3rd edition. Woodbury, NY: Barron's Educational Series.

Kaufer, D. S., C. Geisler, and C. M. Neuwirth. 1989. *Arguing from Sources: Exploring Issues through Reading and Writing*. San Diego: Harcourt Brace Jovanovich. Especially useful for learning to develop arguments in expository writing.

King, Lester S. 1978. *Why Not Say It Clearly: A Guide to Scientific Writing*. Boston: Little, Brown and Co.

Maggio, R. 1987. *The Nonsexist Word Finder: A Dictionary of Gender-Free Usage*. Phoenix: Oryx.

Martin, Phyllis. 1982. *Word Watchers Handbook: A Deletionary of the Most Abused and Misused Words*. New York: St. Martins Press.

Roberts, Philip D. 1987. *Plain English: A User's Guide*. Middlesex, UK: Penguin Books.

Strunk, William, and E. B. White. 1979. *The Elements of Style*, 3rd edition. New York: Macmillan Publishing Co. The best book ever written about writing. If you can afford only one book about writing, this is the one to buy. Read this book.

Temple, Michael. 1978. *A Pocket Guide to Correct Punctuation*. Woodbury, NY: Barron's Educational Series.

Walsh, J. Martyn, and Anna K. Walsh. 1982. *Plain English Handbook: A Complete Guide to Good English*. 8th edition. New York: Random House.

Woolston, Donald C., Patricia A. Robinson, and Gisela Kutzbach. 1988. *Effective Writing Strategies for Engineers and Scientists*. Chelsea, MI: Lewis Publishers.

Writing with a Computer

Hult, C., and J. Harris. 1987. *A Writer's Introduction to Word Processing*. Belmont, CA: Wadsworth, Inc. A practical, start-from-scratch guide to using word-processing in your writing.

Mitchell, Joan P. 1987. *Writing with a Computer*. Boston: Houghton Mifflin.

Scientific and Technical Writing

Alley, Michael. 1987. *The Craft of Scientific Writing*. Englewood Cliffs, NJ: Prentice-Hall.

Bates, Jefferson D. 1988. *Writing With Precision*. Washington: Acropolis Books, Ltd.

Bly, Robert W., and Gary Blake. 1982. *Technical Writing: Structure, Standards, and Style*. New York: McGraw-Hill.

Day, Robert A. 1983. *How To Write and Publish a Scientific Paper*. 2nd edition. Philadelphia: ISI Press. A short, witty book containing much practical advice about preparing a paper for publication.

Gray, P. 1982. *The Dictionary of the Biological Sciences*. New York: Krieger. Gives definitions, spelling, pronunciation, and usage of technical and scientific terms.

Huth, Edward J. 1982. *How To Write and Publish Papers in the Medical Sciences*. Philadelphia: ISI Press.

Huth, Edward J. 1987. *Medical Style and Format: An International Manual for Authors, Editors, and Publishers*. Philadelphia: ISI Press.

King, L. S. 1978. *Why Not Say It Clearly? A Guide to Scientific Writing*. Boston: Little, Brown and Co.

Smith, R. C., W. M. Reid, and A. E. Luchsinger. 1980. *Smith's Guide to the Literature of the Life Sciences*. 9th edition. Minneapolis: Burgess Pub. Co. A detailed guide to scientific literature.

Numbers and Statistics

Baily, N. T. J. 1981. *Statistical Methods in Science*. 2nd edition. San Francisco: Freeman and Co. A good book for beginners.

Budiansky, Stephen, Art Levine, Ted Gest, Alvin P. Sanoff, and Robert J. Shapiro. 1988. "The Numbers Racket: How Polls and Statistics Lie." *U.S. News & World Report* 11 July 1988.

Lippert, H., H. P. Lehmann. 1978. *SI Units in Medicine: An Introduction to the International System of Units with Conversion Tables and Normal Ranges*. Baltimore: Urban & Schwarzengerg. An extensive discussion of SI units and their use.

Moore, Randy. 1989. "Inching Toward the Metric System." *The American Biology Teacher* 51: 213–218.

Doublespeak

Fahey, Tom. 1990. *The Joys of Jargon.* New York: Barron's Educational Series, Inc.

Lutz, William. 1989. *Doublespeak.* New York: Harper Perennial.

Style Manuals

AIP Style Manual. 1990. 4th edition. New York. American Institute of Physics.

American Medical Association, Scientific Publications Division. 1989. *Style Book: Editorial Manual.* 8th edition. Acton, MA: Publishing Sciences Group, Inc.

CBE Style Manual Committee. 1983. *Council of Science Editors Style Manual: A Guide for Authors, Editors, and Publishers in the Biological Sciences,* 5th edition. Washington, D.C.: Council of Science Editors.

Dodd J.S. 1985. *ACS Style Guide: A Manual for Authors and Editors.* Washington, D.C.: Americal Chemical Society.

Huth, Edward J. 1987. *Medical Style & Format.* Philadelphia: ISI Press.

IUA Style Manual. The Preparation of Astronomical Papers and Reports. 1989. G. A. Wilkins. Paris: International Astronomical Union.

Publication Manual of the American Psychological Association. 1994. 4th edition. Washington, D.C.: American Psychological Association.

Suggestions to Authors of the Reports of the United States Geological Survey. 1990. W. R. Hansen. Washington, D.C.: U.S. Government Printing Office.

APPENDIX TWO

Words Frequently Misused by Scientists and Others

You learned in Chapter 3 the importance of always choosing the shortest and most precise word to convey meaning. Carelessly using words not only distracts readers from the author's message, but also discredits the writer. Although poor usage and misspellings can be corrected by an editor or printer before the manuscript appears in a journal, it is the author's job to choose words carefully and accurately to best communicate with readers.

The words listed below are frequently misused by scientists and others. More information about word usage can be found in the reference books listed in Appendix One.

ability
capacity
Ability is the state of being able or the power to do something. *Capacity* is the power to receive or contain.

about
approximately
About indicates a guess or rough estimate (about half full). *Approximately* implies accuracy (approximately 42 m).

absorption
adsorption
Absorption is the *taking up by capillary, osmotic, chemical, or solvent action. Adsorption* is the *taking up by chemical or chemical forces onto the surface of a solid or liquid.*

accuracy
precision
Accuracy refers to the degree of correctness, while *precision* refers to the degree of refinement. Thus, 5.845 is more precise than 5.8, but is not necessarily more accurate.

adequate
Adequate means enough. It is not synonymous with plentiful or abundant.

adapt
adopt
Adapt means *to change* or *adjust to*. *Adopt* means *to accept responsibility for, choose,* or *take from someone else.*

aggravate
The verb *aggravate* means *to intensify or increase.* Many writers incorrectly use *aggravate* to mean to irritate or to annoy.

albumen
albumin
Albumen is the white of an egg. *Albumin* is any of a large class of simple proteins.

aliquot
sample
An *aliquot* is contained an exact number of times in something else. For example, 10 ml is an aliquot of 20 or 30 ml, but not of 25 or 45 ml. A *sample* is a representative part of the whole.

alleviate
Alleviate means to give *temporary relief.* It implies that the underlying problem is still unresolved.

allude
refer
To *allude* to is to refer indirectly. To *refer* is to name.

alternate
alternative
Alternate means *every other one. Alternative* means *another choice; the second of two choices.*

altogether
all together
Altogether means *completely, entirely, whole. All together* means *as a group.*

among
between
Among applies to three or more; *between* applies to two.

amount
number
Amount refers to mass, bulk, or quantity. *Number* applies to a quantity that can be counted.

analogy

An *analogy* is a comparison between two objects or concepts. For example, rain is to snow as water is to ice.

and/or

Avoid this unsightly combination. If you mean *and*, *or*, or *both*, say it.

anthropomorphic

Ascribing human characteristics to nonhuman things.

anticipate

expect

Use *anticipate* when you mean looking forward to something with a foretaste of the pleasure or distress it promises. Use *expect* in the sense of certainty that it will occur. Anticipate a meeting, expect the sun to set.

anxious

eager

Anxious means uneasy, apprehensive. *Eager* is the word to describe earnestly wanting something.

as to whether

Just say *whether*.

as yet

Omit the first word.

bacteria

Plural of bacterium.

basically

An overused and generally unnecessary word.

because of

due to

Because of means *by reason of* or *on account of*. *Due to* means *attributable to*.

believe

think

Believe means *to actively accept as true*. *Think* means *apply mental activity and power in considering a question without necessarily being sure of the answer*.

beside

besides

Beside means *next to*. *Besides* means *in addition to*.

bisect

dissect

Bisect means *to cut into two nearly equal parts*. To *dissect* is *to cut into many parts*.

can

may

Can means *to know how* to or *to be able to*. *May* means *to have permission to* or *to be likely to*.

carry out
perform
Carry out means *to remove from the room. Perform* means to *do*. Scientists do not *carry out* experiments; they *do* experiments.

center around
Impossible. You can only center on.

circadian
diurnal
Circadian is an adjective meaning *about 24 hours. Diurnal* is an adjective meaning *repeated or recurring every 24 hours. Diurnal* also means *chiefly active in daylight hours.*

compare to
compare with
Use *to* when comparing two unlike things. Use *with* when comparing two similar things.

comprise
constitute
Constitute means *to form by putting together. Comprise* means *to contain, include, or consist of.*
   A zoo *comprises* reptiles, birds, and mammals.
   Animals *constitute* a zoo.

connote
denote
Connote implies a meaning beyond the usual, specific meaning. *Denote* indicates the presence or existence of something.

continual
continuous
Continual means *recurring frequently. Continuous* means *without interruption.*

criteria
Criteria is the plural form of the word *criterion.*

dalton
molecular weight
A *dalton* is a unit of mass equal to one-twelfth of the mass of an atom of carbon-12. The *molecular weight* is a ratio of the weight of one molecule of a substance to one-twelfth of the mass of an atom of carbon-12.

data
Data is the plural of *datum.*

decimate
Decimate is from the Latin *decen*, meaning *ten. Decimate* means *to kill one in ten*, as the Romans did to control mutinous troops. Many people use *decimate* incorrectly to mean "the killing of many" or "total destruction."

deductive reasoning
inductive reasoning
Deductive reasoning starts with the observation of general principles and uses these princi-

ples to explain particular details. *Inductive reasoning* uses details to make generalizations. What Conan Doyle often calls deductive reasoning in the Sherlock Holmes stories is actually inductive reasoning.

definite
definitive
Definite means precise or clear. *Definitive* means final or complete. Definite results are exact and unquestionable; definitive results will never be surpassed.

dilemma
A *dilemma* is a situation in which one faces two undesirable alternatives (note the prefix *di-,* implying two, as in *di*alogue).

disinterested
uninterested
Disinterested means *impartial; not influenced by personal or self-interest. Uninterested* means *not interested.* Effective scientists are disinterested in their work.

distinct
distinctive
Distinct means *individual. Distinctive* means *special* or *unique.*

dose
dosage
A *dose* is the amount of a substance to be given at one time. *Dosage* is the regular administration of a substance in some definite amount.

ecology
environment
Ecology is the study of the relationship between organisms and their environment. *Environment* refers to our surroundings.

effect
affect
Effect, when used as a noun, is the result of an action. When used as a verb, *effect* means *to cause* or *to bring about. Affect* is a verb meaning *to influence, to cause a change,* or *cause an effect.*

enable
permit
Enable means *to make possible. Permit* means *to give consent.*

end product
Just say *product.*

essential
Essential means *necessary for the existence of something else.* It is not synonymous with important or desirable.

except
accept
As a preposition, *except* means *other than. Accept* means *to receive* or *to agree with.*

fortunate
fortuitous
Fortunate means *having good luck*. *Fortuitous* means *happening by chance*.

farther
further
Farther refers to distance in space. *Further* applies to additional or advanced degrees or quantities.

fewer
less
Fewer applies to quantities that can be counted. *Less* applies to quantities that must be measured rather than counted. *Fewer* should be used only when an actual count can be made.

genus
Genus is the singular form of *genera*.

homogenous
homogeneous
Use *homogenous* as the adjectival form of *homogeny*—in science, correspondence of organs or parts descending from a common origin. *Homogeneous* means uniform throughout.

incapable
Just say *unable*.

infer
imply
Infer means *to conclude or deduce from an observation or from facts*. *Imply* means *to express indirectly*. A speaker implies, and a listener infers something from what the speaker says.

irregardless
Use *regardless* or *irrespective*, not this bastard mixture of the two. The *ir-* is redundant; it means the same thing as the *-less* on the end of the word. Saying *irregardless* is like saying *irreckless*.

limit
delimit
Limit means *to restrict by fixing limits*. *Delimit* means *to determine or fix limits*.

mitosis
cellular division
Mitosis refers to division of the nucleus. *Cellular division* is division of the cell.

probable
feasible
Probable means *likely to happen*. *Feasible* means *possible*, but not necessarily *probable*.

principal
principle
Principal means *a sum of money* or *a chief person*. *Principle* means *a basic law*.

since
because
Use *since* to refer to time. Use *because* as the conjunction that introduces a reason.
   Since beginning the experiment I have . . .
   The data were duplicated because . . .

small
few
Small refers to the size of an object; *few* refers to the number of items of the same kind.

theory
hypothesis
A *theory* is a broad, integrated, and general concept supported strongly from scientific evidence and useful in predicting a wide range of phenomena. A *hypothesis* is a proposition for experimental or logical testing.

ultimate
penultimate
Ultimate means *last. Penultimate* means *next to last.*

unique
Unique means the only one of its kind. *Unique* cannot be modified and is not synonymous with *unusual, strange,* or *odd.*

APPENDIX THREE

Typical Letter for Requesting Permission to Reproduce Material from Another Source

Date _____

To: Permissions Department

I am writing an article entitled _____

_____ that I will submit for

publication to _____. I would like your permission to include the

following information in my article:

_____

_____

_____

Volume _____ Page(s) _____ Year _____ from the article entitled

_____

_____

written by _____.

    If you grant me permission to use this material, I will credit the authors and

your journal as the source. I am sending a copy of this request to the (publisher or

author).

Sincerely,

_____

(Signature)

Permission Granted:

_____   _____

(Signature)   (Date)

APPENDIX FOUR

Symbols and Abbreviations Used in Science

Term/Unit of Measurement	Symbol or Abbreviation	Term/Unit of Measurement	Symbol or Abbreviation
becquerel	Bq	newton	N
calorie	cal	microliter	μl
centimeter	cm	micrometer	μm
coulomb	C	milliliter	ml
cubic centimeter	cm^3	millimeter	mm
cubic meter	m^3	minute (time)	min
cubic millimeter	mm^3	molar (concentration)	M
day	d	mole	mol
degree Celsius	°C	nanometer	nm
degree Fahrenheit	°F	number (sample size)	N
farad	F	parts per million	ppm
figure, figures	Fig., Figs.	pascal	Pa
gram	g	percent	%
greater than	>	plus or minus	±
hectare	ha	probability	P
height	ht	second (time)	s
hertz	Hz	species (singular)	sp.
hour	h	species (plural)	spp.
joule	J	square centimeter	cm^2
kelvin	K	square meter	m^2
kilocalorie	kcal	square millimeter	mm^2
kilogram	kg	standard deviation	SD
kilometer	km	standard error	SE
less than	<	standard temperature and pressure	STP
liter	l *or* L or liter *to avoid confusing with the numeral 1*	volt	V
		watt	W
logarithm (base 10)	log	weber	Wb
logarithm (base *e*)	ln	weight	wt
mean	$\bar{x}$	year	yr
meter	m		

361

APPENDIX FIVE

Prefixes and Multiples of SI Units

Factor by Which the Unit is Multiplied	SI Prefix	Symbol
$1\ 000\ 000\ 000\ 000\ 000\ 000 = 10^{18}$	exa	E
$1\ 000\ 000\ 000\ 000\ 000 = 10^{15}$	peta	P
$1\ 000\ 000\ 000\ 000 = 10^{12}$	tera	T
$1\ 000\ 000\ 000 = 10^{9}$	giga	G
$1\ 000\ 000 = 10^{6}$	mega	M
$1\ 000 = 10^{3}$	kilo	k
$100 = 10^{2}$	hecto	h
$10 = 10^{1}$	deca	da
$1 = 10^{0}$ (unity)		
$0.1 = 10^{-1}$	deci	d
$0.01 = 10^{-2}$	centi	c
$0.001 = 10^{-3}$	milli	m
$0.000\ 001 = 10^{-6}$	micro	μ
$0.000\ 000\ 001 = 10^{-9}$	nano	n
$0.000\ 000\ 000\ 001 = 10^{-12}$	pico	p
$0.000\ 000\ 000\ 000\ 001 = 10^{-15}$	femto	f
$0.000\ 000\ 000\ 000\ 000\ 001 = 10^{-18}$	atto	a

APPENDIX SIX

Statement of Informed Consent

It is the responsibility of the principal investigator to retain a copy of *each* signed consent form for at least five (5) years beyond the termination of the subject's participation in the proposed activity. Should the principal investigator leave the University, signed consent forms are to be transferred to the Institutional Review Board for the required retention period.

Project Title _____

Grant or Contract No. _____

Principal Investigator _____ Department _____

I consent to the performance upon_____
(myself or name of patient)

of the following treatment or procedure _____

_____

The purpose of the procedure or treatment: _____

Possible alternative methods of treatment: _____

Discomforts and risks reasonably to be expected: _____

Benefits which may be expected: _____

The nature and general purpose of the experimental procedure or treatment and the known risks have been explained to me and I understand them. I understand that any further inquiries I make concerning the procedure or treatment will be answered. I understand that my identify will not be revealed in any publication or document resulting from this research without my permission. I also understand that it is not possible to identify all potential risks in an experimental procedure; however, I believe that reasonable safeguards have been taken to minimize both the known and the potential unknown risks. This authorization is given with the understanding that I may terminate my service as a subject at any time after notifying the project director and without any prejudice. Reasonable and immediate medical attention, as exemplified by the services of the University Student Health Center, will be provided for physical injury caused directly by participating in this protocol. Any financial compensation for such physical injury will be at the option of the University, and decided on a case-by-case basis. In the event questions should arise concerning research-related injury to the subject or subject's rights, please contact the following for additional information.

_____ _____ _____
Principal Investigator or Contact Person   Phone   Signature (Subject's)

GLOSSARY

abstract   A summary of an article, book, or report.

active voice   Writing style in which the subject acts. Example: Charles Darwin wrote *On The Origin of Species*. (See *passive voice*.)

adjective   A word that describes or limits the meaning of a noun or noun phrase. An adjective tells which, what kind of, or how many. Examples: *good* experiment, *red* reagent, *three* meters.

adverb   A word that modifies or expands the meaning of a verb, adjective, or other adverb. An adverb tells how many, when, or where. Adverbs often, but not always, end in -*ly*. Example: work *slowly*.

antecedent   The word, phrase, or clause referred to by a pronoun. Examples: I read Gould's *essay* and liked it. (*Essay* is the antecedent of *it*.) A pronoun must agree with its antecedent in person, number, and gender. Thus, *it* is singular because *essay* is singular.

cliché   An expression that was once fresh but has become dull and stereotyped by overuse. Clichés make people laugh at you, not with you. Example: Avoid clichés *like the plague.*

dangling modifier   A modifier that cannot logically modify any word in a sentence. Example: *Having left in a hurry*, his experiment remained unfinished.

galley proofs   A preliminary reproduction of text made for checking spelling, spacing, type size, styles, format, and related items.

halftone   A photo or printed illustration having a range of tones.

independent clause   A clause that expresses a complete thought and thus can stand alone. Although sentences are independent clauses, most independent clauses are parts of sentences. For example, "He studied elephants in Africa, and he earned money by writing for biology magazines" consists of two independent clauses.

infinitive   The basic form of a verb, usually preceded by *to*. Examples *to work, to study, to write.*

metaphor   a figure of speech containing an implied comparison. Example: Our minds have adjusted to [the nuclear shadow under which we live], as after a time our eyes adjust to the dark. —Jimmy Carter

noun   A word that names things.

passive voice   Writing style in which the subject receives the action. Passive voice usually involves *to be* verbs such as *is, was,* and *were*. Example: The report *was written* by the biology student. (See *active voice*).

pronoun   A word that replaces a noun. Common pronouns include *he, she, they, we, them, I,* and *me*.

redundancy   An unnecessary repetition of meaning. Examples: *advance planning, active participation, present time.*

running head   A title repeated at the top of each page of a book or paper.

subject   The part of a sentence or clause that performs or, in the passive voice, receives the action of the verb.

verb   A word that expresses action or being. Examples: *write, study, is, were.*

INDEX